Falko Dressler and Iacopo Carreras (Eds.)

Advances in Biologically Inspired Information Systems

Studies in Computational Intelligence, Volume 69

Editor-in-chief
Prof. Janusz Kacprzyk
Systems Research Institute
Polish Academy of Sciences
ul. Newelska 6
01-447 Warsaw
Poland
E-mail: kacprzyk@ibspan.waw.pl

Further volumes of this series
can be found on our homepage:
springer.com

Vol. 47. S. Sumathi, S. Esakkirajan
Fundamentals of Relational Database Management Systems, 2007
ISBN 978-3-540-48397-7

Vol. 48. H. Yoshida (Ed.)
Advanced Computational Intelligence Paradigms in Healthcare, 2007
ISBN 978-3-540-47523-1

Vol. 49. Keshav P. Dahal, Kay Chen Tan, Peter I. Cowling (Eds.)
Evolutionary Scheduling, 2007
ISBN 978-3-540-48582-7

Vol. 50. Nadia Nedjah, Leandro dos Santos Coelho, Luiza de Macedo Mourelle (Eds.)
Mobile Robots: The Evolutionary Approach, 2007
ISBN 978-3-540-49719-6

Vol. 51. Shengxiang Yang, Yew Soon Ong, Yaochu Jin Honda (Eds.)
Evolutionary Computation in Dynamic and Uncertain Environment, 2007
ISBN 978-3-540-49772-1

Vol. 52. Abraham Kandel, Horst Bunke, Mark Last (Eds.)
Applied Graph Theory in Computer Vision and Pattern Recognition, 2007
ISBN 978-3-540-68019-2

Vol. 53. Huajin Tang, Kay Chen Tan, Zhang Yi
Neural Networks: Computational Models and Applications, 2007
ISBN 978-3-540-69225-6

Vol. 54. Fernando G. Lobo, Cláudio F. Lima and Zbigniew Michalewicz (Eds.)
Parameter Setting in Evolutionary Algorithms, 2007
ISBN 978-3-540-69431-1

Vol. 55. Xianyi Zeng, Yi Li, Da Ruan and Ludovic Koehl (Eds.)
Computational Textile, 2007
ISBN 978-3-540-70656-4

Vol. 56. Akira Namatame, Satoshi Kurihara and Hideyuki Nakashima (Eds.)
Emergent Intelligence of Networked Agents, 2007
ISBN 978-3-540-71073-8

Vol. 57. Nadia Nedjah, Ajith Abraham and Luiza de Macedo Mourella (Eds.)
Computational Intelligence in Information Assurance and Security, 2007
ISBN 978-3-540-71077-6

Vol. 58. Jeng-Shyang Pan, Hsiang-Cheh Huang, Lakhmi C. Jain and Wai-Chi Fang (Eds.)
Intelligent Multimedia Data Hiding, 2007
ISBN 978-3-540-71168-1

Vol. 59. Andrzej P. Wierzbicki and Yoshiteru Nakamori (Eds.)
Creative Environments, 2007
ISBN 978-3-540-71466-8

Vol. 60. Vladimir G. Ivancevic and Tijana T. Ivacevic
Computational Mind: A Complex Dynamics Perspective, 2007
ISBN 978-3-540-71465-1

Vol. 61. Jacques Teller, John R. Lee and Catherine Roussey (Eds.)
Ontologies for Urban Development, 2007
ISBN 978-3-540-71975-5

Vol. 62. Lakshmi C. Jain, Raymond A. Tedman and Debra K. Tedman (Eds.)
Evolution of Teaching and Learning Paradigms in Intelligent Environment, 2007
ISBN 978-3-540-71973-1

Vol. 63. Wlodzislaw Duch and Jacek Mańdziuk (Eds.)
Challenges for Computational Intelligence, 2007
ISBN 978-3-540-71983-0

Vol. 64. Lorenzo Magnani and Ping Li (Eds.)
Model-Based Reasoning in Science, Technology, and Medicine, 2007
ISBN 978-3-540-71985-4

Vol. 65. S. Vaidya, Lakhmi C. Jain and Hiro Yoshida (Eds.)
Advanced Computational Intelligence Paradigms in Healthcare-2, 2007
ISBN 978-3-540-72374-5

Vol. 66. Lakhmi C. Jain, Vasile Palade and Dipti Srinivasan (Eds.)
Advances in Evolutionary Computing for System Design, 2007
ISBN 978-3-540-72376-9

Vol. 67. Vassilis G. Kaburlasos and Gerhard X. Ritter (Eds.)
Computational Intelligence Based on Lattice Theory, 2007
ISBN 978-3-540-72686-9

Vol. 68. Cipriano Galindo, Juan-Antonio Fernández-Madrigal and Javier Gonzalez
Multiple Abstraction Hierarchies for Mobile Robot Operation in Large Environments, 2007
ISBN 978-3-540-72688-3

Vol. 69. Falko Dressler and Iacopo Carreras (Eds.)
Advances in Biologically Inspired Information Systems, 2007
ISBN 978-3-540-72692-0

Falko Dressler
Iacopo Carreras
(Eds.)

Advances in Biologically Inspired Information Systems

Models, Methods, and Tools

With 132 Figures and 14 Tables

Falko Dressler
Autonomic Networking Group
Department of Computer Science 7
University of Erlangen
Martensstr. 3
91058 Erlangen
Germany
E-mail:- dressler@informatik.uni-erlangen.de

Iacopo Carreras
CREATE-NET
Via Solteri 38
38100 Trento
Italy
E-mail:- iacopo.carreras@create-net.org

ISBN 978-3-642-09176-6   e-ISBN 978-3-540-72693-7
ISSN print edition: 1860-949X
ISSN electronic edition: 1860-9503

This work is subject to copyright. All rights are reserved, whether the whole or part of the material is concerned, specifically the rights of translation, reprinting, reuse of illustrations, recitation, broadcasting, reproduction on microfilm or in any other way, and storage in data banks. Duplication of this publication or parts thereof is permitted only under the provisions of the German Copyright Law of September 9, 1965, in its current version, and permission for use must always be obtained from Springer-Verlag. Violations are liable to prosecution under the German Copyright Law.

Springer is a part of Springer Science+Business Media
springer.com
© Springer-Verlag Berlin Heidelberg 2010

The use of general descriptive names, registered names, trademarks, etc. in this publication does not imply, even in the absence of a specific statement, that such names are exempt from the relevant protective laws and regulations and therefore free for general use.

Cover design: deblik, Berlin

Preface

Technology is taking us to a world where myriads of heavily networked devices interact with the physical world in multiple ways, and at multiple scales, from the global Internet scale down to micro and nano devices. Many of these devices are highly mobile and autonomous, and must adapt to the surrounding environment in a totally unsupervised way.

This vision poses severe challenges to existing approaches to the design and management of ICT systems, since the size of this omni-comprehensive network, in terms of both number of constituent nodes and running services, is expected to exceed by several orders of magnitude that of existing information systems. The resulting large-scale communication system could not be handled according to conventional ICT paradigms, which are not able to accommodate the scale, heterogeneity and complexity of such scenarios. A fundamental research challenge is therefore the design of *robust decentralized computing systems* capable of operating under changing environments and noisy input, and yet exhibit the desired behavior and response time, under constraints such as energy consumption, size, and processing power. These systems should be able to adapt on the short term and evolve on the long one, and they should learn how to react to unforeseen scenarios as well as to display properties comparable to social entities.

Biological systems are able to handle many of these challenges with an elegance and efficiency still far beyond current human artifacts. Most of the desired features have been refined by nature over periods of millions of years, generating living organisms that are able to autonomously repair themselves when damaged, produce emergent behavior, survive despite drastic changes in the environment conditions and evolve over time. All these considerations generated a significant interest in the area of *bio-inspired computing*, i.e. the application of biological principles to the design of human artifacts, with the expectation of reproducing the same observed behavior.

Fully in line with this growing interests in biologically inspired computing, the first international conference on *Bio Inspired mOdels of NEtwork, Information and Computing Systems* (BIONETICS) was successfully organized in Cavalese in December 2006. The aim was to bring together researchers and scientists from several disciplines in computer science and engineering where bio-inspired methods are investigated. The interest in this new research approach has been confirmed by the high number of high quality submissions, and from the high number of conference attendees.

This book collects extended versions of selected outstanding papers originally submitted to BIONETICS 2006. It is structured into four parts, covering different aspects involved in the engineering of future ICT systems:

- Self-Organizing Network Environments
- System Design and Programming
- Sensor and Actor Networks
- Search and Optimization

With this book, we want to address researchers who are interested in bio-inspired computing. According to the broad spectrum addressed by the different book chapters, a variety of biological principles and their application in ICT systems are presented. Based on these information, we want to encourage researchers to initiate interdisciplinary studies especially focusing on biological disciplines.

Erlangen and Trento, *Falko Dressler*
March 2007 *Iacopo Carreras*

Contents

Part I Self-Organizing Network Environments

Bio-inspired Framework for Autonomic Communication Systems
*Sasitharan Balasubramaniam, Dmitri Botvich, William Donnelly,
Mícheál Ó Foghlú and John Strassner* 3

**Towards a Biologically-inspired Architecture for Self-Regulatory
and Evolvable Network Applications**
Chonho Lee, Hiroshi Wada and Junichi Suzuki 21

Biologically Inspired Synchronization for Wireless Networks
Alexander Tyrrell, Gunther Auer and Christian Bettstetter 47

**Bio-Inspired Congestion Control: Conceptual Framework,
Algorithm and Discussion**
Morteza Analoui and Shahram Jamali 63

**Self-Organized Network Security Facilities
based on Bio-inspired Promoters and Inhibitors**
Falko Dressler 81

Part II System Design and Programming

Context Data Dissemination in the Bio-inspired Service Life Cycle
*Carsten Jacob, David Linner, Heiko Pfeffer, Ilja Radusch
and Stephan Steglich* 101

**Eigenvector Centrality in Highly Partitioned Mobile Networks:
Principles and Applications**
*Iacopo Carreras, Daniele Miorandi, Geoffrey S. Canright
and Kenth Engø-Monsen* 123

Toward Organization-Oriented Chemical Programming: A Case Study with the Maximal Independent Set Problem
Naoki Matsumaru, Thorsten Lenser, Thomas Hinze and Peter Dittrich 147

Evolving Artificial Cell Signaling Networks: Perspectives and Methods
James Decraene, George G. Mitchell and Barry McMullin 165

Part III Sensor and Actor Networks

Immune System-based Energy Efficient and Reliable Communication in Wireless Sensor Networks
Barış Atakan and Özgür B. Akan 187

A Bio-Inspired Architecture for Division of Labour in SANETs
Thomas Halva Labella and Falko Dressler 209

A Pragmatic Model of Attention and Anticipation for Active Sensor Systems
Sorin M. Iacob, Johan de Heer and Alfons H. Salden 229

Part IV Search and Optimization

Self-Organization for Search in Peer-to-Peer Networks
Elke Michlmayr 247

A Bio-Inspired Location Search Algorithm for Peer to Peer Networks
Sachin Kulkarni, Niloy Ganguly, Geoffrey Canright and Andreas Deutsch 267

Ant Colony Optimization and its Application to Regular and Dynamic MAX-SAT Problems
Pedro C. Pinto, Thomas A. Runkler and João M. C. Sousa 283

List of Contributors

Özgür B. Akan
Middle East Technical University
Department of Electrical
and Electronics Engineering
Next Generation Wireless
Communications Laboratory
Ankara - Turkey, 06531
akan@eee.metu.edu.tr

Morteza Analoui
Iran University of Science
& Technology
University Road
Tehran - Iran, 16846-13114
analoui@iust.ac.ir

Barış Atakan
Middle East Technical University
Department of Electrical
and Electronics Engineering
Next Generation Wireless
Communications Laboratory
Ankara - Turkey, 06531
atakan@eee.metu.edu.tr

Gunther Auer
DoCoMo Communications Laboratories
Europe GmbH
Landsbergerstrasse 312
Munich - Germany, 80687
auer@docomolab-euro.com

Sasitharan Balasubramaniam
Waterford Institute of Technology
Telecommunication Software
and Systems Group
Carriganore Campus
Waterford - Ireland
sasib@tssg.org

Christian Bettstetter
University of Klagenfurt Mobile
Systems Group Institute of Networked
and Embedded Systems Lakeside
B02 Klagenfurt - Austria, 9020
bettstetter@ieee.org

Dmitri Botvich
Waterford Institute of Technology
Telecommunication Software
and Systems Group
Carriganore Campus
Waterford - Ireland
dbotvich@tssg.org

Geoffrey S. Canright
Telenor R&D
Snaroyveien 30
Oslo - Norway
geoffrey.canright@telenor.com

List of Contributors

Iacopo Carreras
CREATE-NET
Via Solteri 38
Trento - Italy, 38100
iacopo.carreras@
create-net.org

James Decraene
Dublin City University
Research Institute for Networks
and Communications Engineering
Glasnevin
Dublin - Ireland
james.decraene@eeng.dcu.ie

Andreas Deutsch
TU-Dresden
Center for Information Services
and High Performance Computing
Helmholtzstr. 10
Dresden - Germany, 01069
andreas.deutsch@
tu-dresden.de

Peter Dittrich
Friedrich-Schiller-University Jena
Department of Mathematics
and Computer Science
Bio Systems Analysis Group
Ernst-Abbe-Platz 1-4
Jena - Germany, D-07743
dittrich@minet.uni-jena.de

William Donnelly
Waterford Institute of Technology
Telecommunication Software
and Systems Group
Carriganore Campus
Waterford - Ireland
wdonnelly@tssg.org

Falko Dressler
University of Erlangen-Nuremberg
Dept. of Computer Science
Autonomic Networking Group
Martensstr. 3
Erlangen - Germany, 91058
dressler@informatik.
uni-erlangen.de

Kenth Engø-Monsen
Telenor R&D
Snaroyveien 30
N-1331 Fornebu (Norway), Postal Code
kenth.engo-monsen@telenor.
com

Micheál Ó Foghlú
Waterford Institute of
Technology Telecommunication
Software and Systems Group
Carriganore Campus Waterford - Ireland
mofoghlu@tssg.org

Niloy Ganguly
Indian Institute of Technology
Kharagpur
Department of Computer Science
and Engineering
Kharagpur - India, 721302
niloy@cse.iitkgp.ernet.in

Johan de Heer
T-Xchange
Twente University
De Horst (building 20)
P.O. Box 217
Enschede - The Netherlands, 7500 AE
johan.deHeer@txchange.nl

Thomas Hinze
Friedrich-Schiller-University Jena
Department of Mathematics
and Computer Science
Bio Systems Analysis Group
Ernst-Abbe-Platz 1-4
Jena - Germany, D-07743
hinze@minet.uni-jena.de

Sorin M. Iacob
Telematica Instituut
Brouwerijstraat 1
Enschede - The Netherlands, 7523 XC
sorin.iacob@telin.nl

List of Contributors   XI

Carsten Jacob
Fraunhofer Institute for
Open Communication Systems
Kaiserin-August-Allee 31
Berlin - Germany, 10589
carsten.jacob@fokus.
fraunhofer.de

Shahram Jamali
Iran University of Science
& Technology
University Road
Tehran - Iran, 16846-13114
jamali@iust.ac.ir

Sachin Kulkarni
Indian Institute of Technology
Kharagpur
Department of Computer Science
and Engineering
Kharagpur - India, 721302
sachindkulkarni@gmail.com

Thomas Halva Labella
University of Erlangen-Nuremberg
Dept. of Computer Science
Autonomic Networking Group
Martensstr. 3
Erlangen - Germany, 91058
hlabella@ulb.ac.be

Chonho Lee
University of Massachusetts
Department of Computer Science
Morrissey Blvd. 100
Boston, MA - USA, 02125
chonho@cs.umb.edu

Thorsten Lenser
Friedrich-Schiller-University Jena
Department of Mathematics
and Computer Science
Bio Systems Analysis Group
Ernst-Abbe-Platz 1-4
Jena - Germany, D-07743
thlenser@minet.uni-jena.de

David Linner
TU Berlin
Kaiserin-August-Allee 31
Berlin - Germany, 10587
david.linner@tu-berlin.de

Naoki Matsumaru
Friedrich-Schiller-University Jena
Department of Mathematics
and Computer Science
Bio Systems Analysis Group
Ernst-Abbe-Platz 1-4
Jena - Germany, D-07743
naoki@minet.uni-jena.de

Barry McMullin
Dublin City University
Research Institute for Networks
and Communications Engineering
Glasnevin
Dublin - Ireland
barry.mcmullin@dcu.ie

Elke Michlmayr
Vienna University of Technology
Institute of Software Technology
and Interactive Systems
Women's Postgraduate College
for Internet Technologies
Favoritenstrasse 9-11/E188
Vienna - Austria, 1040
michlmayr@wit.tuwien.ac.at

Daniele Miorandi
CREATE-NET
Via Solteri 38
Trento - Italy, 38100
daniele.miorandi@
create-net.org

George Mitchell
Dublin City University
Research Institute for Networks
and Communications Engineering
Glasnevin
Dublin - Ireland
george.mitchell@dcu.ie

Heiko Pfeffer
TU Berlin
Kaiserin-August-Allee 31
Berlin - Germany, 10587
heiko.pfeffer@tu-berlin.de

Pedro Caldas Pinto
Technical University of Lisbon
Department of Mechanical Engineering
Avenida Rovisco Pais
Lisbon - Portugal, 1049-001
pinto.pedro.ext@siemens.com

Ilja Radusch
TU Berlin
Kaiserin-August-Allee 31
Berlin - Germany, 10587
ilja.radusch@tu-berlin.de

Thomas A. Runkler
Siemens AG Corporate Technology
Information and Communications
Otto-Hahn-Ring 6
Munich - Germany, 81730
thomas.runkler@siemens.com

Alfons H. Salden
Almende
Westerstraat 50
Rotterdam - The Netherlands, 3016 DJ
alfons@almende.com

João M. da Costa Sousa
Technical University of Lisbon
Department of Mechanical Engineering
Avenida Rovisco Pais
Lisbon - Portugal, 1049-001
jmsousa@ist.utl.pt

Stephan Steglich
TU Berlin
Kaiserin-August-Allee 31
Berlin - Germany, 10587
stephan.steglich@tu-berlin.de

John Strassner
Motorola Chicago Labs
Schaumburg, IL - USA
john.strassner@motorola.com

Junichi Suzuki
University of Massachusetts
Department of Computer Science
Morrissey Blvd. 100
Boston, MA - USA, 02125
jxs@cs.umb.edu

Alexander Tyrrell
DoCoMo Communications Laboratories Europe GmbH
Landsbergerstrasse 312
Munich - Germany, 80687
tyrrell@docomolab-euro.com

Hiroshi Wada
University of Massachusetts
Department of Computer Science
Morrissey Blvd. 100
Boston, MA - USA, 02125
shu@cs.umb.edu

Part I

Self-Organizing Network Environments

Bio-inspired Framework for Autonomic Communication Systems

Sasitharan Balasubramaniam[1], Dmitri Botvich[1], William Donnelly[1], Mícheál Ó Foghlú[1] and John Strassner[2]

[1] Telecommunication Software and Systems Group,
Waterford Institute of Technology, Carriganore Campus,
Waterford, Ireland
{sasib, dbotvich, wdonnelly, mofoghlu}@tssg.org;
[2] Motorola Chicago Labs, Schaumburg, IL, USA
john.strassner@motorola.com

Abstract – The rapid growth of the Internet has led to tremendous network management complexities. This is largely due to the rapid changes in network technology, traffic behaviour, and service environments, which require communication systems to be more robust, scalable, and adaptive to such changes. Biological systems exhibit these characteristics and have tremendous capabilities to adapt to environmental changes. In this chapter we present a bio-inspired framework for autonomic communication systems, where the framework can be applied to infrastructure or ad hoc networks as well as service environments. Case study applications of the framework for infrastructure network have also been presented.

1 Introduction

The telecommunications network environment has experienced major changes over the last ten years. The changes were driven by a number of factors. The first major change was the introduction of Internet connectivity that revolutionised both communication mechanisms and service creation environment for users (e.g. multimedia services). The second change was the emergence of new broadband wireless communications network technologies (e.g. UMTS, WiMAX) that has enabled users to access information from anywhere at anytime. New network technology developments have also created new applications such as pervasive computing, which supports all aspects of society from leisure through to work activities. Based on these factors, future network infrastructure is expected to support greater volumes of traffic ranging from services such as Voice over IP (VoIP) to more complex multimedia services such as collaborative work environments and remote monitoring applications. The changes of communication networks also extend beyond telecommunication systems or conventional computer networks that support data communication. It is for this reason that ad hoc networks have received particular attention over the last few years. The concepts of ad hoc network were later extended to support networking of micro devices such as sensor networks.

However, as these networks grow in complexity so did new communications management solutions reflecting the need for network autonomy and control. This has lead to the emergence of autonomic management of communication networks and services, whereby network devices and software components are able to exhibit self-governance (self-management, self-organisation, self-learning, etc.) and minimise human intervention. Self-governance can occur at various levels of the communication systems, from infrastructure of core and access networks, to low level sensor networks. In addition, self-governance enables management decisions to be leveraged by human users and operators, making the overall system more efficient.

A number of self-governance principles can be found in nature itself, and one very good example is biological systems. Various organisms have different biological processes that can adapt to environmental changes through mechanisms such as self-management, self-organisation as well as self-learning. Through time biological systems are able to evolve and learn and adapt to changes that environment brings upon living organisms. Our objective is to investigate the various biological processes that exhibit self-governance and combine these different processes into a biological framework that can be applied to various communication systems to realise true autonomic behaviour. Our work has been inspired by other researchers that had applied bio-inspired techniques to various other disciplines. Examples of these applications include applying characteristics of hormone interactions for robot self-organisation [1] or hormone signaling control for Bionode multi-FPGA network nodes [2].

This chapter is organised as follows: Section 2 presents the background and related work on autonomic and bio-inspired communication. Section 3 describes our bio-inspired framework for autonomic communication systems. Section 4 presents a case study describing the application of our bio-inspired framework, and lastly section 5 presents the conclusion.

2 Related Works and Background

2.1 Autonomic Communication Systems

The vision of autonomic communication [3] is to propose a novel technique towards managing and controlling complexity in order to **govern** adaptive behavior. One of the early pioneers to investigate *Autonomic Computing* was IBM [4], who proposed an architecture based on a self-adjusting control loop that monitors, analyses, plans and executes based on the status of the current environment. Various research works have been conducted on applying specific self-governance mechanisms to communication networks. Yagan and Tham [5] proposed a model-free reinforcement learning technique for self-optimising and self-healing capabilities for DiffServ QoS networks. Prehofer and Bettstetter [6] outlined and defined key self-organization design paradigms for future communication systems. The paradigms are based on system-wide adaptive structure with no external or central dedicated control entity, where individual entities interact with each other in a peer-to-peer fashion. The FOCALE (Foundation, Observation, Comparison, Action, and Learning Environment) architecture [7] is based on the concept of transforming any Managed Resource entity (e.g. device, system) into an Autonomic Computing Element (ACE). These ACE's can be uniformly managed and organized into a larger Autonomic Management Domain. Our biological framework can be integrated with the FOCALE architecture, by drawing out key biological analogies and translating them to a set of policies within the Autonomic Manager of the ACE.

2.2 Bio-inspired Communication Systems

The concept of looking to nature for inspirational solutions and applying to communication systems is not new. For instance, Suzuki and Suda [8] have applied the behavior of bee colonies to a bio-networking architecture for autonomous applications, where network applications are implemented as a group of autonomous diverse agents called Cyber-Entity (CE) (e.g. a CE for a web server may contain HTML files). The work was later extended to support evolutionary adaptation using genetic algorithms to evolve CE behaviours and improve their survival fitness in the environment [9]. Leibnitz et al [10] proposed a bio-inspired technique towards self-adaptive routing for overlay networks. Bio-inspired techniques have also been applied towards network security, where Dressler [11] [12] observed the reaction of cells during virus attack and inflammation, and applied this concept to autonomic network security system.

3 Bio-inspired Autonomic Network Management

Inspired by the different bio-inspired applications, we have developed a bio-inspired autonomic network management framework to meet the challenges of next generation communication systems. First, we describe a general development and management cycle of communication systems inspired from a biological development cycle. This is followed by defining key biological analogies used in the framework and mapping them to communication networks and service environments.

3.1 Development and Management Life Cycle

Our objective is to combine and integrate multiple self-governance principles to produce a more robust, scaleable and adaptive communication system. Before we describe the different biological analogies of the framework, we will first describe the biological development cycle that demonstrates how we integrate the various self-governance principles into a single framework. The key candidates for self-governing mechanisms are organised into three main categories, which are self-organisation, self-management, and self-learning. These categories and their organisation in the development cycle are illustrated in Fig. 1.

Biological Development Cycle

The development cycle begins at the formation stage, which is then followed by the system management stage. During the system management stage, the organism will adaptively learn from various transition stages (Fig. 1 illustrates this as environmental and structure changes) that may occur during the lifespan of the organism.

The self-organisation process falls into two stages of the development process, which are the formation and management stages. At the formation stage, biological systems start the growth process from an embryo that develops into a mature structured organism (Ontogeny). During this development phase, the organism slowly forms its internal organ structure and interconnection with various organs, whereby the organism slowly begins various functionalities [13].

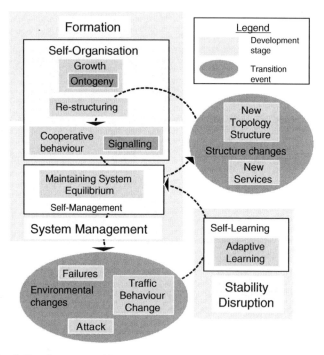

Fig. 1. Development and Management Life Cycle of Bio-inspired System

As the organism matures, the internal organs can perform specific tasks and functions to maintain equilibrium during the life cycle. Example of these equilibrium processes may include the ability to maintain thermoregulation, osmoregulation, or blood glucose homeostasis [13]. At the same time, the self-organisation mechanism can also occur in the management stage, where organs of an organism may cooperatively perform functionalities to support system management. Organisms can also exhibit self-organisation through restructuring mechanisms (e.g. some living organisms may loose specific body parts that they can regenerate).

During the organism's development cycle, a crucial characteristic is the ability for the organism to self-learn. The learning process will be refined and enhanced through experience by noting any disruptions from the environment stability and react by appropriately adapting to re-stabilise the equilibrium. Learning in the biological systems can occur at different time lines, where learning and adaptation can evolve through centuries or learning can be performed during the lifespan of the organism.

Communication Systems Development Cycle

Based on the development cycle described in the previous section, we will describe the mapping from the biological development cycle to the communication system development cycle. Similar to the biological development cycle, the self-organisation concept is applied to the communication system's formation and management stage. In the case of infrastructure network, the self-organisation will largely be the formation of the topology, which will be mainly static. However, in the case of ad hoc or sensor networks, the formation of the topology will be largely dynamic.

The management stage of the communication system will include mechanisms that enable the system to sustain itself. These mechanisms include the ability for the system to balance the internal equilibrium (homeostasis), and rely on the self-organisation mechanism for coordination with other elements to further support sustainable management in the face of environmental changes. An example of this is route organisation to support efficient resource management stability. During the lifespan of the communication system, various disruptions and changes may be encountered that may affect the system's management. Such events may include changes to traffic behaviour in the network or security attacks on the infrastructure. Based on these disruptions, the system must be able to self-learn through these events and gain experience. The management stage may also be disrupted in the event of new services that have been added or a topology re-structuring occurs. Re-structuring of the topology for the network infrastructure maybe due to failures or addition of new nodes or links. This is especially the case for ad hoc networks, when nodes constantly migrate in or out of the current topology. In the case of ad hoc networks, the dynamic changes in topology may also result in role changes of specific devices (e.g. in a piconet, where certain devices may take on a role of master or slave). The re-structuring is similar to living organisms that are injured and must reform limbs (self-healing). The restructuring phase is shown in Fig. 1 as a transition event loop from the self-management process to the self-organisation process. Once the re-structuring process is done, the formation mechanism will progress to the management stage to resume the normal operations.

3.2 Bio-inspired Framework

The framework for our bio-inspired Autonomic Network Management model is illustrated in Fig. 2. The framework is composed of various components that are sub-categorised under self-organisation, self-management, and self-learning. These sub-components represent the different mechanisms described in section 3.1. The Model

Generator evaluates the requirements from each type of communication system through the *Requirements Processing* module and maps this to the *Analogy Mapping* module. The model is then composed using the *Model Composition* module, which maps the composed model to the respective type of communication system.

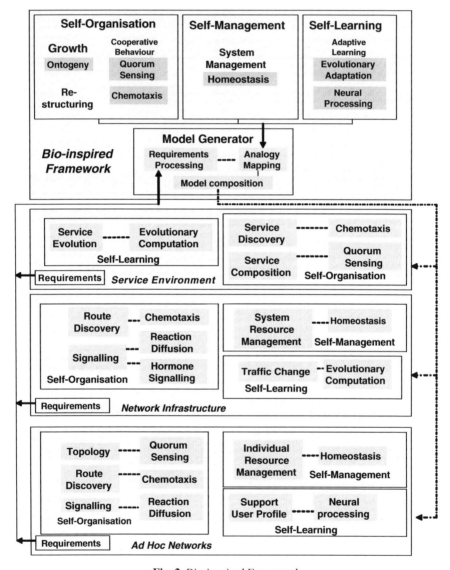

Fig. 2. Bio-inspired Framework

The Analogy Mapping process is performed using a characteristic matching mechanism that allows the salient features of the communication system to be matched to various biological analogies. In Fig. 2, we illustrate how model mapping can be performed for three types of systems – infrastructure and ad hoc networks, and service environments. However, the service environment can't exist independently and must rely on network infrastructure or ad hoc network to transport the content stream. Therefore, Fig. 2 illustrates this by showing the service environments being independently managed above the communication network. A number of characteristics from the different systems are similar and have biological analogy that overlaps.

Network infrastructure requires efficient resource management that must satisfy the business objective of the operator. At the same time, the robustness and scalability issue are a core requirement in network infrastructure. These factors are crucial when supporting changes that require route and resource re-configuration. This factor may be due to changes in traffic behaviour. Based on these characteristics, we matched the homeostasis equilibrium model to the self-management of resources, chemotaxis [14] for self-organisation of dynamic routes, and evolutionary computation for self-learning to support dynamic traffic changes. Chemotaxis is the process that supports micro-organism mobility by sensing and following a chemical gradient. The reaction-diffusion principle is also applied to messaging between the devices [15]. The homeostasis model fits the resource management model because the homeostasis model must maintain an equilibrium level with respect to a high level goal. As described earlier, network infrastructure are required to maintain a specific level of equilibrium in order to support various goals that the systems have. Therefore, the network infrastructure will concentrate largely on management of traffic resources to suit the goals and requirements of an operator, where less emphasis will be put on the self-organisation of nodes because of the static structure of the network topology. This makes sense, because intuitively, it is very undesirable for the organisation of a network infrastructure to vary – this would cause havoc to the devices and services that access it.

On the other hand, ad hoc networks have slightly different requirements as opposed to network infrastructure. Although resource management is a crucial requirement in ad hoc networks, this is largely dependent on peer-peer routing and the ability for nodes to cooperatively determine efficient routes with minimal energy dissipation. Therefore, an important characteristic is the ability for devices to perform specific cooperative tasks. In particular we have focussed on quorum sensing mechanisms [16]. Quorum sensing allows a community of cells to co-ordinate and perform a specific function. The routing mechanisms used in ad hoc network are similar to network infrastructures, where the chemotaxis mechanism can be used. Since ad hoc networks are usually closely associated to specific user needs, an important requirement is the ability to process user requirements, where we have applied a simple neural processing mechanism that has the ability to evaluate user behaviour and profile requirements, and use this information to manage the ad hoc network more efficiently. Since a priority of ad hoc devices is energy efficiency, we have also applied homeostasis mechanism for internal resource management of each individual device.

A crucial requirement in networks and communication systems is supporting content services. Unlike traditional services, current content services are very dynamic and are tailored to meet various user requirements. For instance, the impact of evolving multimedia content has helped drive this dynamic nature. A requirement of content services is the ability for a content service to be composed from or cooperate with other content services to specifically meet the needs of the user. The latter is especially true for ad hoc pervasive services. Therefore, key characteristics include the ability for services to efficiently discover other services, and cooperatively self-organise to meet the user's request. Services are also required to evolve and compose with new services that may address changing user requirements. Therefore, we have mapped the chemotaxis process to enable discovering the right service. For cooperative interaction amongst different services, we have employed quorum sensing. We also have employed evolutionary computation to allow the services to learn, evolve, and compose with new services that are introduced.

4 Bio-inspired Framework Case Studies

In this section, we will demonstrate a case study that illustrates the application of the bio-inspired framework. The case study demonstrates the self-organisation and self-management mechanisms of core network infrastructure.

4.1 Applying Self-Management and Self-Organisation to Autonomic Network Management

As described in the introduction, autonomic network management provides a communication system with the ability to manage and control complexity in order to enhance self-governance. Our architecture is based on three-layer hierarchy, as illustrated in Fig. 3. This is composed of the Business, Systems, and Device layers. In this section, we outline a subset of framework mapping of section 3.2 for autonomic communication in core networks that require self-management and self-organisation. As shown in Fig. 3, the business level codifies business goals, and translates them into a form that can be used by the system and device layers. For self-management mechanism of network resources we applied the blood glucose homeostasis model, while the self-organisation mechanism for de-centralised routing we applied the chemotaxis, reaction diffusion, and hormone signalling model.

Blood Glucose Homeostasis Self-management

Blood glucose homeostasis is one mechanism that living organisms use to maintain system equilibrium. The mechanism relies on monitoring the body of the organism when performing different activities and responding through one or more feedback loops. Fig. 4 illustrates the mapping of the blood glucose homeostasis to the communication network resource management mechanism. The main motivation behind selecting the blood glucose model is based on the requirements of core network infrastructure, where network resources must be managed efficiently in the face of

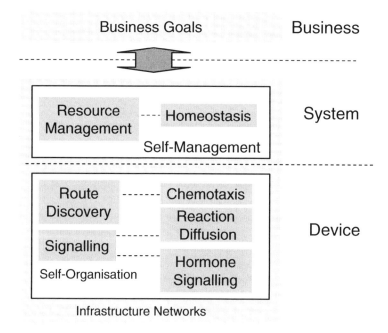

Fig. 3. Application of Self-Organisation and Self-Management to Autonomic Network Management of Core Network Infrastructure

varying traffic intensity to meet the business objectives of network operators. In similar ways, this can be compared to the body controlling the blood glucose balance when subjected to tasks of different intensity. The glucose within the body can be represented as glycogen or fat.

When the body intensity increases, the glucose gets used, and the process of glycogen breakdown for glucose production is performed. In the same way, once the glycogen is used beyond a specific threshold, the breakdown of fat and lipid is performed to obtain glucose. The reversal process could also occur, where if the amount of glucose is increased, the glucose is transformed back into glycogen, and once the glycogen reaches a certain threshold, the remaining glucose is transformed into fat. We apply the same mechanism that the human body uses to manage the different forms of glucose resources to the way networks cope with different traffic types and intensity. The algorithm for self-management of resources is shown in Fig. 5. Our comparison is based on the fact that the way the glycogen is broken down to create glucose is similar to how resources are used to support a particular demand profile for a specific source-destination pair of routers (Fig. 5-line 1). In this case, the demand profile is the average traffic that enters the network, and is compared to an average routine activity intensity a person may exert between specific times. By comparing the traffic profile to the demand profile, the route is streamed through permanent routes established for the source-destination pair (Fig. 5-line 3-6). How-

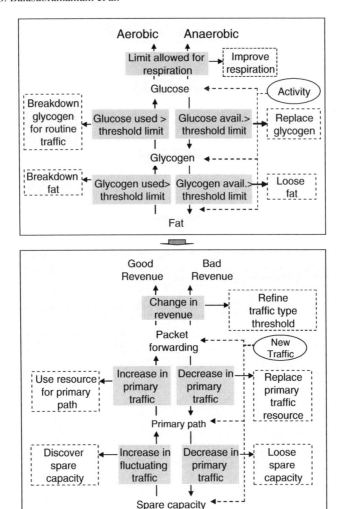

Fig. 4. Blood Glucose Homeostasis Mapping to Core Network Resource Management

ever, in the event that a network may face fluctuating traffic or increase in traffic load beyond the demand profile, the resource manager will start using the spare capacity in the network (Fig. 5-line 8). This process is compared to the human body, when glycogen runs out due to unexpected high intensity activity being performed leading to discovery and breakdown of fat to produce glucose. In the event that this fat is used consistently, the fat quantity is slowly reduced, leading to a fitter body. Our factor for evaluating if a new traffic should join the demand profile is based on time usage (T_{SPARE_USAGE}) and frequency of usage (N_{SPARE_CYLCES}) (Fig. 5-line 9-11). When this principle is applied to network operators for an ISP, the use of spare resources will in turn create a fitter network that translates to higher revenue.

Although the body has the ability to control the volume of fat loss, the opposite effect may also occur that can lead to the body gaining excessive fat. This effect may be caused by changes in the human activity intensity or an increase in resource in-take. When the body becomes unhealthy due to excessive fat, the human body could increase the activity intensity to lower the fat. In such a scenario, the ISP can face situations such as loss of customers. Therefore, the ISP may choose to sell off excess resources in order to maintain network fitness. This is done by comparing the amount of time in reduction of traffic volumes compared to a fixed threshold (Fig. 5-line 12-14). Our future work will consider performing mechanisms to attract more customers such as changing the pricing scheme for different traffic types. An extensive description for the blood glucose management can be found in [17].

1 **for** edge router pair ER_i, ER_j
2 determine bandwidth for demand profile BW_{ER_i, ER_j}
3 **for** new traffic t
4 **if** (new traffic $t_{ER_i, ER_j} \in$ Demand Profile $_{ER_i, ER_j}$)
5 route traffic as permanent stream P_{ER_i, ER_j}
6 add $BWt_{, ER_i, ER_j}$ to used bandwidth BW_{u, ER_i, ER_j}
7 **else if**
8 route traffic as spare capacity P_{SC, ER_i, ER_j}
9 **if** ($time_{P_{SC, ER_i, ER_j}} \geq T_{SPARE\_USAGE}$)
10 **if** (no. of cycles $N_{P_{SC, ER_i, ER_j}} \geq N_{SPARE\_CYLCES}$)
11 add P_{SC, ER_i, ER_j} to demand Profile
12 **if** ($BW_{u, ER_i, ER_j} < BW_{ER_i, ER_j}$)
13 **if** ($time_{BW_{u, ER_i, ER_j} < BW_{ER_i, ER_j}} < T_{GET\_RID\_Spare\_Resource}$)
14 $BW_{ER_i, ER_j} = BW_{ER_i, ER_j} - BW_{u, ER_i, ER_j}$

Fig. 5. Algorithm for resource management

De-centralised Signalling and Route Gradient Calculation

In this section, the mechanism for bio-inspired self-organisation for de-centralised signalling for route management will be discussed. The objective of this process is to allow de-centralised mechanism of route management. Current mechanisms for route management (e.g. OSPF) are reasonably centralised in the sense that each node must have a view of the entire topology. As shown in Fig. 3, our mechanism combines a number of biological mechanisms that includes chemotaxis in combination with signalling mechanisms for self-organisation such as reaction diffusion and hormone signalling. The self-organisation process is performed in phases. The algorithm for the hormone messaging process is shown in Fig. 6. Initially, the hormone messages H, which specify the hop count, are transmitted from the destination to the source H_s (Fig. 6-line 3). The advantage in hormone signals is in its ability to change state as it travels through the blood stream and evaluate the environment condition. Therefore, as the hormone messages travel from node to node, the message deducts the hop count and stores this in each router before being passed on to the next node (Fig. 6-line 4). A back propagation mechanisms is then initiated once the source node has received the final H_f value, to normalize each hormone value to give a final h value, which is represented as (Fig. 6-line 5),

$$h_{j_{(s,d)}} = \frac{((H_s - H_f) + 1) - (H_s - H_j)}{(H_s - H_f) + 1}, \qquad (1)$$

In the event that a node receives multiple hop count values, the node retains the highest value (Fig. 6-line 6-9).

```
1  do while h_n(d,s) ≡ 0 for all nodes n ∈ N
2  for each node n
3      if (H_n(s,d) received from neighbour node m)
4          calculate H_n(s,d) - 1
5          calculate h_n(s,d)
6              if (new h_n(s,d) > old h_n(s,d))
7                  add new hn(s,d) to hop count
8              else
9                  delete new h_n(s,d)
```

Fig. 6. Algorithm for hormone signalling

Once the hormone values are normalized, the nodes can begin the operation of deducing the routes. The algorithm for reaction diffusion messaging and chemotaxis gradient calculation is presented in Fig. 7. During this stage, a reaction-diffusion like mechanism for peer-peer signalling is performed between the devices and its neighbours in order to provide load information to the neighbours. The diffusion mechanism distributes the message packets to each of the neighbours, while the reaction process evaluates the message packets from the neighbours and calculates the load on each node (Fig. 7-line 3-4). Based on information of neighbours' load node as well as the hormone hop count, each node will calculate the gradient weight of each link, where each node n contains i number of links l (Fig. 7-line 5-7). The equation for the weight gradient calculation per node between node n and j for link i is represented as,

$$G_{n,i \to j} = \alpha \Phi_j + \beta l_{n,i \to j} + \gamma H_j, \qquad (2)$$

where Φ, represents the load of neighbour node, and is expressed as,

$$\Phi_n = \frac{\sum_{i}^{I}(l_{n,c,i \to j} - l_{n,i \to j})}{\sum_{i}^{I} l_{n,c,i \to j}}, \qquad (3)$$

and the capacity of each link i of node n is represented as $l_{n,c,i \to j}$. Therefore, as the packets travel through the network, the packets will route hop by hop through the highest gradient weight value. This in turns mimics the chemotaxis behaviour of organism as they migrate towards a food source by sensing a chemical gradient.

```
1  do forever
2    for nodes n∈ N
3      calculate node load Φₙ for n
4      diffuse Φₙ to m where m ∈ neighbours of n
5      if (n receives Φₘ)
6        for link i ∈ Iₙ
7          calculate Gₙ,ᵢ
```

Fig. 7. Algorithm for load node and gradient weight calculation

4.2 Simulation

To illustrate our concepts, we have performed simulations to demonstrate mechanism of route discovery depending on the load of the network, as well as resource management based on the blood glucose model. Our simulation topology is illustrated in Fig. 8, and the input traffic into the network for the three source and destination pairs is shown in Fig. 9. For simplicity, we will not demonstrate the growth and formation stages of the development cycle, but only the self-management and self-organisation mechanism. In particular, we focus on monitoring the behaviour for the source-destination pair S1-D1. Based on numerous tests performed, we chose the following parameters for equation 2: $\alpha = 0.2, \beta = 0.4$, and $\gamma = 0.4$. Before the traffic builds up within the network, the gradient calculation on each node shows the best path for S1-D1 is path 1 (1 – 5 – 6 – 8 – 12). Fig. 10 illustrates the link gradient values for a subset of links. As shown in Fig. 10, at time 0 the gradient value for link

5-6 is higher than 5-4. As the traffic load increases in the network for both data (D) and multimedia (M) streams (path 2 and path 3, Fig. 8 and 9), at time 4 the link gradient for 5-6 reduces and is taken over by link 5-4, which causes the path to diverge to path 4 (1 – 5 – 4 – 10 – 11 – 12, Fig. 8). Therefore, this mechanism demonstrates how the gradient calculation is performed based on the evaluation of link load and the reaction diffusion process of diffusing load messages from the neighbouring nodes. Our simulation also includes tests on the self-management of resources, where Fig. 11 shows the amount of bandwidth consumed by each source and destination pair. The bandwidth consumption at the initial stage is largely from the routine traffic of the demand profile. Therefore, as described in section 4.1.1, this is compared to the body performing normal activity and using only the glycogen stored within the body. At time 13, a new traffic stream enters the network for edge routers S1 – D1. At this point, the resource manager permits the traffic stream to enter the network along the spare capacity links of the network.

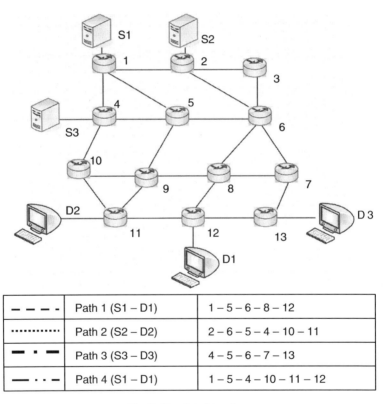

– – – –	Path 1 (S1 – D1)	1 – 5 – 6 – 8 – 12
············	Path 2 (S2 – D2)	2 – 6 – 5 – 4 – 10 – 11
— · —	Path 3 (S3 – D3)	4 – 5 – 6 – 7 – 13
— · · —	Path 4 (S1 – D1)	1 – 5 – 4 – 10 – 11 – 12

Fig. 8. Simulated topology

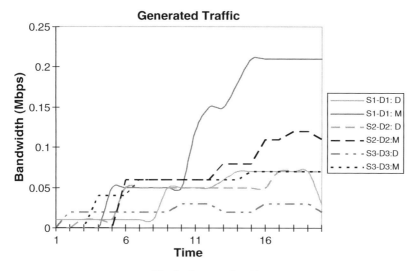

Fig. 9. Generated traffic

The spare capacity traffic is routed between time 14 and 20. According to the algorithm in Fig. 5, the minimum threshold of spare capacity usage $T_{SPARE\_USAGE} = 2$ time units and frequency of usage $N_{SPARE\_CYLES} = 3$ days will permanently include the new traffic stream to the routine demand traffic profile.

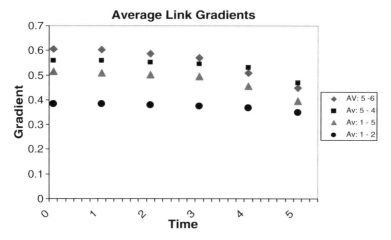

Fig. 10. Average link gradients

This is analogous to the body using the fat constantly, which in turn reduces the fat to transform to glycogen for higher intensity usage. By integrating this,

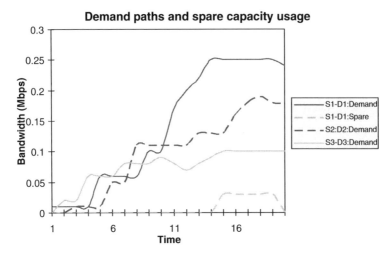

Fig. 11. Self-management of resources

the resource manager can match and compare the revenue of the demand traffic in response to resource subscribed by the ISP.

5 Conclusion

In this chapter, we have presented a bio-inspired framework for autonomic communication systems. The framework is based on integrating three key principle of self-governance (e.g. self-organisation, self-management, self-learning). A number of biological mechanisms meet the characteristic requirements of these self-governance principles. Therefore, we have extracted these principles from a development cycle perspective and combined them together into a framework. The aim of the framework is to apply these diverse mechanisms to different communications systems (e.g. infrastructure, ad hoc, services). We have applied the framework to a case study, which are self-management and self-organisation of network infrastructure. Main elements of the algorithms are illustrated.

Acknowledgement

This work has received support from Science Foundation Ireland via the *Autonomic Management of Communication Networks and Services* programme (grant no. 04/IN3/I4040C).

References

1. Shen W, Salemi B, Will P (2002) Hormone-Inspired Adaptive Communication and Distributed Control for CONRO Self-configurable Robots. IEEE Transaction on Robotics and Automation 18(5)
2. Greensted A, Tyrrell A (2004) An Endocrinologic-Inspired Hardware Implementation of a Multicellular System. In: NASA/DoD Conference on Evolvable Hardware, Seattle, USA
3. Kephart JO, Chess DM (2003) The Vision of Autonomic Computing. IEEE Computer Magazine 1:41-50
4. IBM (2003) An Architectural Blueprint for Autonomic Computing
5. Yagan D, Tham C-K (2005) Self-Optimizing Architecture for QoS Provisioning in Differentiated Services. In: Second International Conference on Autonomic Computing (ICAC' 05), Seattle, WA, USA
6. Prehofer C, Bettstetter C (2005) Self-Organization in Communication Networks: Principles and Paradigms, IEEE Communication Magazine 43(7):78-85
7. Strassner J, Agoulmine N, Lehtihet E (2006) FOCALE – A Novel Autonomic Networking Architecture. In: Latin American Autonomic Computing Symposium (LAACS 2006), Brazil
8. Suzuki J, Suda T (2005) A Middleware Platform for a Biologically Inspired Network Architecture Supporting Autonomous and Adaptive Applications. IEEE Journal on Selected Areas in Communications 23(2)
9. Nakano T, Suda Y (2005) Self-organizing Network Services with Evolutionary Adaptation. IEEE Transaction on Neural Networks 16(5)
10. Leibnitz K, Wakamiya N, Murata M (2006) Biologically inspired Self-Adaptive Multipath routing in overlay networks. Communications of the ACM 49(3)
11. Dressler F (2005) Efficient and Scalable Communication in Autonomic Networking using bio-inspired mechanisms – An Overview. Informatica 29(183):183 – 188
12. Kruger B, Dressler F (2005) Molecular Processes as a Basis for Autonomic Networking. IPSI Transactions on Advanced Research: Issues in Computer Science and Engineering 1(1):43-50
13. Raven P, Johnson G, Losos J, Singer S (2002) Biology, 6th edn, McGraw-Hill
14. Dhariwal A, Sukhatme GS, Requicha AAG (2004) Bacterium-inspired Robots for Environmental Monitoring. In: IEEE International Conference on Robotics and Autonomation, New Orleans, USA
15. Turing AM (1952) The Chemical basis of morphogenesis. Philosophical Transaction of the Royal Society 237:37-72
16. Miller MB, Bassler BL (2001) Quorum sensing in Bacteria. Annual Review in Micro-Biology 55:165-199
17. Balasubramaniam S, Botvich D, Donnelly W, Agoulmine N (2006) Applying Blood Glucose Homeostatic model towards Self-Management of IP QoS Provisioned Networks. In: 6th IEEE International Workshop on IP Operations and Management (IPOM 2006), LNCS, Dublin, Ireland, pp 84-95

Towards a Biologically-inspired Architecture for Self-Regulatory and Evolvable Network Applications

Chonho Lee, Hiroshi Wada and Junichi Suzuki

Department of Computer Science
University of Massachusetts, Boston
{chonho, shu, jxs}@cs.umb.edu

Summary The BEYOND architecture applies biological principles and mechanisms to design network applications that autonomously adapt to dynamic environmental changes in the network. In BEYOND, each network application consists of distributed software agents, analogous to a bee colony (application) consisting of multiple bees (agents). Each agent provides a particular functionality of a network application, and implements biological behaviors such as energy exchange, migration, reproduction and replication. This paper describes two key components in BEYOND: (1) a self-regulatory and evolutionary adaptation mechanism for agents, called iNet, and (2) an agent development environment, called BEYONDwork. iNet is designed after the mechanisms behind how the immune system detects antigens (e.g., viruses) and produces antibodies to eliminate them. It models a set of environment conditions (e.g., network traffic) as an antigen and an agent behavior (e.g., migration) as an antibody. iNet allows each agent to autonomously sense its surrounding environment conditions (i.e., antigens) and adaptively invoke a behavior (i.e., antibody) suitable for the conditions. In iNet, a configuration of antibodies is encoded as a gene. Agents evolve their antibodies so that they can adapt to unexpected environmental changes. iNet also allows each agent to detect its own deficiencies to detect antigen invasions (i.e., environmental changes) and regulate its policy for antigen detection. Simulation results show that agents adapt to changing network environments by self-regulating their antigen detection and evolving their antibodies through generations. BEYONDwork provides visual and textual languages to design agents in an intuitive manner.

1 Introduction

Large-scale network applications such as data center applications and grid computing applications face several critical challenges, particularly *autonomy* and *adaptability*, as they have been increasing in complexity and scale[1]. They are expected to autonomously adapt to dynamic environmental changes in the network (e.g., workload surges and resource extinction) in order to improve user experience, expand applications' operational longevity and reduce maintenance cost [3, 4, 5, 6].

---

[1] For example, Google, Inc. reportedly runs over 450,000 servers in its data centers [1, 2].

Based on an observation that various biological systems have developed the mechanisms necessary to meet the above challenges, the proposed architecture, called BEYOND[2], applies key biological principles and mechanisms to design autonomous and adaptive network applications. In BEYOND, each application is designed as a decentralized group of software agents. This is analogous to a bee colony (application) consisting of multiple bees (agents). Each agent provides a particular functionality of a network application, and implements biological behaviors such as energy exchange, migration, replication, reproduction and death.

This paper focuses on two key components in BEYOND: (1) a self-regulatory and evolutionary adaptation mechanism for agents, called iNet, and (2) an agent development environment, called BEYONDwork. iNet is designed after the mechanisms behind how the immune system detects antigens (e.g., viruses), how it specifically produces antibodies to eliminate them and how it evolves antibodies to react a massive number of antigens. iNet models a set of environment conditions (e.g., network traffic and resource availability) as an antigen and an agent behavior as an antibody. Each agent contains its own immune system (i.e., iNet), and a configuration of the agent's antibodies defines its behavior policy, which determines which behavior to be invoked in a given set of environment conditions. iNet allows each agent to autonomously sense its surrounding environment conditions (i.e., antigen) for evaluating whether it adapts well to the sensed conditions, and if it does not, adaptively invoke a behavior (i.e., antibody) suitable for the conditions. For example, agents may invoke the replication behavior at the network hosts that accept a large number of user requests for their services. This leads to the adaptation of agent availability; agents can improve their throughput. Also, agents may invoke the migration behavior to move toward the network hosts that receive a large number of user requests for their services. This results in the adaptation of agent locations; agents can improve their response time to user requests.

iNet also allows each agent to detect its own deficiencies to detect antigen invasions, i.e., deficiencies to evaluate whether it adapts well to the current environment conditions. Due to the deficiencies, some agents may invoke their behaviors when they have already adapted well to the current environment conditions. Others may invoke no behaviors when they do not adapt to the current environment conditions. With iNet, each agent can regulate its policy for antigen detection so that it can perform its behaviors at the right time. This self-regulation process is intended to avoid the degradation of agent adaptability and the waste of resource consumption incurred by unnecessary behavior invocations.

In iNet, a configuration of antibodies (i.e., behavior policy) is encoded as a gene. Agents evolve their antibody configurations so that the configurations become fine-tuned to the current and even unexpected environment conditions. This evolution process occurs via genetic operations such as mutation and crossover, which alter antibody configurations (genes) during agent reproduction and replication. Agent evolution frees agent developers from anticipating all possible environmental

---

[2] Biologically-Enhanced sYstem architecture beyond Ordinary Network Designs

changes and tuning their agents' antibodies (behavior policies) to the changes at design time. This significantly simplifies the implementation of agents.

Simulation results show that iNet allows agents to autonomously adapt to changing network environments by dynamically regulating their antigen detection and evolving their antibodies through generations.

The second focus of this paper is an application development environment for iNet, called BEYONDwork. BEYONDwork provides visual and textual languages to design agent in an intuitive and easy-to-understand manner. It accepts the visual models and textual programs built with the proposed languages, and transforms them to Java code that are compilable and runnable on a simulator for BEYOND. This code generation enables rapid development and configuration of agents, thereby improving the productivity of agent developers.

2 Design Principles in the BEYOND Architecture

In BEYOND, agents are designed based on the six principles described below.

- **Decentralization:** Inspired by biological systems (e.g., bee colony), there are no central entities to control and coordinate agents in BEYOND so that they can be scalable and simple by avoiding a single point of performance bottlenecks [7] and failures [8] and by avoiding any central coordination in deploying agents [9].
- **Autonomy:** Similar to biological entities (e.g., bees), agents sense their local network environments, and based on the sensed environment conditions, they autonomously behave and interact with each other without any intervention from/to other agents and human users.
- **Emergence:** In biological systems, collective (group) behaviors emerge from local interactions of autonomous entities [10]. In BEYOND, agents only interact with nearby agents. They behave according to dynamic changes in environment conditions such as user demands and resource availability. Through collective behaviors and interactions of individual agents, desirable system characteristics (e.g., load balancing and resource efficiency) emerge in a swarm of agents.
- **Lifecycle:** Biological entities strive to seek and consume food for living. In BEYOND, agents store and expend *energy* for living. Each agent gains energy in exchange for performing its service to other agents or human users, and expends energy to use network and computing resources (e.g., bandwidth and memory). The abundance or scarcity of stored energy affects agent lifecycle. For example, an energy abundance indicates high demand to an agent; thus, the agent may be designed to favor reproduction or replication to increase its availability. An energy scarcity (i.e., an indication of lack of demand) causes death of the agent.
- **Homeostasis:** Biological entities regulate their internal environments to maintain stable conditions (e.g., stable body temperature and blood fluid) even though external environments change. Similarly, in BEYOND, agents strive to maintain the fitness (or the degree of adaptation) to external network environments. When an agent finds that its fitness decreases, it adjusts its antigen detection policy so that it can keep its fitness to dynamic network environments.

- **Evolution:** Biological entities evolve as a species to increase the fitness to the environment across generations. In BEYOND, agents collectively evolve their antibody configurations (behavior policies) across generations. Agents perform this evolution process by generating behavioral diversity and executing natural selection. Behavioral diversity means that different agents possess different antibody configurations (behavior policies). This is generated via mutation and crossover during agent replication and reproduction. Natural selection retains the agents that adapt well to the environment (i.e., the agents that have beneficial/effective behavior policies suitable for the environment) and eliminate the agents that does not adapt to the environment (i.e., the agents that have detrimental/ineffective behavior policies).

3 Agent Structure and Behaviors

Each agent consists of *attributes*, *body* and *behaviors*. Attributes carry descriptive information regarding an agent (e.g., agent ID and energy level). Body implements a functional service an agent provides. For example, an agent may implement a web service in a data center, while another may implement a scientific simulation model in a grid computing system. Behaviors implement the actions inherent to all agents:

- **Migration:** Agents may move between network hosts.
- **Energy exchange and storage:** Agents may gain energy in exchange for providing their services to other agents or users. They may also expend energy for services that they receive from other agents and for resources available at the local network host (e.g., memory space).
- **Replication:** Agents may make their copies in response to higher energy level, which indicates higher demand for the agents. A replicated agent is placed on the host that its parent agent resides on, and it inherits the parent's antibody configuration (behavior policy) as well as the half amount of the parent's energy level. Mutation may occur on the inherited antibody configuration.
- **Reproduction:** Agents may reproduce child agents with other agents (mating partners) running on their local hosts. A child agent is placed on the host that its parents reside on, and it inherits antibody configurations (behavior policies) from both parents through crossover. Each parent gives a child agent the quarter amount of its energy level. Mutation may occur on the antibody configuration of a child agent.
- **Communication**: Agents may communicate with each other for the purposes of, for example, requesting services, exchanging energy units or reproducing child agents.
- **Death**: Agents die due to energy starvation. If an agent cannot balance its energy expenditure with its energy gain, the agent cannot pay for the resources it needs; thus, it dies from lack of energy. When an agent dies, all resources allocated to the agent are released.

Agents expend a certain amount of energy units to invoke each behavior (i.e., behavior cost) except the death behavior.

4 iNet: Agent Adaptation Mechanism in BEYOND

This section overviews how the natural immune system works (Section 4.1) and describes how iNet is designed after the natural immune system (Section 4.2).

4.1 Natural Immune System

The immune system is an adaptive defense mechanism to regulate the body against dynamic environmental changes such as antigen invasions. Through a number of interactions among various white blood cells (e.g., macrophages and lymphocytes) and molecules (e.g., antibodies), the immune system evokes two responses to antigens: *innate* and *adaptive* responses.

In the innate response, the immune system performs self/non-self discrimination. This response is initiated by macrophages and T-cells, a type of lymphocytes. Macrophages move around the body to ingest antigens and present them to T-cells. T-cells are produced in thymus that performs the negative selection. In the negative selection process, thymus removes T-cells that strongly react with the body's own (self) cells. The remaining T-cells are used as detectors to identify foreign (non-self) cells. When a T-cell(s) detects a non-self antigen presented by a macrophage, the T-cell(s) secrete chemical signals to induce the adaptive response.

In the adaptive response, B-cells, another type of lymphocytes, are activated by T-cells. Some of the activated B-cells strongly react to an antigen, and they produce antibodies that specifically kill the antigen. Antibodies form a network and communicate with each other [11]. This immune network is formed with stimulation and suppression relationships among antibodies. With these relationships, antibodies dynamically change their populations and network structure. For example, the population of specific antibodies rapidly increases following the detection of an antigen and, after eliminating the antigen, decreases again.

In order to react a massive number of antigens, the immune system needs to be able to generate a variety of antibodies. A primary repertoire of antibodies is approximately 10^9 using immune genes. B-cells can increase this repertoire further by mutating and recombining immune gene segments so that antibodies can bind an unlimited number of antigens [12].

The immune system regularly encounters anomalies such as immunodeficiency and autoimmunity. Immunodeficiency is a phenomenon that the immune system fails to detect non-self antigens and produce antibodies to eliminate them. Autoimmunity is a phenomenon that the immune system recognizes the constituent self cells as non-self. This results in self-attacks via overreaction of the immune system. When the immune system faces such anomalies, it is alerted with danger signals by cells damaged by the anomalies [13]. Currently, two types of danger signals are known: uric acid [14, 15] and heat shock proteins (HSP) [16, 17]. Uric acid is produced in response to immunodeficiency, and it stimulates macrophages so that T-cells detects non-self antigens properly. This accelerates the production of antibodies. HSP is produced in response to autoimmunity. HSP reforms broken proteins in macrophages and T-cells so that they stop attacking self cells. This suppresses antibody production.

4.2 iNet Artificial Immune System

The iNet artificial immune system consists of the environment evaluation (EE) facility and behavior selection (BS) facility, which implement the innate and adaptive immune responses, respectively (Figure 1). The EE facility allows an agent to continuously sense a set of current environment conditions as an antigen and classify the antigen to self or non-self. A self antigen indicates that the agent adapts to the current environment conditions well, and a non-self antigen indicates it does not. When the EE facility detects a non-self antigen, it activates the BS facility. The BS facility allows an agent to choose a behavior as an antibody that specifically matches with the detected non-self antigen.

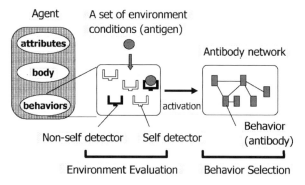

Fig. 1. Design of iNet Adaptation Mechanism

Environment Evaluation Facility

The EE facility performs two steps: initialization and self/non-self classification. The initialization step produces detectors that identify self and non-self antigens. Each antigen is represented as a feature vector (X), which consists of a set of environment conditions, or features, (F_i) and a class value (C):

$$X = (F_1, F_2, ..., F_n, C) \qquad (1)$$

C indicates whether a given antigen (i.e., a set of environment conditions) is self (0) or non-self (1). If an agent senses resource utilization and workload (the number of user requests) on the local host, an antigen is represented as follows.

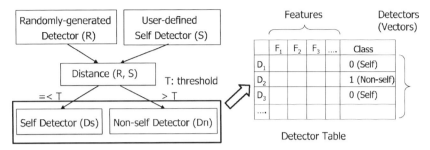

Fig. 2. Initialization Step in the EE Facility

$$X_{current} = ((Low : ResourceUtilization, Low : Workload), 0) \qquad (2)$$

The initialization step in the EE facility is designed after the negative selection process in the immune system (Figure 2). As the immune system randomly generates T-cells first, the EE facility generates detectors (feature vectors) randomly. Then, the EE facility separates the detectors into self detectors, which closely match with self antigens, and non-self detectors, which do not closely match with self antigens. This separation is performed via similarity measurement between randomly generated feature vectors (X) and self antigens (S) that human users supply. After the vector matching, both self and non-self detectors are stored in the detector table (Figure 2)[3].

The second step in the EE facility is self/non-self classification of an antigen (a set of current environment conditions). It is performed with a decision tree built from detectors in the detector table and classifies an antigen into self or non-self[4] The decision tree is built using the information gain technique [18]. First, consider one node as a root of decision tree, and it contains all detectors in the detector table. Then, divide the detectors based on one of feature into two subsets of detectors (Assume that each feature has two distinct values.) Each subset goes to one of two child nodes. If all detectors in the subset have the same class value, then the node becomes a leaf node with the class value; otherwise, divide the subset again based on one of the other features into the subsets. Information gain technique suggests how to select a feature at each dividing step so that the number of paths to leaf nodes and the height of tree can be minimized.

Figure 3 shows an example of decision tree. Each node in the tree specifies which feature (environment condition) is considered. Based on the feature values in a given

---

[3] The immune system removes non-self detectors through negative selection. However, in iNet, both self and non-self detectors are used to perform self/non-self classification.

[4] The reasons for using decision trees as an antigen classifier are implementation simplicity and algorithmic efficiency. Decision trees perform classification much faster than other algorithms such as clustering, support vector machine and Markov model algorithms [18]. The efficiency of classification is one of the most important requirements in iNet because each agent periodically senses and classifies its surrounding environment conditions.

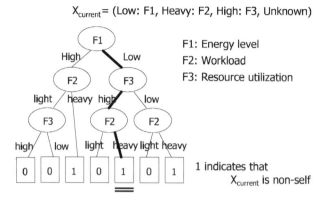

Fig. 3. An Example Decision Tree

antigen, the EE facility travels through tree branches. If the EE facility classifies the antigen to non-self, it activates the BS facility.

Behavior Selection Facility

The BS facility selects an antibody (i.e., agent's behavior) suitable for the detected non-self antigen (i.e., environment conditions). Each antibody consists of three parts: a *precondition* under which it is selected, *behavior ID* and *relationships* to other antibodies. Antibodies are linked with each other using stimulation and suppression relationships. Each antibody has its own concentration value, which represents its population. The BS facility identifies candidate antibodies (behaviors) suitable for a given non-self antigen (environment conditions), prioritizes them based on their concentration values, and selects the most suitable one from the candidates. When prioritizing antibodies (behaviors), stimulation relationships between them contribute to increase their concentration values, and suppression relationships contribute to decrease it. Each relationship has an affinity value, which indicates the degree of stimulation or suppression.

Figure 4 shows a generalized network of antibodies. The antibody i stimulates M antibodies and suppresses N antibodies. m_{ji} and m_{ik} denote affinity values between antibody j and i, and between antibody i and k. m_i is an affinity value between an antigen and antibody i. The concentration of antibody i, denoted by a_i, is calculated with the following equations.

$$\frac{dA_i(t)}{dt} = \left(\frac{1}{N} \sum_{j=1}^{N} m_{ji} \cdot a_j(t) - \frac{1}{M} \sum_{k=1}^{M} m_{ik} \cdot a_k(t) + m_i - k \right) \cdot a_i(t) \quad (3)$$

$$a_i(t) = \frac{1}{1 + exp(0.5 - A_i(t))} \quad (4)$$

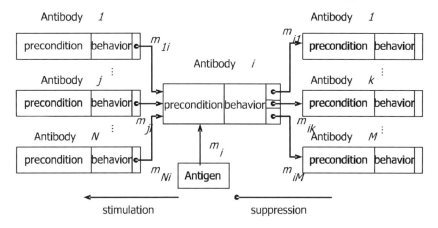

Fig. 4. A Generalized Antibody Network

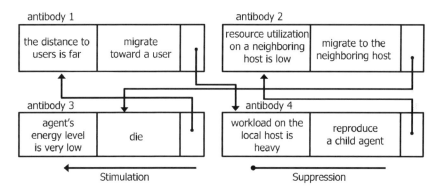

Fig. 5. An Example Antibody Network

In Equation (3), the first and second terms in a bracket denote the stimulation and suppression from other antibodies. m_{ji} and m_{ik} are positive between 0 and 1. m_i is 1 when antibody i is stimulated directly by an antigen, otherwise 0. k denotes the dissipation factor representing the natural death of an antibody. Equation (4) is a sigmoid function used to squash the $A_i(t)$ value between 0 and 1.

Every antibody's concentration is calculated 200 times repeatedly. This repeat count is obtained from a previous simulation experience [19]. If no antibody exceeds a predefined threshold during the 200 calculation steps, the antibody whose concentration value is the highest is selected (i.e., winner-tales-all selection). If one or more antibodies' concentration values exceed the threshold, an antibody is selected based on the probability proportional to the concentration values (i.e., roulette-wheel selection).

Figure 5 shows an example network of antibodies. It contains four antibodies, which represent the migration, replication and death behaviors. Antibody 1 represents the migration behavior invoked when the distance to users is far from an agent. Antibody 1 suppresses Antibody 3 and stimulate Antibody 4. Now, suppose that a (non-self) antigen indicates (1) the distance to users is far, (2) workload is heavy on the local host and (3) resource utilization is low on a neighboring platform. This antigen stimulates Antibodies 1, 2 and 4 simultaneously. Their populations increase, and Antibody 2's concentration value becomes highest because Antibody 2 suppresses Antibody 4, which in turn suppresses Antibody 1. As a result, the BS facility would select Antibody 2.

Evolution of Antibodies

As Section 4.1 describes, the immune system diversifies antibodies by mutating immune genes so that antibodies can react to unanticipated antigens. Similarly, iNet diversifies antibodies via gene operations such as mutation and crossover so that agents can adapt to unanticipated environment conditions. In iNet, each agent encodes and possesses its own antibody configuration (behavior policy) as a set of genes (genotype). The agent genotype consists of the antibody genes, which specify the presence of antibodies, and the affinity genes, which specify relationships among antibodies and their affinity values. When a new agent is born through a replication or reproduction process, it interprets the genes given by its parent(s) and form an antibody network. Figure 6 shows an example genotype and phenotype.

Each agent periodically keeps track of its *fitness*, which quantifies how much it adapts to the the current environment conditions. Agents strive to increase their fitness values by altering their genes through generations. Fitness is calculated as a weighted sum of fitness factors (f_i):

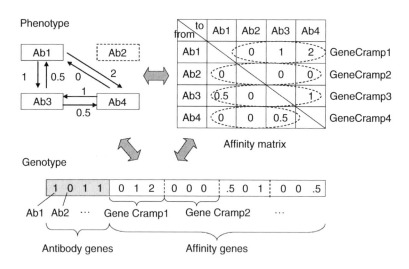

Fig. 6. Phenotype and Genotype of Antibody Network

A Biologically-inspired Architecture for Evolvable Network Applications   31

$$Fitness = \sum w_i \cdot f_i \qquad (5)$$

Currently, iNet considers the following six fitness factors. Each factor value is non-negative between 0 and 1.

- **Response time** (f_1): R is the time for each agent to process a single user request. RT is the total of R and the time for a user request and agent response to travel between a user and agent.

$$f_1 = \frac{R}{RT} \qquad (6)$$

- **Throughput** (f_2): indicates how many user requests agents process.

$$f_2 = \frac{\# \ of \ user \ requests \ processed \ by \ all \ agents}{Total \ \# \ of \ user \ requests} \qquad (7)$$

- **Energy utility** (f_3): indicates the rate of an agent's energy expenditure to its energy gain during its lifetime.

$$f_3 = 1 - \frac{Energy \ expenditure \ during \ lifetime}{Energy \ gain \ during \ lifetime} \qquad (8)$$

- **Load balance** (f_4): indicates how user requests (workload) are distributed over agents. m denotes the number of user requests that an agent processes in a unit time. μ_m denotes the expected average number of user requests that each agent is expected to process. M_{max} denotes the maximum number of user requests that an agent can process in a unit time.

$$f_4 = 1 - \frac{|m - \mu_m|}{M_{max}} \ where \ \mu_m = \frac{Total \ \# \ of \ user \ requests}{Total \ \# \ of \ agents} \qquad (9)$$

- **Resource utilization balance** (f_5): indicates how resource utilization is distributed over hosts. r denotes the resource utilization rate on the local host that an agent resides on. This is measured as the ratio of the amount of resources consumed by agents on the host to the amount of resources available on the host. μ_r denotes the expected average of resource utilization rate over all hosts that agents reside on.

$$f_5 = 1 - |r - \mu_r| \ where \ \mu_r = \frac{Sum \ of \ resource \ utilization \ rate \ on \ all \ hosts}{\# \ of \ hosts \ that \ agents \ resides \ on} \qquad (10)$$

- **Age** (f_6): denotes the lifetime of an agent. S is the total simulation time.

$$f_6 = \frac{Lifetime \ of \ an \ agent}{S} \qquad (11)$$

Upon invoking the reproduction behavior, each agent searches mating partner candidates whose fitness values are higher than the agent's fitness value. The candidates are searched on the local host. If the agent cannot find any candidates, it

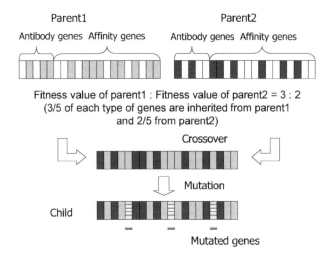

Fig. 7. A Genetic Operations

performs the replication behavior rather than the reproduction behavior. This mating partner selection contributes to increase the population of agents that provide services in higher demand and maintain higher fitness.

In reproduction, two parent agents contribute their genes, via crossover, to a child agent. The amount of their gene contributions follow the ratio of their fitness values. For example, in Figure 7, the fitness value ratio is 3:2 between the parent agent 1 and 2. Thus, the parent agent 1 contributes 60% (3/5) of its genes to a child agent, and the parent agent 2 contributes the rest (2/5). In replication, a parent agent contributes its whole genes to a child agent. In both reproduction and replication, mutation may occur on the genes of a child agent in a certain probability (mutation rate). A reproduced child inherits the quarter amount of energy units from each parent, and a replicated child inherits the half of energy units from its parent.

Self-regulation Process

As Section 4.1 describes, the immune system regulates itself with danger signals when it detects anomalies. Similarly, iNet allows each agent to detect its own deficiencies to recognize antigens, i.e., deficiencies to evaluate whether it adapts well to the current environment conditions. Due to the deficiencies, some agents may invoke their behaviors when they adapt well to the current environment conditions. Others may not when they do not adapt to the current environment conditions. With iNet, each agent can adjust its policies for antigen recognition so that it can perform its behaviors at the right time.

Figure 8 describes the flow of the self-regulation process. Corresponding to danger signals such as Uric acids and Heat shock proteins, each agent responds to two types of signals. *Signal 1* is produced when the current fitness decreases

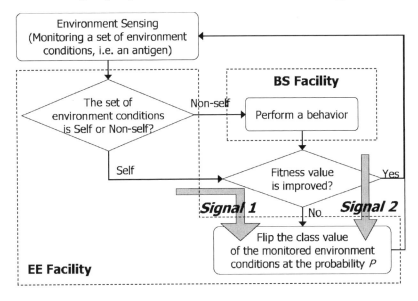

Fig. 8. A Self-regulation Process

by classifying the current environment conditions as self and by not performing any behaviors even though an agent does not adapt well to the conditions (this corresponds to immunodeficiency). *Signal 2* is produced when the current fitness decreases by classifying the current environment conditions as non-self and by performing inappropriate behaviors although there is no necessity to perform behaviors because an agent adapts well to the conditions (this corresponds to autoimmunity).

When an agent receives either of signals, it flips the class value (self ↔ non-self) of the detector, which indicates the miss-classified environment conditions. The strength of the danger signals is represented as a probability, P, that an agent flips the class value. The probability is calculated the weighted sum of the agent's previous fitness, F_{t-1} and the decay of the current fitness, $F_{t-1} - F_t$ as follow:

$$P = \alpha * F_{t-1} + (1 - \alpha) * (F_{t-1} - F_t) \tag{12}$$

5 Simulation Results

This section presents several simulation results to evaluate the autonomous adaptability of agents (network applications). The simulations are carried out on the BEYOND simulator. Figure 9 shows a simulated network as a server farm consisting of network hosts connected in a 10x10 grid topology. BEYOND platform is running on each network host, and each agent implements a web service. Service requests travel from users to agents via user access point. This simulation study assumes that a single (emulated) user runs on the access point and sends service requests to agents. When

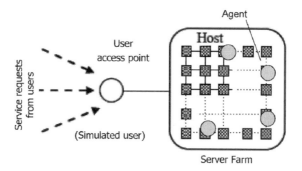

Fig. 9. A Simulated Network

Agent Type	The EE Facility		The BS Facility	
	Tree configuration	Regulation	Antibody network configuration	Evolution
R.ee + R.bs	Random	No	Random	No
R.EE + R.BS	Random	Yes	Random	Yes
M.EE + M.BS	Manual	Yes	Manual	Yes

Fig. 10. Agent Type

a user issues a service request, request messages are broadcasted to search a target agent that can process the issued service requests.

In order to investigate how the self-regulation (the EE facility) and evolution process (the BS facility) impact the adaptability of agents, three different types of agents, described in Figure 10, are evaluated. (1) *R.ee+R.bs*, an agent with a randomly configured tree in the EE facility and a randomly configured antibody network in the BS facility does not perform a regulation and evolution process, (2) *R.EE+R.BS*, an agent with a randomly configured tree and a randomly configured antibody network does dynamically perform a regulation and evolution process, and (3) *R.ee+R.bs*, an agent with a manually tuned tree and a manually tuned antibody network; and does also dynamically perform a regulation and evolution process.

At the beginning of simulations, four agents are randomly deployed on the network. This simulation generates a random workload for web service agents as described in figure 11. The workload trace is designed based on a daily request rate for the www.ibm.com site in February, 2001 [20]. The request rate peaks to about 5,500 requests per min in the morning and 9,000 requests per min in the evening.

Figure 12 shows how agents autonomously adapt their population to workload changes. When agents receive requests, they start to provide their service for users and gain more energy from users. Agents (M.EE+M.BS) successfully adapt their population in timely manner. For example, at 3:00, 6:00, and 15:00, when the workload surges, they increase their population by performing replication or reproduction

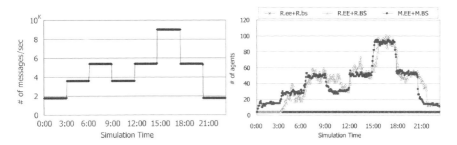

Fig. 11. Workload **Fig. 12.** Population of Agents

behavior; subsequently, at 9:00 and 18:00, when the workload drops, they immediately decrease their population by performing death behavior. On the other hand, agents (R.ee+R.bs) did not perform any behaviors because their EE facility classified environment conditions (e.g., workload is high) as self and also their self-regulation process is never executed; so they could not adapt their population. However, agents (R.EE+R.BS) dynamically regulate their policies for environment evaluation in the EE facility so that they perform behaviors to adapt their population. Especially, after 3:00, agents start to update the EE facility as a response to danger signals; subsequently, although they could not immediately reduce their population when the workload drops at 9:00, they adjust the EE facility and successfully perform death behavior at about 11:00. In addition, agents (R.EE+R.BS) may invoke inappropriate behaviors (not suitable for the current environment conditions) in the early stage of simulation due to a randomly configured antibody network. For example, between 3:00 and 6:00, the plotted line for agent population is swinging. This implies that some agents keep invoking death behaviors inappropriately. But, in the late stage of simulation, the plotted line is almost following that of manually configured agents (M.EE+M.BS). It follows that agents tried to dynamically adjust their antibody network by performing reproduction behavior. They reproduce children having the adapted antibody network by which appropriate behaviors are selected in timely manner.

Figure 13 shows how agents autonomously adapt response time for a user. At the beginning of simulation, response time becomes very high because only four agents process 2,000 requests a minute and a distance between the agent and users is long. However, after agents (M.EE+M.BS) accumulate enough energy from users and start migrating towards users or replicating themselves, they rapidly decrease response time. For example, at 3:00, 6:00, and 15:00, the response time dramatically spikes (up to about 7 sec) due to the workload surges, but the agents reduce the response time quickly by adapting their population as described in Figure 12. On the other hand, agents (R.ee+R.bs) cannot reduce their response time at all because they could not perform any behaviors. However, when agents (R.EE+R.BS) recognize that the EE facility wrongly evaluated the environment conditions (i.e., receive danger signals) at 3:00, they tried to regulate evaluation policies in the EE facility. In

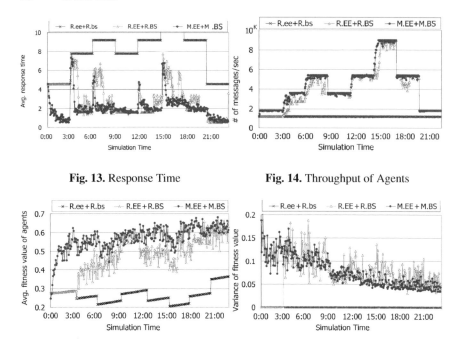

Fig. 13. Response Time

Fig. 14. Throughput of Agents

Fig. 15. Average Fitness Value

Fig. 16. Variance of Fitness Value

addition, the reproduced children keep adjusting their antibody network to perform behaviors suitable for the current environment conditions in timely manner.

Figure 14 also shows how three different types of agents dynamically adapt their throughput. It is measured as the number of responses that users receive a minute from agents. Agents (M.EE+M.BS) successfully maintain high throughput by dynamically adapting their locations and population through migration and reproduction behaviors while agents (R.ee+R.bs) could not improve their throughput because the agents did not migrate or replicate at all. Until 3:00, agents (R.EE+R.BS) also could not improve their throughput. However, after 3:00, the agents regulate the behavior invocation by dynamically updating a tree in the EE facility, and the reproduced children adjust an antibody network in the BS facility to invoke appropriate behaviors in timely manner. As the result, they increase their throughput (i.e., tried to reply all user requests in timely manner).

Figure 15 shows the average fitness value of agents (i.e., the degree of adaptation to the environment) as described in Section 4.2. Agents (M.EE+M.BS) dynamically improve their fitness value to about 0.6-0.7 while other agents (R.ee+R.bs) could not improve the fitness value (although the fitness value slightly increases because of energy utility (f3) and age (f6) factors). On the other hand, agents (R.EE+R.BS) keep trying to improve their fitness value after 3:00. In early stage of simulation, because their iNet configuration is not optimized yet, the plotted line for fitness value is swinging (i.e., the fitness value easily drops) compared to that of manually configured agents (M.EE+M.BS). Some of agents with high fitness value might die unexpectedly. However, in the late stage of simulation, the trace of their fitness value eventually close in that of manually configured agents. It follows that self-regulation and evolution process contribute for agents to autonomously improve their adaptability by dynamically tuning their iNet configurations.

Finally, Figure 16 shows the variance of agents' fitness values; that is, how the fitness values are spread around the average. The variance for agents (M.EE+M.BS) has gradually converged. The lower variance implies that every agents' fitness values are close to each other. Together with the results in figure 15, figure 16 explains that most agents improve their fitness values at the same time. This concludes that the optimal or adapted antibody network is successfully spread out to other surviving agents by evolution; thus, agents adapts to the environment conditions well through generations. Agents (R.EE+R.BS) also reduce the variance gradually. However, the plotted line is unstable compared to that of manually configured agents. Some agents still posses non-adapted antibody network and then invoke inappropriate behaviors. This may kill agents with high fitness value and make the spread speed of the adapted iNet configuration slow.

6 BEYONDwork: Agent Development Environment in BEYOND

BEYONDwork is an application development environment for iNet. It provides two visual modeling languages and a textual programming language for configuring environment conditions, detectors and behavior policies. Figure 17 shows the iNet configuration process with BEYONDwork. BEYONDwork consists of four facilities:

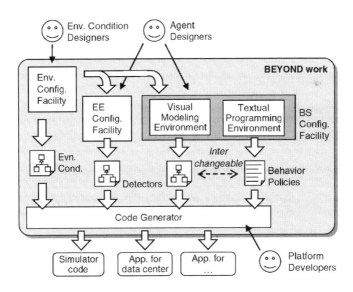

Fig. 17. iNet Configuration with BEYONDwork

environment configuration facility, EE configuration facility, BS configuration facility and code generator. The environment configuration facility allows environment condition designers to configure environment conditions with a visual language. The EE configuration facility allows agent designers to configure a set of detectors to identify self and non-self antigens (see Section 4.2) based on environment conditions configured in the environment configuration facility. The BS configuration facility allows agent designers to configure their agents' behavior policies (antibody configuration) with visual or textual languages. Both languages have the same level of expressiveness, and the artifacts of the languages (models and programs) are transparently translatable with each other. Agent designers can configure the behavior policy of each agent through the use of either language.

Once environment conditions, detectors and a behavior policy are complete in the form of visual models or textual programs, the code generator transforms them to compilable source code by following a transformation rule between the languages and source code. The transformation rules are implemented by platform developers, who know the details of platform technologies. (e.g., operating systems, middleware, simulators and programming languages) Through changing one transformation rule to another, the code generator can generate source code that are compatible with different deployment environments such as simulators and real networks. Environment condition designers and agent designers do not have to write different models/programs for the same agent running on different platform technologies. This flexible code generation feature improves the productivity of agent designers. Currently, BEYONDwork supports Java code generation for a simulator in BEYOND.

Figure 18 shows the environment configuration facility. As Figure 18 illustrates, the visual language visualizes an environment condition as a rectangle. Each rectangle can contain arbitrary number of rounded rectangles representing categories of a corresponding environment condition. For example, in Figure 18, the `LocalWorkload` environment condition has two categories, `HEAVY` and `LIGHT`. Also, each category specifies a condition to classify a corresponding environment

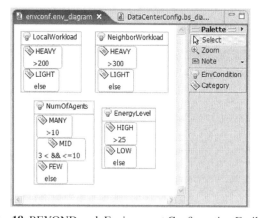

Fig. 18. BEYONDwork Environment Configuration Facility

condition. In iNet, each environment condition is supposed to have one representative value, e.g., workload value or the number of agents, and the representative value is used to identify the category of a corresponding environment condition. For example, in Figure 18, the `LocalWorkload` environment condition is classified as `HIGH` when its representative value exceeds 200, otherwise classified as `LIGHT`.

The details of representative values, e.g., how and when to obtain the value, are hidden from environment condition designers and agent designers. Platform developers implement such details on skeleton code generated by the code generator. For example, the following is a fragment of Java source code generated from the `LocalWorkload` environment condition in Figure 18. By implementing the `getRepValue` method, e.g., returns a request rate, CPU load, or the summation of them, a representative value of an environment condition is retrieved and evaluated against conditions specified by each category. The environment configuration facility is implemented on Eclipse Graphical Modeling Framework (GMF)[5]. The transformations from visual models and Java source code are implemented with a model-code transformation engine in openArchitectureware[6].

```
class LocalWorkload extends EnvCondition{
  enum Category{ HIGH, LOW };
  public Category evaluate(){
    double repValue = getRepValue();
    if( repValue > 200 ){ return Category.HIGH; }
    else{ return Category.LOW; }
  }
  private double getRepValue(){
    // TODO: platform developers add code here
  }
}
```

Figure 19 shows the EE configuration facility appears as one of windows in BEYONDwork, located below the BS configuration facility (Figure 20). Each column in the table represents an environment condition configured in the environment configuration facility (Figure 18) and each row represents a detector. The facility allows agent designers to add and remove detectors, and configure detectors by selecting the categories of each environment condition. For example, in Figure 18, the `NumOfAgents` environment condition is configured to have three categories, `MANY`, `MID` and `FEW`, and cells in the corresponding column in Figure 19 allows agent designers to select its value from `MANY`, `MID` and `FEW`. From a set of detectors, BEYONDwork automatically create a decision tree and deploys it in agents (see Section 4.2).

Figure 20 and 21 show the visual modeling and textual programming environments in the BS configuration facility, respectively. As Figure 20 illustrates, the visual language visualizes an antibody as a rounded rectangle. Each rectangle consists of three compartments: (1) the name and the initial concentration of an antibody,

---

[5] www.eclipse.org/gmf/
[6] www.openarchitectureware.org

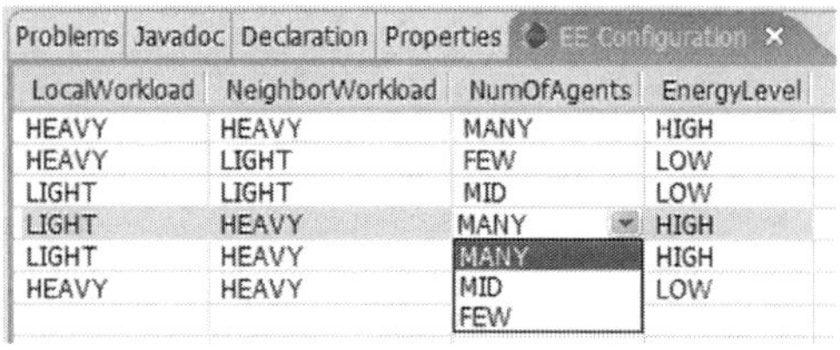

Fig. 19. BEYONDwork EE Configuration Facility

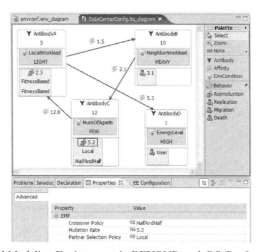

Fig. 20. Visual Modeling Environment in BEYONDwork BS Configuration Facility

(2) an environment condition to which an antibody reacts, and (3) an agent behavior and its properties. For example, in Figure 20, `AntibodyA`'s initial concentration value is 5, and it represents the reproduction behavior. The behavior is invoked when `LocalWorkload` is high. A stimulation/suppression relationship between antibodies is visualized as an solid arrow between rounded rectangles. Each arrow has value, which represents affinity value of a relationship. As Figure 20 demonstrates, the visual language supports all the concepts in antibody configurations as built-in model elements, and agent designers can configure antibodies (agent behavior policies) in an intuitive and rapid manner. The BS configuration facility is implemented on GMF and openArchitectureWare as well as the environment configuration facility.

In the textual language (Figure 21), each antibody is defined with the built-in keyword `antibody`. The program in Figure 21 and the model in Figure 20 define the semantically same antibody configuration. As Figure 21 shows, the textual programming environment in BEYONDwork shows built-in keywords in boldface, automatically performs a syntax check, and reports syntax errors while antibody designers configure antibodies. In Figure 21, a syntax error is reported as a cross mark. (The textual language does not support keyword `energylevel` but `EnergyLevel` because of the environment conditions defined in Figure 18.) The textual programming environment in the BS configuration facility is implemented on Eclipse. The

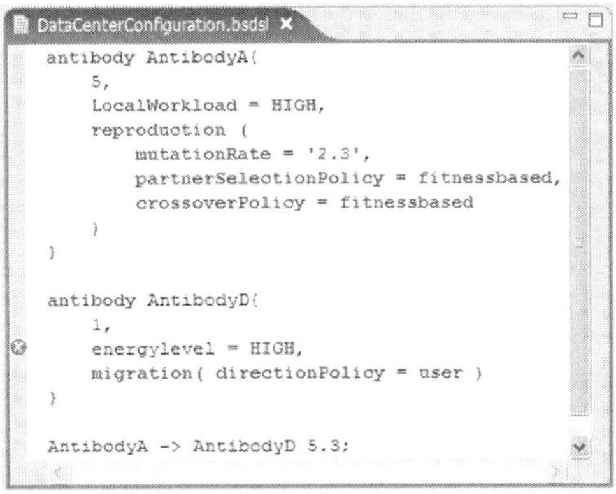

A Biologically-inspired Architecture for Evolvable Network Applications   41

```
antibody AntibodyA{
    5,
    LocalWorkload = HIGH,
    reproduction (
        mutationRate = '2.3',
        partnerSelectionPolicy = fitnessbased,
        crossoverPolicy = fitnessbased
    )
}

antibody AntibodyD{
    1,
    energylevel = HIGH,
    migration( directionPolicy = user }
}

AntibodyA -> AntibodyD 5.3;
```

Fig. 21. Textual Programming Environment in BEYONDwork BS Configuration Facility

transformations from textual programs and Java source code are implemented with a model-code transformation engine in openArchitectureware.

The following is a fragment of Java source code generated from the textual program in Figure 21.

```
void setupAntibodiesOfINet(){
  Antibody antibodyA =
    new Antibody( "AntibodyA", 5, LocalWorkload.HIGH,
      new Reproduction(
        2.3, CROSSOVER.FITNESSBASED, PARTNER.FITNESSBASED ) );
  Antibody antibodyD =
    new Antibody( "AntibodyD", 1, EnergyLevel.HIGH,
      new Migration( DirectionPolicy.USER ) );

  ImmuneNetwork inet = getImmuneNetwork();
  inet.add( antibodyA );
  inet.add( antibobyD );
  antibodyA.addAffinity( antibodyD, 5.3 );
}
```

The BS configuration facility allows agent designers to not only configure an antibody configuration (behavior policy) from scratch, but also investigate and fine tune existing antibody configurations in running agents. In iNet, antibody configurations are evolved automatically via genetic operations (see Section 4.2). The BS configuration facility helps agent designers to understand evolved antibody configurations by showing it in a visual manner, and experienced agent designers can fine-tune them by hand.

Without the BS configuration facility, agent designers need to know the details on how to implement agents in Java (e.g., how to define new agents, where to implement antibody configuration code, and which iNet API to use.) For example, agent designers need to define a new class extending the `Agent` class provided by a simulator in BEYOND. Also, as the above code fragment shows, they need to write the `setupAntibodiesOfINet()` method using iNet API in order to configure the agent's antibodies. The visual and textual languages hide these implementation details and allow agent designers to focus on the design of antibody configurations. In addition, compared with the Java code shown above, a model or program in the BS configuration facility is easier to read and understand.

7 Related Work

This paper describes several extensions to the prior work on iNet [19, 21]. [19] does not investigate the iNet evolutionary mechanism. Thus, agent designers needed to manually and carefully configure antibodies in their agents at design time. In contrast, the iNet evolutionary mechanism allows agents to autonomously adjust their antibody configurations at runtime; it does not require manual antibody configurations. [21] describes preliminary simulation results of the iNet evolutionary mechanism; however, it does not investigate the languages in BEYONDwork as well as the self-regulatory mechanism in the iNet EE facility.

The Bio-Networking Architecture [22] is similar to BEYOND in that it applies biological principles and mechanisms to allow network applications to autonomously adapt to dynamic environmental changes in the network. However, its adaptation engine is different from iNet. While iNet is designed after immune responses, [22] employs a simple weighted sum calculation for behavior selection. Although [22] has an evolutionary mechanism that dynamically adjusts weight values in the weighted sum calculation, agent designers still need to manually define a weighted sum equation for each behavior and configure a threshold value for each weighted sum equation. In contrast, iNet requires no manual configuration work for agent designers.

Artificial immune systems have been proposed and used in various application domains such as anomaly detection [23] and pattern recognition [24]. [23] focuses on the generation of detectors for self/non-self classification and improves the negative selection process of the artificial immune system. [24] focuses on the accuracy for the matchmaking of an antigen and antibody. Unlike those work, this paper proposes an artificial immune system to improve autonomous adaptability of network applications. To the best of our knowledge, this work is the first attempt to apply an artificial immune system to this domain.

In addition, some research work [25] using artificial immune systems extend their work with the concept of danger signals. [25] proposes the mechanism to detect misbehaving nodes as antigens based on event sequences of routing process in ad hoc network. Danger signals contribute to reduce the number of false positives (i.e., the system evaluates a correctly working node as a misbehaving node) by dynamically updating the definition of normal event sequences (self). On the other hand, iNet self-regulation process allows agents to respond false positives as well as false negatives (i.e. the system cannot catch unknown non-self antigens).

BEYONDwork provides visual and textual languages to configure iNet, i.e., configuring environment conditions, detectors and behavior policies. The work of the languages in BEYONDwork is parallel to the existing research on domain specific languages (DSLs) [26]. The languages are considered as DSLs focusing on directly capturing the concepts and mechanisms specific to a particular problem domain. There are several DSLs to model biological systems such as biochemical networks for simulating and understanding biological systems (e.g., [27, 28]). However, the objective of the languages in BEYONDwork is different from theirs; languages in BEYONDwork aim to model biological (immunological) mechanisms for building autonomous and adaptive network applications. This work is the first attempt to investigate a DSL for biologically-inspired networking.

8 Conclusion

This paper describes the BEYOND architecture, which applies biological principles and mechanisms to design evolvable network applications that autonomously adapt to dynamic environmental changes in the network. This paper proposes two key components in BEYOND: (1) a self-regulatory and evolutionary adaptation mechanism for agents, called iNet, and (2) an agent development environment, called BEYONDwork. iNet allows each agent to autonomously sense its surrounding environment conditions (i.e., antigens) and adaptively invoke a behavior (i.e., antibody) suitable for the conditions. iNet also allows each agent to detect its own deficiencies to recognize antigens and regulate its policies for antigen recognition. Agents evolve their antibodies so that they adapt to unexpected environmental changes. Simulation results show that agents adapt to changing network environments by self-regulating their antigen recognition and evolving their antibodies through generations. In addition, BEYONDwork provides visual and textual languages to configure iNet in an intuitive and easy-to-understand manner. It accepts the visual models and textual programs built with the proposed languages, and transforms them to Java code that are compilable and runnable on a simulator for BEYOND. This code generation enables rapid development configuration of agents, thereby improving the productivity of agent developers.

References

1. D. F. Carr, "How google works," *Baseline*, July 2006.
2. J. Markoff and S. Hansell, "Google's not-so-very-secret weapon," *International Herald Tribune*, June 2006.
3. P. Dini, W. Gentzsch, M. Potts, A. Clemm, M. Yousif, and A. Polze, "Internet, grid, self-adaptability and beyond: Are we ready?" *Proc. of IEEE International Workshop on Self-Adaptable and Autonomic Computing Systems*, August 2006.
4. R. Sterritt and D. Bustard, "Towards an autonomic computing environment," *Proc. of IEEE International Workshop on Database and Expert Systems Applications*, September 2003.
5. J. Rolia, S. Singhal, and R. Friedrich, "Adaptive internet data centers," *Proc. of International Conference on Advances in Infrastructure for Electronic Business, Science, and Education on the Internet*, July 2000.

6. S. Ranjan, J. Rolia, E. Knightly, and H. Fu, "Qos-driven server migration for internet data centers," *Proc. of International Workshop on Quality of Service*, May 2002.
7. N. Minar, K. H. Kramer, and P. Maes, "Cooperating mobile agents for dynamic network routing," in *Software Agents for Future Communications Systems*, A. Hayzelden and J. Bigham, Eds. Springer, June 1999.
8. R. Albert, H. Jeong, and A. Barabasi, "Error and attack tolerance of complex networks," *Nature*, July 2001.
9. G. Cabri, L. Leonardi, and F. Zambonelli, "Mobile-agent coordination models for internet applications," *IEEE Computer*, February 2000.
10. S. Camazin, J. L. Deneubourg, N. R. Franks, J. Sneyd, G. Theraula, and E. Bonabeau, *Self Organization in Biological Systems*. Princeton University Press, May 2003.
11. N. K. Jerne, "Idiotypic networks and other preconceived ideas," *Immunological Review*, 1984.
12. C. Berek, "Somatic hypermutation and b-cell receptor selection as regulators of the immune response," *Transfusion Medicine and Hemotherapy*, 2005.
13. P. Matzinger, "The danger model: A renewed sense of self," *Science*, April 2002.
14. K. R. Jerome and L. Corey, "The danger within," *The New England Journal of Medicine*, 2004.
15. W. R. Heath and F. R. Carbonel, "Immunology: Dangerous liaisons," *Nature*, October 2003.
16. B. Goldman, "White paper: Heat shock proteins' vaccine potential from basic science breakthroughs to feasible personalized medicine," *Antigenic*, July 2002.
17. W. A. Fenton and A. L. Horwich, "Chaperonin-mediated protein folding: Fate of substrate polypeptide," *Quarterly Reviews of Biophysics*, May 2003.
18. T. Mitchell, *Machine Learning*. McGraw-Hill, 1997.
19. C. Lee and J. Suzuki, "Biologically-inspired design of autonomous and adaptive grid services," *Proc. of International Conference on Autonomic and Autonomous Systems*, July 2006.
20. J. Chase, D. Anderson, P. Thakar, A. Vahdat, and R. Doyle, "Managing energy and server resources in hosting centers," *Proc. of the Eighteenth Symposium on Operating Systems Principles*, October 2001.
21. C. Lee and J. Suzuki, "An immunologically-inspired adaptation mechanism for evolvable network applications," *Proc. of the Fourth IEEE Consumer Communications and Networking Conference*, January 2007.
22. T. Nakano and T. Suda, "Self-organizing network services with evolutionary adaptation," *IEEE Transactions on Neural Networks*, September 2005.
23. F. A. Gonzalez and D. Dasgupta, "Anomaly detection using real-valued negative selection," *Genetic Programming and Evolvable Machines*, 2003.
24. L. N. de Castro and J. I. Timmis, "Artificial immune systems: A novel paradigm to pattern recognition," in *Artificial Neural Networks in Pattern Recognition*, J. M. Corchado, L. Alonso, and C. Fyfe, Eds. University of Paisley, UK, 2002.
25. S. Sarafijanovic and J.-Y. L. Boudec, "An artificial immune system approach with secondary response for misbehavior detection in mobile ad-hoc networks," *IEEE Transactions on Neural Networks, Special Issue on Adaptive Learning Systems in Communication Network*, April 2005.
26. G. Cook, "Domain-specific modeling and model-driven architecture," in *The MDA journal: Model Driven Architecture Straight from the Masters*. Meghan-Kiffer Press, December 2004, ch. 5.

27. M. Hucka, A. Finney, B. Bornstein, S. Keating, B. Shapiro, J. Matthews, B. Kovitz, M. Schilstra, A. Funahashi, J. Doyle, and H. Kitano, "Evolving a lingua franca and associated software infrastructure for computational systems biology: The systems biology markup language (sbml) project," *Systems Biology Journal*, June 2004.
28. F. Kolpakov, "Biouml - framework for visual modeling and simulation biological systems," in *International Conference on Bioinformatics of Genome Regulation and Structure*, July 2002.

Biologically Inspired Synchronization for Wireless Networks

Alexander Tyrrell[1], Gunther Auer[1] and Christian Bettstetter[2]

[1] DoCoMo Euro-Labs, Wireless Solutions Lab, 80687 Munich, Germany
{tyrrell, auer}@docomolab-euro.com
[2] University of Klagenfurt, Networked and Embedded Systems, Mobile Systems Group, 9020 Klagenfurt, Austria
lastname@ieee.org

Summary Fireflies exhibit a fascinating phenomenon of spontaneous synchronization that occurs in nature: at dawn, they gather on trees and synchronize progressively without relying on a central entity. The present chapter reviews this process by looking at experiments that were made on fireflies and the mathematical model of Mirollo and Strogatz, which provides key rules to obtaining a synchronized network in a decentralized manner. In this article challenges related to the implementation in ad hoc networks are addressed. In particular, the effects of transmission delays and the constraint that a node cannot receive and transmit at the same time are studied. A novel delay tolerant synchronization scheme, derived from the original firefly synchronization principle is presented. Simulation results show that an accuracy limited only by propagation delays is retained.

1 Introduction

On riverbanks in South-East Asia male fireflies gather on trees at dawn, and emit flashes regularly. Over time, synchronization emerges from a seamingly chaotic situation, which makes it seem as though the whole tree is flashing in perfect synchrony. This phenomenon forms an amazing spectacle, and has intrigued scientists for several hundred years [3]. Over the years, two fundamental questions have been studied. Why do fireflies synchronize? And how do they synchronize?

The first question led to many discussions among biologists. In all species of fireflies, emissions of light serves as a means of communication that helps female fireflies distinguish males of its own species. While male as well as female fireflies emit light, the response of male fireflies to emissions from females is different in each species. For some species such as *Pteroptyx cribellata* and *Pteroptyx malaccae*, males synchronize their blinking. The reason behind this spontaneous synchronization remains subject of controversial discussions. Several hypothesis exist: either it could accentuate the males' rhythm or serve as a noise-reduction mechanism that helps them identify females [15]. This phenomenon could also enable small groups of males to attract more females, and act as a cooperative scheme.

Although the reason behind synchronization is not fully understood, fireflies are not the only biological system displaying a synchronized behavior. This emergent pattern is present in heart cells [16], where it provides robustness against the death of one or more cells, and in neurons, where it enables rapid computation [9]. Among humans, synchronization also occurs. For example, women living together tend to synchronize their menstrual periods [21], and people walking next to each other on the street tend to step foot in synchrony.

In biological systems distributed synchronization is commonly modeled using the theory of coupled oscillators [25]. For fireflies, an oscillator represents the internal clock dictating when to flash, and upon reception of a pulse from other oscillators, this clock is adjusted. Over time, synchronization emerges, i.e. pulses of different oscillators are transmitted simultaneously. Synchronization in populations of coupled oscillators lies within the field of nonlinear dynamics. A theoretical framework for the convergence to synchrony in fully-connected mesh networks was published by Mirollo and Strogatz in [14].

An interesting question is whether firefly synchronization can be applied to wireless systems. The behavior of synchronized fireflies is reminiscent of time slots required to deploy a slotted Medium Access Control (MAC) protocol in telecommunication networks: all participating nodes in the system are aligned in time and emit synchronously, light in the case of fireflies and packets for nodes in wireless systems. To this end, all nodes need to agree on the beginning and end of a time slot. Naturally, deploying a slotted MAC protocol requires nodes to be synchronized, and yields higher throughput than unslotted ones as it reduces the number of collisions [20].

In centralized systems, the time slot structure is dictated by a master node. However in distributed networks without a predefined leader, a centralized synchronization scheme cannot be applied. Therefore a natural target application of the Mirollo and Strogatz synchronization model of [14] is ad hoc and sensor networks.

One of the first papers to apply the Mirollo and Strogatz model [14] to wireless networks was [8]. It utilized the characteristic pulse of Ultra-Wide-Band (UWB) to emulate the synchronization process of pulse-coupled oscillators, and included effects such as channel attenuation and noise. To lift the restriction of using UWB pulses and apply the model of [14] to wireless systems, delays need to be taken into account. Both models of [14] and [8] assume that fireflies form a fully-connected mesh network and communicate through pulses of infinitesimal small duration. However pulses are hardly considered for communications in a wireless environment, because they are difficult to detect.

To reflect more realistic effects such as message delay and loss, [24] proposed to synchronize using a low-level timestamp on the MAC layer. The principle is similar to the original firefly synchronization scheme, in the way that each node adjusts its clock when receiving such a timestamp. Because timestamps need to be exchanged, the approach of [24] tries to avoid the ideal case of the Mirollo and Strogatz model where all nodes transmit simultaneously. This case creates too many collisions, which prevails nodes from exchanging timestamps and thus from synchronizing.

However, from a physical layer perspective, all nodes transmitting synchronously a common word can help a faraway receiver synchronize and communicate with the rest of the network, because it receives the sum of all transmitted powers (known as the reachback problem) [10]. Hence, unlike data transmission, a synchronization process where all nodes transmit the same word is not affected by collisions, in a similar way to flooding.

The proposed delay tolerant firefly synchronization strategy combats transmission and processing delays by modifying the intrinsic behavior of a node. The resulting synchronization algorithm presented in Section 4 is entirely based on the physical layer. This has several advantages over [24]: the exchange of timestamps is avoided, collisions are in fact a benefit to the scheme, and the time to reach a synchronized state is shorter because there is no random backoff. This synchronization strategy has also been extended to connected mesh networks in [23] and takes into consideration constraints from the MAC layer.

Before introducing the synchronization strategy adapted to wireless networks, Section 2 looks at experiments that have been made in order to comprehend the biological synchronization mechanism. In particular, we focus on experiments made on fireflies and how their flashing instants are affected by external stimuli, providing interesting insights to compare nature with mathematical models. The mathematical model of Mirollo and Strogatz [14] is then presented in Section 3. This model serves as a basis for the synchronization algorithm of Section 4, and in the remainder, this model is referred to as the "MS model." The proposed delay tolerant firefly synchronization algorithm of Section 4 is evaluated through simulations, to show that synchronization is always obtained in finite time, independent of the initial condition, with an accuracy upper bounded by propagation delays.

2 Experiments on Fireflies

Early hypotheses had difficulties explaining the firefly synchronization phenomenon. For example, Laurent in 1917 dismissed what he saw and attributed the phenomenon to the blinking of his eyelids [12]. Others argued that synchrony was provoked by a single stimulus received by all fireflies on the tree [1]. However the presence of a leading firefly or a single external factor is easily dismissed by the fact that not all fireflies can see each other and fireflies gather on trees and progressively synchronize. The lack of a proper explanation lasted until the 1960s and is mostly due to a lack of experimental data.

Among early hypotheses, Richmond [19] stated in 1930 what came very close to the actual process: "Suppose that in each insect there is an equipment that functions thus: when the normal time to flash is nearly attained, incident light on the insect hastens the occurrence of the event. In other words, if one of the insects is almost ready to flash and sees other insects flash, then it flashes sooner than otherwise. On the foregoing hypothesis, it follows that there may be a tendency for the insects to fall in step and flash synchronously."

This statement identifies that synchronization among fireflies is a self-organized process, and fireflies influence each other: they emit flashes periodically, and in return are receptive to the flashes of other males.

To understand this process, a set of experiments was conducted by Buck *et al.* [2]. These experiments concentrated on the reaction of a firefly to an external signal depending on when this light is received. Naturally a firefly emits light periodically every 965 ± 90 ms [2], and the external signal changes this natural period. For the experiments, the firefly was put in a dark room and was restrained from seeing its own flashes. Stimuli were made by guiding 40 ms signals of light from a glow modulator lamp into the firefly's eye via a fiber optics. Responses were recorded and are shown on Fig. 1.

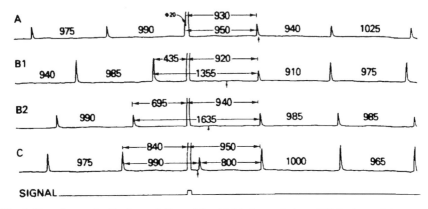

Fig. 1. Experiments by Buck *et al.* (from [2] with kind permission of ©Springer Science and Business Media). Delays are expressed in ms

From Fig. 1, when an external signal is emitted, three different responses are identified:

- In response A, the artificial signal occurs only 20 ms after the firefly's spontaneous flash. As the following response from the firefly occurs at a normal time of 950 ms, the signal has not modified the natural response. This behavior corresponds to a refractory period: during this time, the potential of the flash regains the "resting" position and no modification of the internal clock is possible.
- In responses B1 and B2, the signal inhibits the response of the firefly: instead of emitting light after about 960 ms, it delays its response until 920 ms and 940 ms after receiving the signal. Thus successive flashes occur 1355 ms and 1635 ms apart, which is far more than the natural period.
- In response C, the artificial signal occurs 150 ms before the natural flashing, and does not have any incidence on this flash. This is due to a processing delay in the central nervous system of a firefly, which is equal to about 800 ms [2]. Therefore the external signal influences the following flash, which is advanced by 150 ms.

Thus the external signal exhibits an excitatory effect on the response and brings the firefly to flash earlier.

From these experiments the modified behavior of the firefly depended only on the instant of arrival of the external signal. The responses display both inhibitory (responses B1 and B2) and excitatory (response C) couplings depending on the instant the external flash is perceived. Furthermore a refractory period placed after emission is also present (response A).

In all cases, the external flash only altered the emission of one following flash, and in the following period, nodes regained their natural period of about one second. Variating the amplitude of the input signal yielded similar results.

For more insights into the phenomenon of firefly synchronization, Chapter 10 in [4] provides a history of studies on fireflies, including early interpretations, and analyzes different experiments including the one presented in this section.

These experiments have helped mathematicians modeling fireflies. However proving that synchrony occurs when both inhibitory and excitatory couplings are present has, to our knowledge, not been done yet. Therefore we will concentrate on the existing model of [14], which considers excitatory coupling and no delays, in order to derive a synchronization algorithm suited for wireless systems.

3 Mathematical Model

The internal clock of a firefly, which dictates when a flash is emitted, is modeled as an oscillator, and the phase of this oscillator is modified upon reception of an external flash. In general this type of oscillator is termed *relaxation oscillator*. It is not represented by a sinusoidal form but rather by a series of pulses. A review of this class of oscillators and their implications can be found in Chapter 1 of [18]. Examples include Van der Pol oscillators and integrate-and-fire oscillators [14]. There is however no general model describing this class of oscillators.

In the remainder, we focus on integrate-and-fire oscillators, which are also termed "pulse-coupled oscillators". They interact through discrete events each time they complete an oscillation. The interaction takes the form of a pulse that is perceived by neighboring oscillators. This model is used to study biological systems such as heart cells, neurons and earthquakes [9]. This section describes how time synchronization is achieved in a decentralized fashion in a system of N pulse-coupled oscillators.

Differential Equations

Each oscillator i, $1 \leq i \leq N$, is described by a state variable x_i, similar to a voltage-like variable in a RC-circuit, and its evolution and interactions are described by a set of differential equations [5]:

$$\frac{dx_i(t)}{dt} = -x_i(t) + I_0 + \sum_{\substack{j=1 \\ j \neq i}}^{N} J_{i,j} \cdot P_j(t) \qquad (1)$$

where I_0 controls the period of an uncoupled oscillator and $J_{i,j}$ determines the coupling strength between oscillators. When $x_i = x_{th}$, where x_{th} is the state variable threshold, an oscillator is said to 'fire': at this instant, its state is reset to zero, and it emits a pulse, which modifies the state of other coupled oscillators. The coupling function P_j is defined as a train of emitted pulses:

$$P_j(t) = \sum_m \delta\left(t - \tau_j^{[m]}\right) \tag{2}$$

where $\tau_j^{[m]}$ represents the mth firing time of oscillator j and $\delta(.)$ is the Dirac delta function.

As (1) is not solvable in closed-form for arbitrary N, the mathematical demonstration of [14] describes the system as two states to show that synchrony is always achieved independently of initial conditions. In the free running evolution state, no interaction occurs and nodes simply maintain a variable dictating their next emission, and in the phase adjustment state, a node transmits and all receivers adjust their internal clock.

Free Running Evolution

Instead of describing the system by a voltage-like state variable, each oscillator is described by a phase function ϕ_i which linearly increments from 0 to a phase threshold ϕ_{th} during the free running evolution:

$$\frac{d\phi_i(t)}{dt} = \frac{\phi_{th}}{T} \tag{3}$$

When $\phi_i(t) = \phi_{th}$, a node resets its phase to 0. The phase function is similar to the state variable of (1), in the sense that it periodically fires. The state variable can be written as a function of ϕ_i through the *firing map* $x_i(t) = f(\phi_i(t))$. Variable x_i describes a voltage-like evolution, whereas ϕ_i evolves linearly over time and is adjusted differently upon reception of a pulse.

If not coupled to any other oscillator, i.e. it is isolated, the phase function naturally oscillates and fires with a period equal to T. Fig. 2(a) plots the evolution of the phase function during one period in this case.

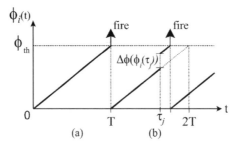

Fig. 2. Time evolution of the phase function (a) in the free running state, (b) when receiving one pulse at τ_j

The phase function encodes the remaining time until the next firing, which corresponds to an emission of light for a firefly. For a firefly, the natural period T is equal to about one second. In order to simplify the analysis, the MS model does not encompass the clock jitter of ± 90 ms that is experimentally observed (Fig. 1).

For considerations about clock jitter and frequency adjustment, a different oscillator model was proposed by Ermentrout [6], where oscillators have different frequencies. In the following, we consider that all nodes have the same dynamics, i.e. clock jitter is considered negligible.

Phase Adjustment

Mirollo and Strogatz analyze spontaneous synchronization phenomena and derive a theoretical framework based on pulse-coupled oscillators for the convergence of synchrony [14]. The goal of the synchronization algorithm is to align all internal counters ϕ_i, so that all nodes agree on a common firing instant. To do so, the phase function is adjusted upon reception of an external pulse.

When coupled to others, an oscillator is receptive to the pulses of its neighbors. Coupling between nodes is considered instantaneous, and when a node j ($1 \leq j \leq N$) fires at $t = \tau_j$, all nodes adjust their phase as follows:

$$\phi_j(\tau_j) = \phi_{\text{th}} \rightarrow \begin{cases} \phi_j(\tau_j) \rightarrow 0 \\ \phi_i(\tau_j) \rightarrow \phi_i(\tau_j) + \Delta\phi(\phi_i(\tau_j)) \text{ for } i \neq j \end{cases} \quad (4)$$

Fig. 2(b) plots the time evolution of the phase when receiving a pulse at time τ_j. To simplify notations, the parameter m in (2) is dropped, which is coherent with Fig. 1 where the external signal affected only one following flashing.

By appropriate selection of $\Delta\phi$, a system of N identical oscillators forming a fully-connected mesh network is able to synchronize their firing instants within a few periods [14]. The phase increment $\Delta\phi$ is determined by the Phase Response Curve (PRC). For their mathematical demonstration, Mirollo and Strogatz derive that synchronization is obtained whenever the firing map f is concave up, and the return map $\phi_i(t) + \Delta\phi(\phi_i(t)) = g(x_i(t) + \epsilon)$, where ϵ is the amplitude increment, is its inverse [14]. The resulting operation $\phi_i(t) + \Delta\phi(\phi_i(t)) = f^{-1}(f(\phi_i(t)) + \epsilon)$ yields the PRC, and is a piecewise linear function:

$$\phi_i(\tau_j) + \Delta\phi(\phi_i(\tau_j)) = \min\left(\alpha \cdot \phi_i(\tau_j) + \beta, 1\right)$$
$$\text{with } \begin{cases} \alpha = \exp(b \cdot \epsilon) \\ \beta = \frac{\exp(b \cdot \epsilon) - 1}{\exp(b) - 1} \end{cases} \quad (5)$$

where b is the dissipation factor of the state variable x_i. Both factors α and β determine the coupling between oscillators and are identical for all nodes. The threshold ϕ_{th} is normalized to 1 without loss of generality.

It was shown in [14] that if the network is fully-connected as well as $\alpha > 1$ and $\beta > 0$ ($b > 0$, $\epsilon > 0$), the system always converges to a state in which all oscillators will fire as one. This convergence is independent of initial conditions, and the time to synchrony is inversely proportional to α.

As it can be observed in Fig. 2(b), detection of a pulse shortens the current period and causes an oscillator to fire early, because $\Delta\phi(\phi) > 0, \forall \phi$. Compared to the experimental data of Fig. 1, the MS model exhibits only excitatory coupling, and no refractory period is present. However the main features of the experiments of Buck *et al.* are encompassed in this model: nodes do not need to distinguish the source of the synchronization pulse, and adjust their current phase upon reception of a pulse. The synchronization scheme relies on the instant of arrival of a pulse and receivers adjusting their phases when detecting this pulse.

4 Adaptation to Wireless Systems

Applying the MS synchronization model to distributed wireless systems appears very attractive. It has the advantage that phenomena like interference and collisions are not problematic. First, there is no need for a receiver to identify the source of emission. Second, two pulses emitted simultaneously may superimpose constructively, which helps a faraway receiver synchronize. This type of spatial averaging has been shown to beneficially bound the synchronization accuracy to a constant, making the algorithm scalable with respect to the number of nodes [10].

To properly develop a synchronization scheme for wireless systems, different delays need to be taken into account. Within the field of nonlinear oscillators, it is known that when a delay occurs, even when constant between all nodes, a system of pulse-coupled oscillators becomes unstable, and is never able to synchronize [7]. The algorithm needs to account for propagation delays, so that the system remains stable. Moreover *long synchronization words* need to be considered, and a receiver typically requires some decoding delay to properly identify that a synchronization message was transmitted. As these delays affect the achievable accuracy, we modify the intrinsic behavior of the MS algorithm, so that the the only source for impairments are propagation delays, and evaluate the performance of the novel scheme through simulations.

4.1 Synchronization through Pulses

In wireless systems, even when considering communication through pulses, a propagation delay dependent on the distance between nodes occurs. Define $T_0^{(i,j)}$ the propagation delay between two nodes i and j then the pulse of i influences j not instantly, but after time $T_0^{(i,j)}$. If this causes j to reach the threshold and transmit a pulse, then i in turn also increments its phase after $T_0^{(i,j)}$, which can cause it to fire, and so on. If more than two nodes are present in the system, nodes continuously fire.

To avoid this unstable behavior, a refractory period of duration T_{refr} is added after firing [13]: after transmitting its pulse, a node i stays in a refractory state, where $\phi_i(t) = 0$ and no phase increment is possible, and then goes back into the listen state where its phase follows:

$$\frac{d\phi_i(t)}{dt} = \frac{\phi_{\text{th}}}{T_{\text{Rx}}} \quad \text{when } i \text{ in listen state} \tag{6}$$

where $T_{\text{Rx}} = T - T_{\text{refr}}$ is the duration of the listen state. The refractory state is also observed in case A in the experimental data of Fig. 1.

Stability in the case of propagation delays is maintained if echoes are not acknowledged, which translates to a condition on T_{refr}:

$$T_{\text{refr}} > 2 \cdot T_0^{[\text{max}]} \tag{7}$$

where $T_0^{[\text{max}]}$ is the maximum propagation delay between two nodes in the network. With the introduction of the refractory state, the accuracy of the synchronization scheme is equal or smaller to the maximum propagation delay [13].

4.2 Synchronization through Long Synchronization Words

The previous scheme implies that nodes communicate through pulses and that a receiver is able to immediately detect a single pulse of infinitely small width, and no decoding is done by the receiver. In a wireless environment solitary pulses are hardly used alone as they are virtually impossible to detect. More realistically a synchronization word of certain duration is used. In the MS model, nodes do not need to distinguish between emitters. Therefore a *common synchronization word* is broadcasted by all nodes when firing.

The synchronization word can be chosen from a variety of schemes:

- a sequence of pulses,
- a Pseudo-Noise (PN) sequence [17],
- the 802.11 preamble [11].

In all these cases the synchronization word has a certain duration T_{Tx}.

During the transmission of this word, a node is unable to receive. This constraint is due to limitations on the Radio Frequency (RF) part of transceivers.

Due to processing delays and to clearly identify a transmitted synchronization message, some decoding delay T_{dec} is allowed at the receiver. The decoding delay T_{dec} is fixed and known by all nodes. It denotes the delay between the instant transmission of the synchronization word finishes and the instant the phase function is incremented. As the processing time varies depending on the hardware implementation, T_{dec} should be overestimated to account for the slowest receiver.

In the remainder, propagation delays are considered negligible in order to focus on the effects of finite length synchronization words and the decoding delay. This assumption is valid when considering Wireless LAN settings in an ad hoc scenario. If the maximum operation range is 50 m, the propagation delay is $T_0^{[\text{max}]} = \frac{50 \text{ m}}{c} \approx 0.17\,\mu\text{s}$. In comparison, the preamble of an 802.11 frame, which serves as the synchronization word, has a duration of $T_{\text{Tx}} = 8\,\mu\text{s}$ [11]. Noise at the receiver is also neglected to emphasize the effect of delays on the original scheme.

Impact of Delays

When node j fires, it enters a transmit state: the synchronization word $x(t)$ passes through a shaping filter and starts being emitted. It then propagates through a channel $h(t)$ before being received by node i, which collects the incoming signal through a matched filter [17]. As $x(t)$ is the same for all nodes, node i detects it by correlating the received signal with the known message. The output of the correlation detector is given by [17]:

$$\Lambda_i(t) = \int_0^{T_{\text{Tx}}} x(\tau) \cdot y_i^*(t-\tau) d\tau \qquad (8)$$

where $y_i(t)$ is the incoming signal at node i. With the above assumptions, node i receives the sum of transmitted synchronization words $y_i(t) = \sum_j x(t-\tau_j)$.

Choosing a PN sequence as the synchronization word, its autocorrelation ideally results in a peak T_{Tx} after j fired:

$$\rho(t) = \int_0^{T_{\text{Tx}}} x(\tau) \cdot x^*(t-\tau) d\tau = \delta(t - T_{\text{Tx}}) \qquad (9)$$

The resulting correlation metric of node i, which is used for deciding a phase increment, is:

$$\Lambda_i(t) = \sum_j \rho(t-\tau_j) \\ = \begin{cases} \sum_j \delta(t-\theta_j) & \text{for } T_{\text{Tx}} < t < \tau_i \\ 0 & \text{elsewhere} \end{cases} \qquad (10)$$

where $\tau_i \leq T$ is the next firing instant for node i and $\theta_j = \tau_j + T_{\text{Tx}}$. The observation interval is reduced by T_{Tx}, because node i is transmitting in $[0, T_{\text{Tx}}]$.

From (10), when several nodes transmit, the output of the correlation detector produces a series of peaks when $t = \theta_j$, in a similar way to the received pulses in the MS model. Thus several transmissions are distinguishable and several phase increments occur. If nodes fire and transmit synchronously, peaks superimpose constructively.

The metric $\Lambda_i(t - T_{\text{dec}})$ is observed at the output of the detector, and a phase increment is decided when node i is in listen state and T_{dec} after a peak is observed:

$$\phi_i(t) \rightarrow \begin{cases} \phi_i(t) + \Delta\phi(\phi_i(t)) & \text{for } \Lambda_i(t - T_{\text{dec}}) > \Lambda_{\text{th}} \text{ and } t \geq T_{\text{Tx}} + T_{\text{refr}} \\ \phi_i(t) & \text{elsewhere} \end{cases} \qquad (11)$$

where Λ_{th} is the decision threshold.

As an example of the system model described previously, Fig. 3 plots the output of the correlation detector $\Lambda_i(t)$ and the corresponding phase function of node i during one period. The synchronization sequence of duration $T_{\text{Tx}} = 0.2 \cdot T$ is transmitted at $t = \tau_j$ and $t = \tau_k$ corresponding to the firing instants of nodes j and k, and is collected by the receiver. This random sequence is known by the receiver, and results in two peaks with a duration of one sample T_{Tx} after nodes j and k fired. The peak at $\theta_k = \tau_k + T_{\text{Tx}}$ is not acknowledged, because node i is in transmit state at $t = \theta_k + T_{\text{dec}}$, where T_{dec} is fixed to $0.12 \cdot T$. On the other hand, the receiver increments its phase at $t = \theta_j + T_{\text{dec}}$.

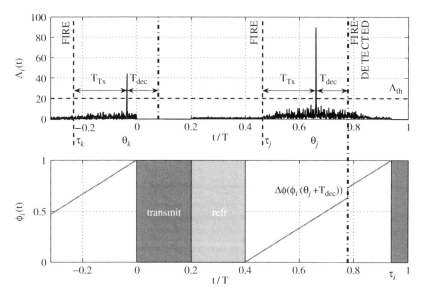

Fig. 3. Example of the output of the correlator and a delayed phase increment when considering a random sequence as the synchronization word

From (11) the delay from the instant a transmitter starts emitting to the instant a receiver has completely received and decoded a message is equal to:

$$T_{\text{del}} = T_{\text{Tx}} + T_{\text{dec}} \tag{12}$$

The delay in the coupling by T_{del} and the reduced observation interval are the most significant differences from the MS model, which assumes an infinitely short transmission time and no decoding delay [14]. This delay has two consequences.

First, the coupling between nodes is always delayed by T_{del}, whereas it was considered instantaneous in the MS model. From the theory of coupled oscillators, it is known that delays impact heavily on the synchronization process and its stability [26]. For pulse-coupled oscillators, it has been shown that the system becomes unstable in the presence of delays [22], but this model does not account for the transmission and refractory periods, where no coupling is possible. In our model, as coupling is only possible when nodes are in listen state, stability issues are prevented by adjusting T_{refr}, which is found through simulations.

Second, the last message that can be fully received and decoded (and therefore have an incidence on the phase function) occurs T_{del} before firing, i.e. at $t = \tau_i - T_{\text{del}}$. Hence part of transmitted synchronization messages cannot be heard, which divides a period of duration τ_i starting at the previous firing instant into three intervals:

$$U_{\text{Tx}} = [0, T_{\text{Tx}}] \tag{13}$$

$$U_{\text{obs}} = [T_{\text{Tx}}, \tau_i - T_{\text{del}}] \tag{14}$$

$$U_{\text{nc}} = [\tau_i - T_{\text{del}}, \tau_i] \tag{15}$$

Within U_{Tx} node i transmits and its receiver is switched off, and within U_{nc} the node is listening but remains uncoupled to emitted synchronization messages due to the total delay.

Due to the fact that node i is not coupled to nodes firing in U_{nc}, a *deafness* of duration T_{del} appears in which nodes cannot listen to the network. Within this deafness, no mutual coupling between nodes can occur, which implies that the attainable synchronization accuracy is lower bounded by T_{del}. For transmission techniques where the time for one symbol block, T_{Tx}, cannot be assimilated as a pulse, such an accuracy is clearly unacceptable. Therefore, there is a need to modify the synchronization strategy.

4.3 Compensating Delays

In order to regain high accuracy, we propose to combat transmission delays by modifying the intrinsic behavior of a node: after firing, a node *delays* its transmission of the synchronization word. This approach is similar to the one observed in the experiments of fireflies on Fig. 1 where responses to the external are not instantaneous. More precisely, in response C of Fig. 1, the advance in flashing is not effective immediately upon reception of a signal, but occurs in the following period. Interestingly, in responses B1 and B2, the firefly drops the flash subsequent to the stimulus and adjusts its phase for the following flash, which in fact closely resembles the proposed modified behavior of a node.

The waiting delay is chosen to be:

$$\begin{aligned} T_{wait} &= T - T_{del} \\ &= T - (T_{Tx} + T_{dec}) \end{aligned} \quad (16)$$

where T denotes the synchronization period. With this approach, receivers increment their phases exactly T seconds after a transmitter fired.

This scheme modifies the natural oscillatory period of a node, which is now equal to $2 \cdot T$. Nodes are coupled only if they can hear each other during T_{Rx}, which is the time during which the phase function linearly increments over time according to (6). This time is reduced by the waiting, transmitting, and refractory states, and is now equal to:

$$\begin{aligned} T_{Rx} &= 2 \cdot T - (T_{wait} + T_{Tx} + T_{refr}) \\ &= T + T_{dec} - T_{refr} \end{aligned} \quad (17)$$

To summarize the modified behavior of a node, Fig. 4 represents the four successive states of a node: wait, transmit, refr and listen when $T_{wait} = 0.75 \cdot T$, $T_{Tx} = 0.2 \cdot T$, $T_{dec} = 0.05 \cdot T$, $T_{refr} = 0.2 \cdot T$ and $T_{Rx} = 0.85 \cdot T$. A node is represented as a marker that circles around the phase diagram linearly over time and counterclockwise. Using this diagram, N nodes can be represented on the same circle, which helps analyzing the dynamical evolution of the system. One full rotation of a marker corresponds to a period $2 \cdot T$.

For a system of N oscillators, all firings instants are initially distributed over a period of $2 \cdot T$, i.e. all markers are randomly uniformly distributed around the circular

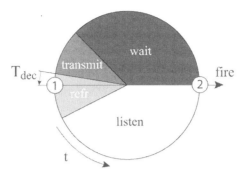

Fig. 4. Phase diagram of a system of N nodes following the novel synchronization strategy. Two groups of oscillators form, spaced exactly T apart

state machine representation. Each oscillator follows the same rules of waiting before transmitting. When a message is successfully received during listen, the marker representing this node abruptly shifts its position towards the diametrically opposed state on the representation of Fig. 4. Over time the oscillators split into two groups diametrically opposed on the state machine representation, each group firing T seconds apart and helping each other synchronizing.

The formation of two groups is a necessary requirement for maintaining high accuracy, because nodes that transmit almost simultaneously cannot hear each other (deafness while transmitting). Therefore each group helps the other to synchronize by transmitting T seconds after the other.

With this new transmitting strategy the accuracy of synchronization is no longer limited by T_{del}. Successful synchronization is therefore declared when firing instants are spread over a time interval that is equal or smaller than the maximum propagation delay.

4.4 Simulation Results

Deriving a thorough mathematical demonstration that synchrony is reached when each node follows the simple rules of waiting before transmitting is a task of formidable difficulty, as proving synchrony when no delays are present is already difficult. The behavior of the system lies within the field of nonlinear dynamics, and a complete description and analysis does not seem easily reachable. Therefore we rely on simulation results to evaluate our synchronization scheme.

To verify the validity of the synchronization scheme, Monte Carlo simulations are carried out. Fig. 5 plots the cumulative distribution function (cdf) of the time to synchrony normalized to the period duration T, denoted T_{sync}, for several values of $T_{Tx} = \{0.1\ T, 0.45\ T, 0.7\ T\}$ for a system of $N = 30$ nodes forming a fully-connected mesh network. Refractory and decoding delays are fixed to $T_{refr} = 0.35\ T$ and $T_{dec} = 0.15\ T$, and coupling is set to $\alpha = 1.2$ and $\beta = 0.01$. Nodes are able to perfectly distinguish each transmitted synchronization word, e.g. by using a long random sequence as the synchronization word. Successful synchronization is declared if two groups of oscillators firing T seconds apart form, and time to synchrony is defined as the time needed for a system to synchronize starting from a random condition. Initial conditions correspond to a scenario where initially all state variables are randomly distributed around the phase diagram of Fig. 4. The number of realizations of initial conditions is set to 10^4.

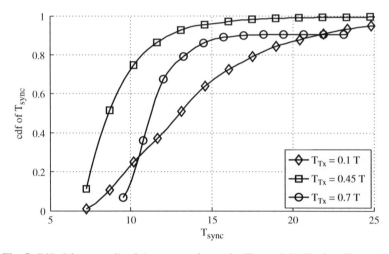

Fig. 5. Cdf of the normalized time to synchrony for $T_{\text{refr}} = 0.35\ T$ when T_{Tx} varies

The fastest time to synchrony is obtained for $T_{\text{Tx}} = 0.45\ T$, and for this setting, synchronization is obtained in less than 15 periods for 95% of initial conditions. In all simulations nodes synchronize in finite time, and the longest case took around $25\ T$ to reach synchrony.

Increasing the transmission time to $T_{\text{Tx}} = 0.7\ T$ shows a different type of behavior. In only 10% of simulations, the system synchronizes in less than $10\ T$, whereas this rate was over 70% for $T_{\text{Tx}} = 0.45\ T$. However the curve grows very rapidly until $15\ T$, where a plateau is reached for higher synchronization times. The cause behind this plateau is that as the transmit time T_{Tx} is large and nodes cannot receive and transmit simultaneously, there exists a set of initial conditions for which synchronization is never reached due to this deafness. The size of this set represents about 10% of initial conditions.

For a lower value of T_{Tx}, i.e. $T_{\text{Tx}} = 0.1\ T$, the time to synchrony is much higher than in other cases and the cdf grows linearly with T_{sync}, and synchronization is obtained in 90% of cases within $22\ T$. However synchronization always happens in bounded time (limit not shown on the figure) and no plateau appears. The reason for this high time to synchrony comes from relatively high listening time compared to the time taken to receive a synchronization message. This means that in early stages of the synchronization process, nodes receive many messages and adjust their phase more often than when $T_{\text{Tx}} = 0.45\ T$ or $T_{\text{Tx}} = 0.7\ T$.

To limit the number of messages that can be received and limit the instability in initial stages, the length of the refractory period T_{refr} is increased. Fig. 6 plots the cdf of the time to synchrony for $T_{Tx} = 0.1\ T$ and $T_{refr} = \{0.35\ T, 0.45\ T, 0.6\ T\}$ using the previous settings of Fig. 5.

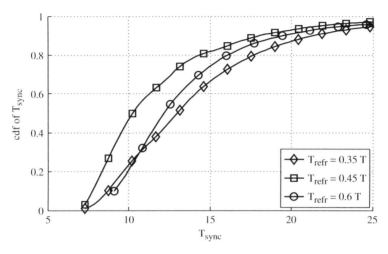

Fig. 6. Cdf of time to synchrony for $T_{Tx} = 0.1\ T$ when T_{refr} varies

Increasing the refractory time to $0.45\ T$ does improve the performance of the system, and synchronization is reached within $18\ T$ in 90% of initial conditions, instead of $22\ T$ for $T_{refr} = 0.35\ T$. Further increasing the refractory period does not bring further performance improvement and for a large value of T_{refr}, e.g. $T_{refr} = 0.6\ T$ on Fig. 6, the performance degrades for a time to synchrony lower than $20\ T$. Further performance improvement for low values of T_{Tx} is possible by increasing the coupling parameters α and β.

5 Conclusions

Fireflies provide an amazing spectable with their ability to synchronize using simple rules: each node maintains an internal clock dictating when to emit, and in return, this clock is adjusted when receiving. These synchronization rules are particularly simple and well suited for a deployment in ad hoc networks. However they are not directly applicable when accounting for transmission delays and the fact that a node cannot receive and transmit simultaneously.

To regain a level of accuracy that is upper bounded by the propagation delay, the intrinsic behavior of nodes was modified to compensate for transmission and decoding delay. Thanks to this modification, communication through pulses is no longer required, and accurate synchrony of oscillators is possible. The simplicity and generality of the synchronization scheme makes its implementation very appealing.

If all nodes cooperate, synchrony can be reached within 20 periods. Once nodes have agreed on a common time scale, they are then able to communicate in a synchronous manner using a slotted medium access protocol, benefiting from fewer collisions and higher throughput.

Acknowledgements

A preliminary version of this chapter was presented at the 3rd workshop of the ESF-funded scientific program MiNEMA, Leuven, Belgium, February 2006. The authors thank for the support of the European Science Foundation (ESF).

References

1. Blair K (1915) Nature 96:411–415
2. Buck J, Buck E, Case J, Hanson F (1981) Journal of Comparative Physiology A 144:630–633
3. Buck J (1988) The Quarterly Review of Biology 63:265–289
4. Camazine S, Deneubourg J-L, Franks N, Sneyd J, Theraulaz G, Bonabeau E (2001) Self-Organization in Biological Systems. Princeton
5. Campbell S, Wang D, Jayaprakash C (1999) Neural Computation 11:1595–1619
6. Ermentrout B (1991) Journal of Mathematical Biology 29:571–585
7. Ernst U, Pawelzik K, Geisel T (1967) Physical Review Letters 74:1570–1573
8. Hong Y-W, Scaglione A (2003) Proc. IEEE Conference on Ultra Wideband Systems and Technologies 2003
9. Hopfield J, Herz A (1995) Proceedings of the National Academy of Sciences 92:6655–6662
10. Hu A, Servetto S (2006) IEEE Transactions on Information Theory 52:2725–2748
11. IEEE 802.11 (1999) Wireless LAN Medium Access Control (MAC) and Physical Layer (PHY) Specifications. IEEE
12. Laurent P (1917) Science 45:44
13. Mathar R, Mattfeldt J (1996) SIAM Journal on Applied Mathematics 56:1094–1106
14. Mirollo R, Strogatz S (1990) SIAM Journal on Applied Mathematics 50:1645–1662
15. Otte D, Smiley J (1977) Biology of Behaviour 2:143–158
16. Peskin C (1975) Mathematical Aspects of Heart Physiology. New York: Courant Institute of Mathematical Sciences
17. Proakis J (1995) Digital Communications. McGraw-Hill
18. Ramírez-Ávila G.M. (2004) Synchronization Phenomena in Light-Controlled Oscillators. Université Libre de Bruxelles, Belgium
19. Richmond C (1930) Science 71:537–538
20. Roberts L (1975) ACM SIGCOMM Computer Communication Review 5:28–42
21. Strogatz S, Stewart I (1993) Scientific American 269:68–74
22. Timme M, Wolf F, Geisel T (2002) Physical Review Letters 89:154105
23. Tyrrell A, Auer G, Bettstetter C (2006) Proc. International Conference on Wireless Communications, Networking, and Mobile Computing 2006
24. Werner-Allen G, Tewari G, Patel A, Welsh M, Nagpal R (2005) Proc. Conference on Embedded Networked Sensor Systems 2005
25. Winfree A (1967) Journal of Theoretical Biology 16:15–42
26. Yeung M, Strogatz S (1999) Physical Review Letters 82:648–651

Bio-Inspired Congestion Control: Conceptual Framework, Algorithm and Discussion

Morteza Analoui and Shahram Jamali

Iran University of Science & Technology (IUST), Tehran, Iran
{analoui, jamali}@iust.ac.ir

Abstract. We believe that future network applications will benefit by adopting key biological principles and mechanisms. This work propose that bio-inspired algorithms are best developed and analyzed in the context of a multidisciplinary conceptual framework that provides for sophisticated biological models and well founded analytical principles. We outline such a framework here, in the context of bio-inspired congestion control (BICC) models, and show that relations of those Internet entities that involved in congestion control mechanisms is similar to population interactions such as predator-prey. This similarity motivates us to map the predator-prey model to the Internet congestion control mechanism and design a bio-inspired congestion control scheme. The results show that using appropriately defined parameters, this model leads to a stable, fair and high performance congestion control algorithm.

Key words: Communication Networks, Congestion Control, Biology, Predator-Prey

1 Introduction

Biologically inspired approaches have already proved successful in achieving major breakthroughs in a wide variety of problems in information technology (IT). A more recent trend is to explore the applicability of bio-inspired approaches to the development of self-organizing, evolving, adaptive and autonomous information technologies, which will meet the requirements of next-generation information systems, such as diversity, scalability, robustness, and resilience. These new technologies will become a base on which to build a networked symbiotic environment for pleasant, symbiotic society of human beings in the 21st century. The central aim of this paper is to obtain methods on how to engineer congestion control algorithms, which have similar high stability and efficiency as biological entities often have. Several examples are available in the area of computer technology, where biological concepts are considered as models to imitate. Each one of these examples focuses on a different biological aspect and applies it to solve or to optimize a specific technological problem. The most known examples are swarm intelligence, evolutionary or genetic algorithms, and the artificial immune system. The adapted mechanisms find application in computer networking for example in the areas of network security [1, 2], pervasive computing, and sensor networks [3].

Previous congestion control research has been heavily based on measurements and simulations, which have intrinsic limitations. There are also some theoretical frameworks and especially mathematical models that can greatly help us understand the advantages and shortcomings of current Internet technologies and guide us to design new protocols for identified problems and future networks [4]. The steady-state throughput of TCP Reno has been studied based on the stationary distribution of congestion windows, e.g. [5, 6, 7, 8]. These studies show that the TCP throughput is inversely proportional to end-to-end delay and to the square root of packet loss probability. Padhye [9] refined the model to capture the fast retransmit mechanism and the time-out effect, and achieved a more accurate formula. This equilibrium property of TCP Reno is used to define the notion of TCP–friendliness and motivates the equation-based congestion control [10]. Misra [11, 12] proposed an ordinary differential equation model of the dynamics of TCP Reno, which is derived by studying congestion window size with a stochastic differential equation. This deterministic model treats the rate as fluid quantities (by assuming that the packet is infinitely small) and ignores the randomness in packet level, in contrast to the classical queuing theory approach, which relies on stochastic models. This model has been quickly combined with feedback control theory to study the dynamics of TCP systems, e.g., [13, 14], and to design stable AQM (Active Queue Management) algorithms, e.g., [15, 16, 17, 18, 19]. Similar flow models for other TCP schemes are also developed, e.g., [20, 21] for TCP Vegas, and [22, 23] for FAST TCP. The analysis and design of protocols for large-scale network have been made possible with the optimization framework and the duality model. Kelly [24, 25] formulated the bandwidth allocation problem as a utility maximization over source rates with capacity constraints. A distributed algorithm is also provided by Kelly [25] to globally solve the penalty function form of this optimization problem. This algorithm is called the primal algorithm where the sources adapt their rates dynamically, and the link prices are calculated by a static function of arrival rates. Low and Lapsley [26] proposed a gradient projection algorithm to solve its dual problem. It is shown that this algorithm globally converges to the exact solution of the original optimization problem since there is no duality gap. This approach is called the dual algorithm, where links adapt their prices dynamically, and the users' source rates are determined by a static function.

In our previous works [27, 28] a multidisciplinary conceptual framework has been proposed that provides principles for designing and analyzing bio-inspired congestion control algorithms. We proposed that the biological population control approaches such as predator-prey is susceptible for mapping to the congestion control problem in the Internet. This paper is an extended version of [27] and attempts to examine how this predator-prey interaction can be adapted in order to apply and gain a new solution to the dynamics of the Internet. The goal is to show how this model can address the scalability, stability, the fairness and the performance of the communication networks. We design a bio-inspired equation-based congestion control mechanism that has two significant features. First, it can inherent some intrinsic characteristics of biology such as stability and robustness. Second, since we use the mathematical models of biology then it provides us a theoretic and mathematical framework that we can benefit from its facilities in analysis and development of the proposed model.

Section 2, briefly presents a conceptual framework for Biologically Inspired Congestion Control (BICC) and also explains analogy between the biological environment and the communication networks. Section 3 presents a methodology for applying the predator-prey mathematical model to the Internet congestion control scheme. Section 4 presents an illustrative example for the proposed algorithm. The implementation consideration for the proposed algorithm will be discussed in section 5 and we conclude in section 6 with future works.

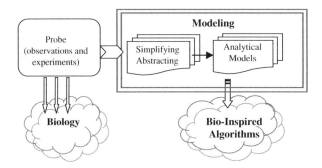

Fig. 1. An outline conceptual framework for a bio-inspired computational domain

2 Conceptual Framework: Internet as an Ecosystem Analogy

The first step in bio-inspired computation should be to develop more sophisticated biological models as sources of computational inspiration, and to use a conceptual framework to develop and analyze the computational metaphors and algorithms. We believe that bio-inspired algorithms are best developed and analyzed in the context of a multidisciplinary conceptual framework that provides for sophisticated biological models and well-founded analytical principles. Figure 1 illustrates a possible structure for such a conceptual framework. Here probes are used to provide a (partial and noisy) view of the complex biological system. This limited view, build and validate simplifying abstract representations and models of the biology. These biological models are used to build and validate analytical computational frameworks. Validation may use mathematical analysis, benchmark problem, and engineering demonstrators. We borrow some mathematical models from biology and use these frameworks for designing and analyzing bio-inspired algorithms applicable to congestion control problem, contain as much or as little biological realism as appropriate. This is necessarily an interdisciplinary process, requiring collaboration between (at least) biologists, mathematicians, and computer scientists to build a complete framework.

2.1 Internet Ecology

Consider a network with a set of k source nodes and a set of k destination nodes. We denote $S = \{S_1, S_2, \ldots, S_k\}$ as the set of source nodes with identical round-trip propagation delay (*RTT*), and $D = \{D_1, D_2, \ldots, D_k\}$ as the set of destination nodes. Our network model consists of a bottleneck link from LAN to WAN as shown in Figure 2 and uses a window-based algorithm for congestion control. The bottleneck link has capacity of B packet per *RTT*. The congestion window (W) is a sender-side limit on the amount of data the sender can transmit into the network before receiving an acknowledgment (*ACK*). We assume that all connections are long-lived and have unlimited bulk data for communication.

We propose that, this network can be imagined as an ecosystem that connects a wide variety of habitats such as routers, hosts, links and operating systems (OS), and etc. We can consider this network from congestion control viewpoint and assume that there are some species in these habitats such as "Congestion Window"(W), "Packet Drop"(P), "Queue"(q) and "link utilization" (u). The size of these network elements refers to their population size in Internet ecosystem. Figure 3 shows the typology of Internet ecosystem from congestion control perspective. In this ecosystem the species are interacting and hence, the population sizes of the species are affected. In general, there is a whole web of interacting species, which makes structurally complex communities.

Let the population of W in source S_i be W_i (congestion window size of connection i). It is clear that if the population size of this species is increased, then the number of sent packet would be inflated. Hence, in order to control the congestion in the communication networks the population size of W (all of the W_is) must be controlled. This means that *the population control problem in the nature can be mapped to the congestion control problem in the communication networks*. We can use the natural population control tactics for this purpose. Nature uses many tactics such as predation, competition, parasites and etc. to control the population size of species. In this paper a methodology is proposed to use the predation tactic to control the population size of W species.

2.2 Predation

In this work the predation tactic is used to design congestion control algorithms in communication networks. So, according to the proposed conceptual framework, it is needed to build and validate simplifying abstract representations and models of the predation [30]:

Predation: This interaction refers to classical predators that kill individuals and eat them. We can summarize this interaction as below:

(1) In the absence of predators, prey would grow exponentially. **(2)** The effect of predation is to reduce the prey's growth rate. **(3)** In the absence of prey, predators

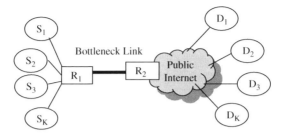

Fig. 2. General Network Model

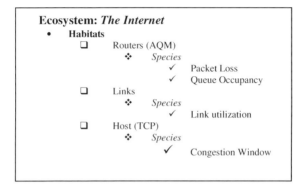

Fig. 3. Internet Ecosystem Typology

will decline exponentially. **(4)** The prey's contribution to the predator's growth rate is proportional to the available prey as well as to the size of the predator population. **(5)** A prey's carrying capacity puts a ceiling (i.e., a limit from outside) on the prey population oscillations, and so this puts a limit on the predator population oscillations and the ability of the predator to drive prey oscillations.

The Lotka-Volterra predator-prey model [30, 31] describes interactions between predator and prey species. Since we are considering two species, the model will involve two equations, one, which describes how the prey population changes, and the second, which describes how the predator population changes. For concreteness, let us assume that the preys in our model are rabbits, and that the predators are foxes. If we let r and f represents the number of rabbits and foxes, respectively, then the Lotka-Volterra model is:

$$\frac{dr}{dt} = ar - brf \qquad (1)$$

$$\frac{df}{dt} = crf - hf \qquad (2)$$

Where the parameters are defined by:
a is the natural growth rate of rabbits in the absence of predation. b is the death rate per encounter of rabbits due to predation. c is the efficiency of turning predated rabbits into foxes. h is the natural death rate of foxes in the absence of food (rabbits).

3 Congestion Control Based on Predator-Prey Model

This section describes two different issues. First it discusses on similarities of current Internet congestion control mechanism (TCP/AQM) and predator-prey interaction. Second, it outlines a methodology to design a new congestion control mechanism that has inspired from predator-prey interaction.

3.1 Similarities of TCP/AQM and Predator-Prey

In order to clarify the similarity between TCP/AQM congestion control mechanism and predator-prey interaction, we use two different approaches for explaining the TCP/AQM running on a network: descriptive exposition and mathematical models.

Descriptively, TCP/AQM can be explained as follows. **(1)** In the absence of packet drop (P), congestion window (W) would grow. **(2)** On the occurrence of a packet drop, congestion window size would decline. **(3)** Incoming packet rate contribution to packet drop rate growth is proportional to available traffic intensity, as well as, the packet drop rate itself. **(4)** In the absence of a packet stream, packet drop rate will decline. **(5)** Bottleneck bandwidth is a limit for packet rate (carrying capacity). We see that this behavior is so similar to the predator-prey interaction.

For mathematical comparison of TCP/AQM and predator-prey models we use the mathematical models of TCP/AQM. In [4] there are some mathematical models that explain the TCP/AQM. For example equations set (3)–(4) shows the fluid model of TCP/AQM, where p_l is packet drop probability at link l and $x_i = W_i/T_i$ is rate of source i. It is clear that there are considerable similarities between equations sets (1)–(2) and (3)–(4).

$$x_i(t+1) = x_i(t) + \frac{\gamma_i}{T_i} \left\langle \alpha_i - x_i(t) \sum_l R_{li} p_l(t) \right\rangle \qquad (3)$$

$$p_l(t+1) = p_l(t) + \frac{1}{c_l} \left\langle \sum_i R_{li} x_i(t) - c_l \right\rangle \qquad (4)$$

A similar discussion can also be carried out to show that the interaction of W and q (queue size in congested router) is also similar to the predator–prey interaction. These similarities are another motivation to use predator-prey model to design congestion control scheme.

3.2 Congestion Control Inspiring by Predator-Prey

Assume that there are two species P and q that can prey W species and can control its population size. Note that P and q species are not exactly the packet drop count and the queue size. At this point only consider that they are predator species living in the congested router and can control the population size of W. Later we will discuss about their interpretation in the network context. Since according to the Figure 2,

W contains k species $(W_1, W_2, \ldots, W_k)$, so suppose that P have also k species $(P_1, P_2, \ldots, P_k)$ in the congested router. P_i can prey all the W individuals but there can be a weighting preference for P_i to prey W_i several times more significant to all other $W_j (j \neq i)$ individuals.

To specify a congestion control system, these questions must be answered: (i) How do the sources adjust their rates based on their aggregate prices (the TCP algorithm)? (ii) How do the links adjust their packet mark rates based on their aggregate and local information (the AQM algorithm)? We can in general postulate a dynamic model of the form

$$\dot{W}_i = F(W_j, Price_j) \qquad i, j = 1, 2, \ldots, k$$
$$\dot{Price}_i = Y(Price_i, W_j) \qquad i, j = 1, 2, \ldots, k$$

In this model $Price_i$ can be calculated based on P_is and q, hence this dynamic model can be rewritten as follows:

$$\dot{W}_i = F(W_i, P_j, q) \qquad i, j = 1, 2, \ldots, k$$
$$\dot{P}_i = G(P_i, W_j) \qquad i, j = 1, 2, \ldots, k$$
$$\dot{q}_i = H(W_j, q) \qquad j = 1, 2, \ldots, k$$

Since we adopted predator–prey interaction for population control of W, hence, the generalized Lotka–Volterra predator–prey system is used to drive F, G and H. In this model $k+1$ species $P_1, P_2, \ldots, P_k$ and q predate and control the population size of the species $W_1, W_2, \ldots, W_k$. This deliberation leads to the following bio-inspired distributed congestion control algorithms (BICC) in which Eq. (5) explains the evolution regime of congestion windows size and (6) and (7) explain the evolution of its predators i.e. P_is and q:

$$\frac{dW_i}{dt} = W_i \left\langle h_i - \sum_{j=1}^{k} b_{ij} P_j - r_i q \right\rangle \qquad i = 1, \ldots, k \qquad (5)$$

$$\frac{dP_i}{dt} = P_i \left\langle \sum_{j=1}^{k} c_{ij} W_j - d_i \right\rangle \qquad i = 1, \ldots, k \qquad (6)$$

$$\frac{dq}{dt} = \left\langle \sum_{j=1}^{k} e_j W_j - m \right\rangle \qquad (7)$$

Where the parameters are defined by: h_i is the growth rate of W_i in the absence of P_is and q. b_{ji} is the decrement rate per encounter of W_i due to P_j. r_i is the decrement rate per encounter of W_i due to q. d_i is the decrement rate of P_i in the absence of W_is. c_{ij} is the efficiency of turning predated W_j into P_i. e_j is the efficiency of turning predated W_j into q and m is set to $Min(B, q + W_i)$. We set time step of this system to RTT i.e. $dt = RTT$, by this setting RTT is included in the model.

3.3 Discussion on Behavior and Parameters Setting

BICC is a unified source/AQM system that addresses both the source and the AQM algorithms in the network model by Eqs. (5)–(7). Eq. (5) refers to source algorithm and Eqs. (6)–(7) describe the AQM algorithm. Assuming packets are probabilistically marked, the AQM part of BICC must address three issues: (*i*) How the congestion is quantified. (*ii*) How the congestion quantity is embedded in the probability function. (*iii*) How the probability is fed back to the sources.

Any source i that implements BICC, maintains and updates a variable that we call it $Price_i$, as a congestion measure:

$$Price_i = \sum_{j=1}^{k} b_{ij} P_j + r_i q \tag{8}$$

According to (5) $Price_i$ is used to adjust sending rate at the sources:

$$\frac{dW_i}{dt} = W_i \langle h_i - Price_i \rangle$$

When the numbers of active sources increases or source rates are increased, P_is and q grow, pushing up $Price_i$ and hence marking probability. This sends a stronger congestion signal to the sources which then reduce their rates. When the source rates are too small, the P_is and q will be small, pushing down $Price_i$ and raising source rates, until eventually, the $Price_i$ is driven to equilibrium. In equilibrium, the queue size, the window size and the P_is are stabilized and hence, the following conditions are held:

$$\frac{dW_i}{dt} = W_i \langle h_i - Price_i \rangle = 0$$

$$\frac{dP_i}{dt} = P_i \left\langle \sum_{j=1}^{k} c_{ij} W_j - d_i \right\rangle = 0 \quad where\ i = 1, \ldots, k$$

$$\frac{dq}{dt} = \left\langle \sum_{j=1}^{k} e_j W_j - m \right\rangle = 0$$

These conditions lead to Eqs. (9)–(11) (note that if W_is and P_is was zero then the system would be unstable):

$$\sum_{j=1}^{k} b_{ij} P_j + r_i q = h_i \tag{9}$$

$$\sum_{j=1}^{k} c_{ij} W_j = d_i \quad where\ i = 1, 2, \ldots, k \tag{10}$$

$$\sum_{j=1}^{k} e_j W_j = m \tag{11}$$

Eqs. (9)–(11) depict the fact that the steady state behavior of the BICC is significantly affected by the choice of the control parameters such as b_{ij}s, c_{ij}s, e_js, h_is and d_is. Other characteristics such as stability and convergence time are also affected by these parameters. In what follows, we discuss how a good parameter configuration can help us to achieve better results.

We use the following settings for the parameters of the proposed model in which k is the number of sources sharing congested link and can be estimated by some techniques as in [32, 33].:

$$\forall i,j \quad b_{ij} = c_{ij} \tag{12}$$

$$\forall i,j \quad for\ j \neq i \quad c_{ii} = c_1 = 0.9\ and\ c_{ij} = c_2 = \frac{1-c_1}{k}$$

$$hence\ c_{ii} \succ\succ c_{ij}\ and\ \sum_{j=1}^{k} c_{ij} = 1 \tag{13}$$

$$\forall i \quad h_i = 0.25, d_i = d = 0.9B/k, r_i = r = 0.02\ and\ e_i = 1 \tag{14}$$

We expect that, these settings lead to the following results:

- According to the (10), (13) and (14), due to symmetry of c_{ij} coefficients, all of the W_is will be equal to the d in equilibrium. This means that the proposed scheme satisfy max-min fairness [34, 35].
- By summing up Eqs. (10), aggregate traffic on the bottleneck link can be computed as follows:

$$\sum_{i=1}^{k}\sum_{j=1}^{k} c_{ij}W_j = kd \implies \sum_{i=1}^{k} W_i = 0.9B \tag{15}$$

Eq. (15) shows that in equilibrium the bottleneck link is utilized around 90 percent. Since $\sum Wj = 9B \leq B$, then in the steady-state the input rate is less than true link capacity and hence, according to Eq. (7), q approaches to zero. Note that we have used the coefficient of 0.9 conservatively to avoid queue inflation.
- The ratio of c_{ij}/c_{ii} indicates the degree of influence of the connection j that share the same bottleneck link with connection i. To converge window size i to a positive value despite its share of bandwidth d_i it is necessary to satisfy the condition $0 < c_{ij}/c_{ii} < 1$. Furthermore, smaller c_{ij}/c_{ii} leads to faster convergence. On the other hand according to the mathematical biology, in order to establish a stably operated ecology effect of self-inhibitive action must be larger than inhibitive action by others [36]. With these backgrounds any c_{ii} must be several times larger than any c_{ij}. These conditions can be held by (13).
- According to the formal definition of queue dynamics any e_i must be set to 1. Degree of influence of queue length on $Price_i$ is indicated by r_i. We set r_i to 0.02. These are reflected in (14).
- According to (6) P_i refers to rate mismatch i.e. difference between input rate and target capacity (d_i). P_i can also be interpreted as an unfairness measure

for source i i.e. P_i is incremented if source i uses more than its fair share of bandwidth (d_i) and is decremented otherwise. q refers to queue mismatch and is positive if there is any waiting packet in queue. Hence the $Price_i$ is positive when the input rate exceeds the link capacity or there is excess backlog to be cleared and is zero otherwise. This is so similar to REM active queue management algorithm [37] in which $Price$ is updated, periodically or asynchronously, based on rate mismatch and queue mismatch.
- According to (8) equilibrium $Price_i$ will be equall to h_i. Since it is desireable that equilibrium $Price_i$ be small, hence h_i must be set to small non-zero values. As mentioned zero value for P_i and W_i leads to instablity of network so, h_i can not be set to zero. We set h_i to 0.25.

3.4 Scalability

As mentioned $Price_i$ is computed in any source that passes through congested node. According to (8) each source needs to obtain all of the P_is and also q. Since, P_is can only be computed in the router then the congested router must keep track of per flow P_is in order to communicate them to the sources. It is clear that when the number of connections is increased, the load on the router is increased too. Hence this scheme cannot be scalable by number of connections. In order to solve this problem we propose a procedure in which any source i itself computes P_i and the congested routers are responsible only to compute and communicate aggregated information to the sources:
By using Eq. (13) we can rewrite Eq. (6) as follows:

$$\frac{dP_i}{dt} = P_i \left\langle c_1 W_i + c_2 \sum_{j \neq i} W_j - d_i \right\rangle$$

$$= P_i \left\langle (c_1 - c_2) W_i + \frac{(1 - c_1) \sum_j W_j - 0.9B}{k} \right\rangle \quad (16)$$

We assume that c_1 is a global variable in the network that is set to 0.9 and k is large enough to regard $c_1 \gg c_2$. Hence, Eq. (16) can be rewritten as follows:

$$\frac{dP_i}{dt} = P_i \left\langle c_1 W_i + \frac{(1 - c_1) \sum_j W_j - 0.9B}{k} \right\rangle = P_i \langle c_1 W_i + P_g \rangle \quad (17)$$

$$P_g = \frac{(1 - c_1) \sum_j W_j - 0.9B}{k} \quad (18)$$

According to Eq. (18) it is clear that P_g can only be computed in the router. Eq. (17) says that if P_g is computed in the router and is communicated to all of the sources that share it then any source i can compute P_i. In other words the router doesn't have to compute all of the P_is and only computes P_g by using local and aggregated information.

The router also monitors the queue length (q) of the congested link and disseminates it to all of the sources that share the congested link.

Any source i that receives P_g computes new P_i by using Eq. (17). Since $c_1 \gg c_2$, then by an acceptable estimation (if we assume that all of the P_is are equal that take place around equilibrium) Eq. (5) can be rewritten as follows:

$$\frac{dW_i}{dt} = W_i \langle h_i - P_i - r_i q \rangle \tag{19}$$

Eq. (19) says that if source i obtains P_g and q then it can compute its new window size.

This procedure leads to the following results: First, since BICC uses only local and aggregated information in the router and in the sources, so it is scalable by complexity and number of flows. Second, the loads of routers don't increased by number of flows. Third, the coefficient of P_i in (17) shows that dP_i/dt is increasing function of the current P_i. This means that any source is a learning entity and its reaction to congestion measure depends on the current status of network. For example in the case of heavy congestion that can be realized by large P_is, small rates mismatch or a small queue length mismatch leads to a strong congestion feedback in the sources.

4 Illustrative Example

In this section we discuss about the behavior of proposed model through of an example. We use a four-connection network, as given in figure 2, and assume the network with a single bottleneck link of capacity 50 pkts/RTT (for example if $RTT = 20$ ms and packet size $= 1500$ Byte then this capacity will refer to capacity of 30 Mbps), shared by 4 sources. All other links have bandwidth of 100 pkt/RTT. Suppose that all flows are long-lived, always are active and have the same end-to-end propagation delay.

According to Eqs. (5)–(7) and (12)–(14), Eqs. (20)–(28) describe the congestion control regime of the example network. Considering the following initial state we use Matlab 7.1 to solve Eqs. (20)–(28) and simulate the network behavior:
$P_1(0) = P_2(0) = P_3(0) = P_4(0) = 0.1, q(0) = 1, W_1(0) = 1, W_2(0) = 2, W_3(0) = 4, W_4(0) = 6$

$$dW_1/dt = W_1(0.25 - 0.9P_1 - 0.033P_2 - 0.033P_3 - 0.033P_4 - 0.02q) \tag{20}$$

$$dP_1/dt = P_1(0.9W_1 + 0.033W_2 + 0.033W_3 + 0.033W_4 - 11.25) \tag{21}$$

$$dW_2/dt = W_2(0.25 - 0.033P_1 - 0.9P_2 - 0.033P_3 - 0.033P_4 - 0.02q) \tag{22}$$

$$dP_2/dt = P_2(0.033W_1 + 0.9W_2 + 0.033W_3 + 0.033W_4 - 11.25) \tag{23}$$

$$dW_3/dt = W_3(0.25 - 0.033P_1 - 0.033P_2 - 0.9P_3 - 0.033P_4 - 0.02q) \tag{24}$$

$$dP_3/dt = P_3(0.033W_1 + 0.033W_2 - 0.9W_3 + 0.033W_4 - 11.25) \tag{25}$$

$$dW_4/dt = W_4(0.25 - 0.033P_1 - 0.033P_2 - 0.033P_3 - 0.9P_4 - 0.02q) \quad (26)$$
$$dP_4/dt = P_4(0.033W_1 + 0.033W_2 - 0.033W_3 + 0.9W_4 - 11.25) \quad (27)$$
$$dq/dt = W_1 + W_2 + W_3 + W_4 - min(q + W_1 + W_2 + W_3 + W_4, 50) \quad (28)$$

Figures (4)–(7) shows the simulation results. Figure 4 illustrates the behavior of the sources sharing the bottleneck link and shows the time curves of the congestion windows sizes. Figure 5 illustrates the behavior of the congested router and shows the evolution of P_is. The throughputs of bottleneck link, has been given in Figure 6.a. This throughput refers to aggregate loads of the sources on the bottleneck link. In Figure 6.b we can find the queue size of congested router and finally Figure 7 shows the phase-plane trajectories of (W_i, P_i) that are used for stability analysis.

We discuss about these results from two points of view: equilibrium and dynamic.

4.1 Equilibrium

The following metrics are used for equilibrium analysis:

Fairness: The congestion control is not satisfactory if it does not ensure a fair sharing of network resources. There are several ways of defining and reaching fairness in a network, each one leading to a different allocation of link capacities [34, 35]. We are talking here about fairness in the sharing of the bandwidth of the bottleneck link regardless of the volume of resources consumed by a connection on the other links of the network. This kind of fairness is called in the literature the max-min fairness. Average of W_is and P_is are summarized in table 1. This table shows that in spite of inequality of initial states, all sources enjoy same amount of bandwidth in the bottleneck link. We can see also that all of the P_is are equal.

Table 1. Average throughput and Pi of sources in BICC

Traffic sources	Mean of W_i	Mean of P_i
Source 1	11.26	0.2342
Source 2	11.26	0.235
Source 3	11.26	0.2363
Source 4	11.26	0.2366

Network Throughput and Utilization: It is a clear goal of most congestion control mechanisms to maximize throughput, subject to application demand and to the constraints of the other metrics. In most occasions, it might be sufficient to consider the aggregated throughput only. According to Figure 6.a, simulation results show that the average of aggregated traffic on the congested link is *45.05 pkt/RTT*. This throughput refers to over *90%* utilization on the bottleneck link that is good enough.

Equilibrium Queues: Figure 6.b shows that the queue size stabilizes to zero in the steady state. The small size of queue length, leads to low jitter and low delay for those connections that share this queue capacity.

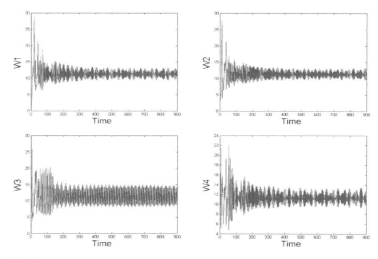

Fig. 4. Evolution of congestion window size of the sources (W_1, W_2, W_3, W_4)

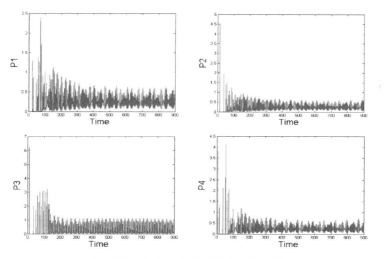

Fig. 5. Evolution of P_is (P_1, P_2, P_3, P_4)

Packet Drop: If we set the queue capacity of the congested router around 30 packets then there won't be any packet loss in this scenario.

4.2 Dynamics

Stability: Equilibrium considerations are meaningful if the system can operate at or near this point. For this reason we pose as a basic requirement, the stability of the equilibrium point. Ideally, we would seek global stability from any initial condition, but at the very least, we should require local stability from a neighborhood. This objective is sometimes debated, since instability in the form of oscillations could

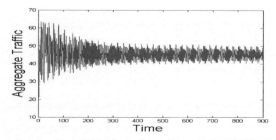

a. Evolution of aggregated traffic on the bottleneck link ($W_1+W_2+W_3+W_4$)

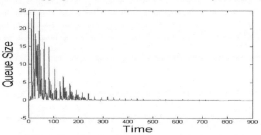

b. Evolution of queue length in congested router (q)

Fig. 6. Aggregated traffic and queue trace

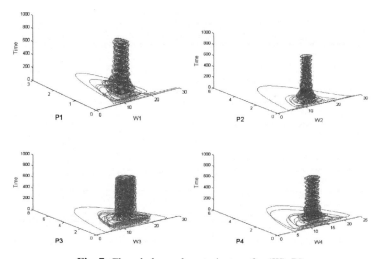

Fig. 7. Closed phase plane trajectory for (W_i, P_i)

perhaps be tolerated in the network context, and might be a small price to pay for an aggressive control. Typically, analysis of linear stability of differential equations can be carried out using community matrix and its eigenvalues examination [38, 39]. But in order to simplify we use the numerical solution of (20)-(28) and use phase-plane approach for stability analysis in terms of minimizing oscillations of queuing delay or of throughput. Figure 7 shows the closed (W_i, P_i) phase-plane trajectories. First note that the direction of time arrows in Figure 7 is clockwise. This is reflected in

Figure 4 and Figure 5. We can see that all of the trajectories converge around the point (11.26, 0.235), and so have a plausible stability near this point.

Figures 6.a and 6.b show stability of aggregated traffic and queue size. Figure 6.a shows that the aggregated traffic has minimizing oscillation regime and converges to 45.05 pkt/RTT. Figure 6.b shows that the queue length of congested router approaches to zero and its evolution has a stable regime.

Speed of Convergence: Convergence times concern the time for convergence to fairness between an existing flow and a newly starting one. The simulation results show slow response to the traffic demand. Furthermore, the windows, queue and P_is are not extremely smooth in the steady state.

5 Implementation Issues

During each RTT congested router monitors the incoming packet rate (ΣW_i) and uses (18) to compute new P_g. It also monitors its queue size and computes q. Then P_g and q are communicated to all of the sources that go through congested link. To communicate P_g and q router they are embedded in a probability function. For dissemination of P_g an ECN bit [40] of each packet arriving at the congested link would be marked with probability P_{pg}:

$$p_{pg} = 1 - \phi^{p_g} \quad where \quad \phi > 1 \tag{29}$$

According to (18), P_g is non positive always, so we have $0 \leq p_{pg} \leq 1$. Once a packet is marked, its mark is carried to the destination and then conveyed back to the source via acknowledgement. Source i estimate the end-to-end marking probability P_{pg} by the fraction of its packets marked, and computes P_g by inverting the (29):

$$p_g = \log_\phi (1 - p_{pg}) \tag{30}$$

With a same mechanism, for dissemination of q an additional ECN bit would be marked at the congested link with probability $p_q = 1 - \phi^{-q}$ where $\phi > 1$ and then estimated in the sources. Note that q is positive always, so we have $0 \leq p_q < 1$. The following algorithms summarize the implementation process.

BICC Algorithm: Hybrid Approach
  Congested router's algorithm:
  At time RTT, 2RTT, 3RTT,... congested router:
  1- Receives W_s packet from all of the sources $s \in S$ that goes through bottleneck link l.
  2- Computes the new P_g and q for all of the sources that use link l:

$$P_g = \frac{(1-c_1)\sum_j W_j - 0.9B}{k}$$

$$\frac{dq}{dt} = \left[\sum_{j=1}^{k} e_{ij} w_j - m\right]$$

3- *Computes* marking probability p_{pg} and p_q through

$$p_q = 1 - \phi^{-q}, \; p_{pg} = 1 - \phi^{p_g} \; where \, \phi > 1$$

4- *Uses ECN bit 1 and ECN bit 2 to communicate the new p_{pg} and p_q for all the sources that use link l.*

Source i's algorithm:
At time RTT, 2RTT, 3RTT,… source i:
1- *Receives from the congested router marked packets and extracts p_{pg} and p_q from marking statistics.*
2- *compute) P_g and q and then P_i* :

$$p_g = \log_\Phi(1 - p_{pg})$$

$$q = -\log_\Phi(1 - p_q)$$

$$\frac{dP_i}{dt} = P_i \langle c_1 W_i + P_g \rangle$$

3- *Choose a new window size for the next period:*

$$\frac{dW_i}{dt} = Wi \langle h_i - p_i - r_i q \rangle$$

The exponential form of the marking probability is critical in a large network where the end-to-end marking probability for a packet that traverses multiple congested links from source to destination depends on the link marking probability at every link in the path. When and only when, individual link marking probability is exponential in its link price, this end-to-end marking probability will be exponentially increasing in the *sum* of the link prices at all the congested links in its path. This sum is a precise measure of congestion in the path. Since it is embedded in the end-to-end marking probability, it can be easily estimated by sources from the fraction of their own packets that are marked, and used to design their rate adaptation [37].

6 Conclusion Remarks

This paper proposed that the population control mechanisms of the nature are applicable to design congestion control schemes in communication networks. Toward this idea, we have used a model based on predator-prey interaction to design a congestion control mechanism in communication networks. We used general predator-prey model and observed that with some consideration on parameters, this model leads to a stable, fair and high performance congestion control algorithm.

Although the simulation results indelicate the credibility of proposed model, but a number of avenues for future extensions remains. First our model can be enhanced to account realistic models of predator-prey. Second the model can be further extended to incorporate the effects of short flows. Third, this work needs some analytical foundations. Fourth, with mathematical characterization of network objective such as fairness, stability and etc. we can use mathematical rules for definition of parameters of purposed model to achieve well-designed communication network.

Acknowledgment

This work is done under grant of Iran Telecommunication Research Center (ITRC).

References

1. P. D'haeseleer, S. Forrest, An Immunological Approach to Change Detection: Algorithms, Analysis and Implications, *IEEE Symposium on Security and Privacy*, Oakland, CA, USA, 1996.
2. S. Hofmeyer, An Immunological Model of Distributed Detection and Its Application to Computer Security, University of New Mexico, 1999.
3. F. Dressler, Efficient and Scalable Communication in Autonomous Networking using Bio-inspired Mechanisms – An Overview, informatica 29, 2005.
4. J. Wang, A Theoretical Study of Internet Congestion Control: Equilibrium and Dynamics, PhD thesis, University of Caltech, 2005.
5. S. Floyd, Connections with multiple congested gateways in packet-switched networks part 1: One-way traffic, *Computer Communications Review*, 1991.
6. T. V. Lakshman and U. Madhow, The performance of TCP/IP for networks with high bandwidth-delay products and random loss, *IFIP Transactions*, 1994.
7. T. Ott, J. Kemperman, and M. Mathis, The stationary behavior of ideal TCP congestion avoidance, 1998.
8. M. Mathis, J. Semke, J. Mahdavi, and T. Ott, The macroscopic behavior of the TCP congestion avoidance algorithm, *Computer Communication Review*, 1997.
9. J. Padhye, V. Firoiu, D. Towsley, and J. Kurose, Modeling TCP Reno performance: A simple model and its empirical validation, *IEEE/ACM Transactions on Networking*, 2000.
10. M. Handley, S. Floyd, J. Padhye, and J. Widmer, TCP Friendly Rate Control (TFRC): Protocol specification, RFC 3168, Internet Engineering Task Force, 2003.
11. V. Misra, W. Gong, and D. Towsley, Stochastic differential equation modeling and analysis of tcp-window size behavior, 1999.
12. V. Misra, W. Gong, and D. Towsley, Fluid-based analysis of a network of AQM routers supporting TCP flows with an application to RED, *ACM Sigcomm*, 2000.
13. C. Hollot, V. Misra, D. Towsley, and W. Gong, A control theorietic analysis of RED, *IEEE Infocom*, 2001.
14. S. H. Low, F. Paganini, J. Wang, and J. C. Doyle, Linear stability of TCP/RED and a scalable control, *Computer Networks Journal*, 2003.
15. J. Aweya, M. Ouellette, and D. Y. Montuno, A control theoretic approach to active queue management, *Computer Networks,* 2001.
16. C. Hollot, V. Misra, D. Towsley, and W. Gong, On designing improved controller for AQM routers supporting TCP flows, *IEEE Infocom*, 2001.

17. K. B. Kim and S. H. Low, Analysis and design of aqm for stabilizing tcp, Technical Report, 2002.
18. H. Zhang, L. Baohong, and W. Dou, Design of a robust active queue management algorithm based on feedback compensation, *ACM Sigcomm*, 2003.
19. S. Ryu, C. Rump, and C. Qiao, Advances in active queue management (AQM) based TCP congestion control, *Telecommunication System*, 2004.
20. H. Choe and S. H. Low, Stabilized Vegas, *IEEE Infocom*, April 2003.
21. S. H. Low, L. Peterson, and L.Wang, Understanding Vegas: a duality model, *Journal of ACM*, 2002.
22. C. Jin, D. X. Wei, and S. H. Low FAST TCP: motivation, architecture, algorithms, performance, *IEEE Infocom*, 2004.
23. J. Wang, D. X. Wei, and S. H. Low, Modeling and stability of FAST TCP, *IEEE Infocom*, 2005.
24. F. Kelly, Charging and rate control for elastic traffic, *European Transactions on Telecommunications*, 1997.
25. F. P. Kelly, A. Maulloo, and D. Tan, Rate control for communication networks: Shadow prices, proportional fairness and stability, *Journal of Operations Research Society*, 1998.
26. S. H. Low and D. E. Lapsley, Optimization flow control I: basic algorithm and convergence, *IEEE/ACM Transactions on Networking*, 1999.
27. M. Analoui, Sh. Jamali, A Conceptual Framework for Bio-Inspired Congestion Control In Communication Networks, IEEE/ACM BIONETICS, 2006.
28. M. Analoui, Sh. Jamali, Inspiring by predator-prey interaction for Congestion Control In Communication Networks, CSICC2006, Iran, Tehran, 2006.
29. M. Analoui, Sh. Jamali, Bio-Inspired Congestion Control: Conceptual Framework, Case Study and Discussion, Springer CI series, 2007 (to appear).
30. S. Elizabeth, John A. Rhodes, Mathematical Models in Biology: An Introduction, Cambridge press, 2003.
31. Lotka, A., Elements of Physical Biology, Williams and Wilkins, Baltimore,1925.
32. H. Ohsaki, Y. Mera, M. Murata, and H. Miyahara, Steady state analysis of the RED gateway: stability, transient behavior, and parameter setting, *ACM SIGMETRICS*, 2000.
33. T. J. Ott, T. V. Lakshman, and L.Wong, SRED: Stabilized RED, IEEE INFOCOM'99, 1999.
34. F. Kelly, Mathematical Modeling of the Internet, Mathematics Unlimited-2001 and Beyond, Springer-Verlag, Berlin, 2001.
35. M. Analoui, Sh. Jamali, TCP Fairness Enhancement Through a parametric Mathematical Model, CCSP2005, IEEE International Conference, 2005.
36. M. Murata, Biologically Inspired Communication Network Control, International Workshop on Self-* Properties in Complex Information Systems, 2004.
37. S. Athuraliya, V. H. Li, S. H. Low, and Q. Yin, REM: active queue management. *IEEE Network*, 2001.
38. J.D. Murray, Mathematical Biology: I. an Introduction, Third Edition, Springer press, 2002.
39. George f. Simmons, differential equations (with applications and historical notes), McGraw-Hill Inc., 1972.
40. K. Ramakrishna, S. Floyd, and D. Black, The addition of explicit congestion notification (ECN) to IP, RFC 3168, Internet Engineering Task Force, September 2001.

Self-Organized Network Security Facilities based on Bio-inspired Promoters and Inhibitors

Falko Dressler

Autonomic Networking Group
Dept. of Computer Science 7,
University of Erlangen, Germany
dressler@informatik.uni-erlangen.de

Summary. Self-organization techniques based on promoters and inhibitors has been intensively studied in biological systems. Promoters enable an on-demand amplification of reactions to a particular cause. This allows to react quickly with appropriate countermeasures. On the other hand, inhibitors are capable of regulating this uncontrolled amplification by suppressing the reaction. In this paper, we demonstrate the applicability of these mechanisms in a network security scenario consisting of network monitoring elements, attack detection, and firewall devices. Previous work identified most existing detection approaches as not suitable for high-speed networks. This problem can be alleviated by separating the methodologies for network monitoring and for subsequent data analysis. In this paper, we present an adaptation algorithm that allows to manage the individual configuration parameters in order to optimize the overall system. We show the advantages of self-regulating techniques based on promoters and inhibitors that lead to maximized security and that gracefully degrade in case of overload situations. We created a simulation model to verify the algorithms. The results of the conducted simulations encourage further studies in this field.

1 Introduction

Network security facilities usually include mechanisms for attack detection and appropriate countermeasures. If employed in high-speed networks, a third component is added in order to cope with the steadily increasing amount of data and the very high bandwidths in nowadays backbone networks: network monitoring. All these components interoperate in a distributed environment. Driven by network security demands, network monitoring methods and techniques have been developed and standardized by several organizations, first of all by the IETF (Internet Engineering Task Force). Figure 1 depicts a typical scenario. Monitoring probes are used to obtain traffic statistics and to capture selected packets. This information is forwarded to associated attack detection systems, e.g. intrusion detection systems (IDS), which in turn analyze the data and close the loop by configuring firewall systems to counteract identified attacks. The same figure also introduces possible extensions to distributed attack detection by interconnecting autonomous IDS systems to share information about ongoing attacks and legitimate traffic. In this scenario, the efficiency of the security mechanisms strongly depends on the performance of all involved components. In the following, we concentrate on attack detection as one of the major applications of network monitoring, especially DDoS (distributed denial-of-service) attacks are concerned [3, 20]. Such attack detection entities rely on the quality of received monitoring data, i.e. their timeliness, correlation, and completeness. The most important issue is to prevent the attack detection system from becoming overloaded by monitoring probes, which are sending too much information. Primarily, two reasons lead to this objective. We need to prevent the detection system from becoming a target itself as well as to increase the availability of the overall system. In a distributed attack detection environment, each subsystem can perform the detection autonomously. Nevertheless, a single overloaded system can miss packets of a primary attack accompanied by a large amount of meaningless packets.

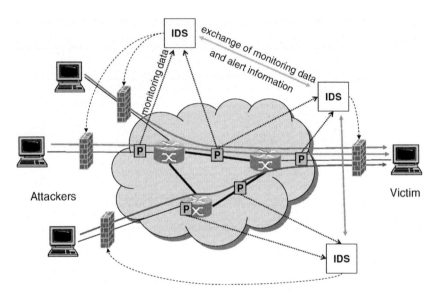

Fig. 1. Network security scenario consisting of monitoring probes, associated attack detection systems, and firewall devices

Searching for methods available to reduce the amount of monitoring data, two solutions can be found: flow monitoring and packet sampling. Flow monitoring [1] provides statistical measures of packet flows described by common properties, e.g. by the IP-5-tuple [21]. Depending on the specification of the flow characteristics, the necessary bandwidth to transport all monitoring data can be successfully reduced at cost of granularity and content. On the other hand, packet sampling is used for the selection and analysis of a few selected packets [13]. Filter are used for deterministic packet selection while statistical operations allow to select packets based on content-independent measures. All these mechanisms must be properly configured in order to achieve an optimal result. In summary, the scalability of the overall system strongly depends on three parameters: the current network load, the amount of flows that was successfully identified to be malicious or legitimate, and the configuration of the involved systems. Similar observations hold for the reliability in terms of overload prevention of subsystems, e.g. monitoring probes or attack detection systems.

Turning to nature, we figured out several commonalities between the structure of organisms and computer networks [9]. This is also true for the cellular signaling pathways and data communication, thus promising high potentials for computer networking in general and adaptive network security in particular [18]. Based on these investigations, we started to adapt mechanisms known from molecular biology to enable a self-organized operation and control in network security scenarios.

We developed and analyzed an adaptive mechanism [8, 6] for reducing the amount of monitoring data to adjust the load of involved attack detection systems. The primary goal is to monitor as much packet data as possible in order to achieve most accurate detection results. A threshold is given by the processing capacity of the involved systems, thus, an upper bound is defined. Technically, this threshold depends on the many parameters including the form and amount of monitoring data. In order to achieve this goal, we created two separate feedback loops as inspired by similar solutions found in nature [7]. These feedback loops represent promoter / inhibitor functions, i.e. they either stimulate monitoring probes to send dare, i.e. higher quality data, or they suppress this amplification effect if the detection modules approach their maximum capacity.

The following general objectives are addressed with our adaptive re-configuration scheme. *Self-maintenance* relies on the adaptation of configuration parameters depending on the environmental conditions. The autonomously working entities must be able to adapt to changing environmental conditions. *Self-healing* mechanisms respond to system failures. For example, on-demand re-configuration is required in the case of resource shortages. *Self-optimization* refers to the overall detection quality. This can be achieved by exchanging information about identified attacks or suspicious network connections and also by statistically forwarding parts of collected data packets and network statistics to neighboring probes.

We created appropriate simulation models to analyze the scalability of the developed approaches. Basically, we implemented the behavior of monitoring, firewall, and attack detection systems. In order to allow practically significant simulations and to easily compare different configurations, we used previously monitored data for trace-driven input modeling. We studied the configuration and possible adaptation of individual subsystems to increase the scalability and reliability of the overall system. It turned out that the dynamic reconfiguration depending on the current network behavior is possible without any global control, i.e. we achieve an optimized system behavior at all times.

The rest of the paper is organized as follows. Sections 2 and 3 reflect the state-of-the-art in network monitoring and present an overview to the investigates biological mechanisms, respectively. The self-organizing adaptive control scheme is depicted in section 4. The used simulation model including the discussion of selected results is presented in section 5. Some conclusions in section 6 summarize the paper.

2 Network Monitoring and Attack Detection

2.1 High-speed Network Onitoring

In this section, a general overview to monitoring solutions is provided. Due to the fact that available bandwidths grow much faster than the processing speed of the monitoring probes and subsequent analyzers, solutions have been developed that allow reducing the processing requirements at the analysis stage. The primary idea behind all these concepts is to split the monitoring and the subsequent analysis into two independent tasks and to discard as much data as possible at the monitoring stage. The first concept that was developed in this context is flow monitoring. The key idea is to store information about flows and the corresponding statistics instead of individual packets. Thereby, a flow is defined as a unidirectional connection between two end systems as defined by common properties, e.g. the IP 5-tuple (protocol, source IP address, destination IP address, source port, destination port). Using flow monitoring, a single measurement data set contains information of one up to several thousand individual packets. For the transmission of the monitoring data to an analyzing system, a special protocol was developed: Netflow.v9 [4]. Its successor, the IPFIX (IP flow information export) protocol provides sufficient information for a distributed deployment [23, 5]. Even if this methodology works well under normal conditions (usual connections consist of about 7.7 packets per flow [19]), there is a major problem during DDoS attacks. Typical attacks are using forged IP addresses, different ones for each attack packet, which results in the creation of individual flows per packet. Thus, in such an attack situation, flow monitoring does not scale well, i.e. it tends to overflow the connection between the monitoring probe and the intrusion detection system (regardless of the computational expense at the analysis). To cope with this problem, recently an aggregation mechanism was introduced [12, 10] that allows to aggregate individual flows into so called meta-flows. This aggregation mechanism allows a free scaling of the amount of monitoring data and provides the basic functionality to build adaptive self-optimizing flow-based accounting solutions.

An additional problem is based on the basic principle of flow monitoring: the loss of payload information. For intrusion detection reasons, this information is often required and, therefore, the applicability of flow monitoring is limited. To support the selection of single but complete packets and transporting them to an analyzer, PSAMP (packet sampling) was developed [14, 26]. It allows the free combination of filters and samplers. Filters are used for deterministic packet selection based on matching fields in the IP packet. Samplers are statistical algorithms that select packets using particular sampling algorithms, e.g. count-based. PSAMP therefore allows to monitor and to export complete packets providing sufficient information for intrusion detection. Additionally, the sampling algorithms, filters, and parameters can be freely defined and re-configured allowing a full-adaptive behavior. The monitoring can be optimized to provide all required data to the analyzer and no more than it is able to process.

To conclude the overview to network monitoring solutions, some strengths and disadvantages of flow accounting and packet sampling are summarized: Flow monitoring provides strong data compression in general. Unfortunately, during ongoing attacks, the number of flow records easily reaches the number ob observed packets and there is no possibility for signature-based attack detection due to the missing payload information. In case of packet sampling, the reduction depends only on the sampling algorithm, i.e. on then current configuration. The methodology-inherent loss of packet information makes the usage of such approaches in accounting and charging scenarios difficult. Therefore, even more functionality is required to cope with the growing amount of network traffic. In the next section, we discuss an approach for adaptively reconfiguration of involved components in the network security scenario.

2.2 CATS - Attack Detection using Cooperating Autonomous Detection Systems

The objective of this section is to describe an approach for attack detection using cooperative autonomous detection systems. This system, CATS [11], is one of the first approaches that provide an architecture clearly split into the mentioned three parts: monitoring, analysis, counteracting. Additionally, aspects of distributed operation are included into the monitoring part as well as into the analyzing part.

The architecture of an individual detection system is described in [11]. It consists of an outer part for network monitoring and an inner part for detection. The network monitoring part is responsible for capturing packets and flow statistics from the network, either directly using a connected network interface, or by employing monitoring probes and the standardized protocols IPFIX and PSAMP. This part also performs necessary preprocessing of the gathered data, such as packet filtering or generation of statistical flow measurements needed by the detection part. It is further divided into a layer for packet monitoring and sampling and a layer for statistical measurements. The detection part is divided into two detection engines, one providing statistical anomaly detection and the other applying knowledge-based detection mechanisms. The required packet data and statistical measures are provided by the network monitoring part.

The main reason for separating the network monitoring part and the detection part is to allow for a multi-hierarchy monitoring environment for capturing packets and flow statistics. The metering NSLP protocol [15] can be employed for the configuration of the monitoring environment. This allows for deploying one detection system that analyzes data monitored at different points of the network. Furthermore, a detection system can become itself a source of information to other detection systems by exporting monitoring data.

In the following subsections, the network monitoring part and the detection part of the detection system are described in more detail. This and additional information on CATS can be found in [11].

Packet Monitoring and Sampling Layer

The architecture of our detection system allows two ways to capture packet data from the network: by using a directly connected NIC, and by employing PSAMP exporters, which send the collected information in a standardized way. The packet monitoring and sampling layer is responsible for capturing of packet data received via NICs or PSAMP. Moreover this layer may preprocess the packet data. Filters or sampling algorithms may be applied to reduce the amount of packets being further processed. Within the detection system, the collected packet data is used for two purposes. First, it can be directly passed on to the detection part in order to look for known attack signatures. Secondly, it can be forwarded to the statistical measurement layer that generates flow statistics from the packet data. Additionally, the detection system can export packet data to other detection systems using PSAMP.

Statistical Measurement Layer

The statistical measurement layer generates statistical flow measures based on the packet data received by the packet monitoring and sampling layer, and the flow statistics received via IPFIX. Examples for statistical measures are the number of bytes and packets per flow or per aggregate, the number of connections per time, and the number of similar connections. The resulting statistical measures build the basis for further anomaly detections. For instance, an unusually high connection rate may indicate a distributed denial of service attack where typically each connection consists of only a single packet.

The statistical measurement layer does not only provide the data for the local detection mechanisms. It may also export the generated flow statistics via IPFIX. Using the terms of IPFIX, this corresponds to the functionality of an exporter or concentrator.

Attack Detection

In the detection system, we integrate two separate, independently working detection engines - an anomaly detection engine and a knowledge-based detection engine - in order to achieve high detection rates. The detection of an attack results in the generation of an event that is combined with additional information for characterizing the attack. This information can be exchanged with other detection systems in order to improve the detection performance. On the other hand, it can be used to trigger appropriate countermeasures.

The anomaly detection works on statistical data received from the lower statistical measurement layer. This detection process is looking for unusual behavior without any precognition. It compares long-time behavior to short-time behavior and maintains different profiles, e.g. per destination, aggregate, and others. Potential techniques are statistical tests, neural networks, and Bayes networks. The architecture of our autonomous detection system allows to integrate a variety of other detection algorithms. The knowledge-based approach represents the second main pillar of our detection engine. This engine searches the packet stream for known signatures and misbehaviors. Open-source tools such as Snort [24, 2] and Bro [22], which are widely used in the Internet community, build the basis for this part of the detection.

3 Biological Background

In an interdisciplinary team we are identifying appropriate mechanisms in cell biology and to adapt them to networking technology with the focus on self-organization based on adaptive feedback loops. A structural comparison of organisms and computer networks depicts that both show high similarities. Also, the communications between the systems, the signal transduction pathways, follow the same requirements [9]. Here, a specific regulation mechanism is discussed.

Some organs such as the kidney do play a central role on physiological functions and dysfunctions of the organism. For example a descent of arterial blood pressure below a critical value which will have many negative consequences for the whole body is monitored by a small population of cells in the filtration unit of the kidney [25]. As an answer to this information, these cells produce a protein (renin) which has the function to initiate a cascade of conversions and activations, respectively, of another constitutive but quiescent protein (angiotensinogen) produced by the liver and distributed in several organs. The conversion of this protein to a shorter one (now called angiotensin I) is the first step to form the right answer for solving the initial problem. Further proteins are necessary for the formation of this final answer. The protein ACE (angiotensin converting enzyme) further modulates this protein, angiotensin I by cleaving it into the short and potent protein angiotensin II. This protein represents the final answer which now has many effects on different cells in different organs in order to increase the blood pressure to normal level. On the one hand, angiotensin II stimulates the production of further protein signals in the adrenal gland, which is e.g. a hormone called aldosterone. This protein in turn stimulates the retention of Na+ ions in the kidney which finally has consequences for the blood volume regulation. On the other hand, angiotensin II stimulates the contraction of smooth muscle cells surrounding blood vessels within the kidney. Finally, angiotensin II also activates the production of the hormone vasopressin in the adenohypophysis in the brain which finally plays a role in the blood volume regulation. All these effects enhance the blood pressure in the whole body. The complete procedure is depicted in figure 2.

Looking at one of the target cells of angiotensin II in the kidney or smooth muscle cells, the protein binds to certain receptors on the cell surface. This binding induces an intracellular signal transduction cascade that finally results in the aforementioned actions to increase the blood pressure. A molecular negative feedback mechanism finishes the whole cellular reaction. If all receptor are bound by angiotensin II, the reaction is blocked which in turn also blocks the primary conversion of angiotensinogen to angiotensin II in the way that the initial renin secretion is blocked. Therefore, this mechanism describes a very effective remote and local control of the blood pressure which plays a central role in the body.

In summary, renin is a promoter for the development of angiotensin II, which in turn works as an inhibitor for the production of renin. A smooth self-regulation is the result of this feedback loop [17]. We will describe the application of this methodology for the envisioned network security scenario in the next section.

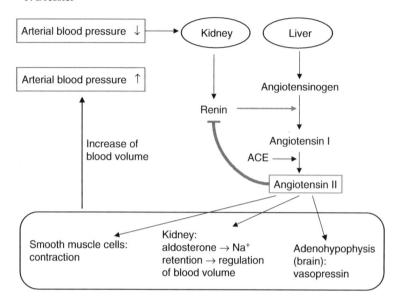

Fig. 2. Overview of the regulation of blood pressure (signaling cascades including a molecular negative feedback mechanism)

4 Adaptive Control Scheme

Our developed adaptive control scheme depends on the current network behavior, i.e. on the observed traffic as well as on the current state of the individual subsystems. The primary intention of this approach is to prevent overload situations. In this section, we outline the used models and the application of previously depicted biological promoter/inhibitor principles. The developed model is the basis for the conducted performance measures.

4.1 Modeling the Security Solution

The overall goal is to adapt the rate of packets sent to the attack detection system. The following three methods can be used to regulate the rate of monitored data: *Compression/encoding* – Monitored packet data can be encoded in a way reducing the number of transmitted bytes to the minimum. This is done using IPFIX and associated aggregation mechanisms. *Statistical sampling* – In many cases statistically chosen packets can be successfully used to determine possible attacks or to identify legitimate traffic. The PSAMP framework specifies a number of sampling algorithms that fulfill the requirements of anomaly detection systems. *Blacklists/whitelists* – Usually, blacklists represent hosts involved in an attack and whitelists represent legitimate traffic. Blacklists in packet filtering systems, i.e. firewalls, represent a functionality having two advantages. First, the packets are prevented from reaching the systems under attack and, secondly, these packets also no longer reach the monitoring system. Therefore, the data rate being sent by the monitoring systems to the attack detection systems is reduced. Additionally, whitelists are used at the monitoring probes to reduce the amount of data transmitted to the analyzing systems.

Adaptive flow aggregation has been addressed by Hu et al. [16], proposing to adapt the aggregation level dynamically according to the available resources of the monitoring probe. Although this approach avoids flow loss in DDoS attack situations as described above, it does not take into account that arbitrarily defined flow aggregates usually do not meet the requirements of the analyzer.

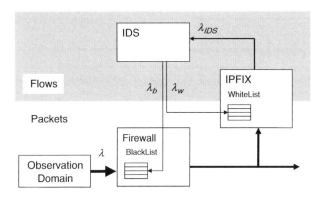

Fig. 3. Basic model for adaptive re-configuration including the main components and the control and data flows

In figure 3, a model is shown that represents the considered architecture. On the lower half, the packet-oriented monitoring part is shown. The observation domain reflects the network behavior. All data packets regardless of their content are received by the overall system. In a first step, a firewall component is used to filter all packets that belong to previously detected attack flows. A blacklist is involved here that can be configured dynamically by an IDS system. In a subsequent step, the monitoring probe is used to gather packet information, invoke flow accounting and/or packet sampling algorithms, and to transmit the monitored data sets to the attack detection system. We introduce a whitelist, also configured by the IDS, representing all legitimate flows. Therefore, appropriate filters can be used at the monitoring probe to reduce the amount of data packets to be processed. Finally, an IDS system is used to process the monitoring data. All well-detected flows, either attack or legitimate, are communicated to the firewall or monitoring systems, respectively. The depicted parameters as used in figure 3 represent possible measures to adapt the system behavior. The total arrival rate λ can be decomposed into attack traffic λ_b, legitimate traffic λ_w, and unknown data flows $\lambda - \lambda_b - \lambda_w$. λ_{IDS} is a system parameter describing the maximum data rate that can be processed by the attack detection system. Finally, λ_{bi} is the arrival rate of 'black' packets belonging to the i-th flow, i.e. representing the possibility to optimize the system behavior depending on particular attack flows.

4.2 Biologically Inspired Promoters and Inhibitors

As previously mentioned, bio-inspired methodologies can be used to create appropriate feedback loops for adapting the system parameters. In our system, we want to adapt the parameters of the monitoring environment depending on the load of the detection system and of the current network behavior. Usually, two kinds of feedback loops are used in combination: positive feedback for short-term amplification and negative feedback for long-term regulation. Both loops are depicted in figure 3. The intrusion detection reports detected attacks to the firewall that in turn blocks this traffic and, therefore, reduces the number of packets that have to be monitored. Additionally, the detection system reports legitimate traffic to the monitor. This monitor stops reporting on packets belonging to these flows and, therefore, reduces the number of packets that have to be analyzed. Obviously, both configurations cannot be permanent. Sources sending legitimate traffic might begin to send attack packets at any time. Also, previously attacking machines might become 'corrected' and should not be starved by our firewalls.

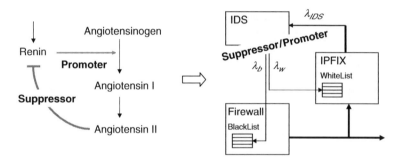

Fig. 4. Adapting the biological model to network monitoring; promoter and suppressor mechanisms are needed

In contrast to other system that dynamically configure firewall rules based on attack detection results, we introduce two other basic methods:

(i.) direct configuration of filters at the monitoring probes corresponding to legitimate data flows and
(ii.) associated timeout values to each blacklist and whitelist entry corresponding to the current overall system load.

The adaptation of biological promoters and inhibitors is depicted in figure 4. The self-regulating process that amplifies the production of angiotensin II by the production of renin is reflected by the measurement of λ_b and λ_w. If there are resources available at the attack detection system, the amplification effect is initiated by decreasing the associated timeouts for entries in the backlists and whitelists. Similarly, the suppressing reaction, i.e. angiotensin II inhibits the production of renin, is modeled by modifying the timeouts in the opposite direction in case of overload situations. Thus, a methodological approach can be developed that is based on the parameterization being adapted to the current situation in the network. The attack detection system can communicate its current load to the monitoring probes. These can adapt a number of parameters, usually timeouts, based on the number of packets received from the network, the number of packets reported to the detection systems, and the current load of the analyzers.

4.3 Mathematical Modeling

Finally, the adaptation is done using the following formulas for calculating appropriate timeouts. TO_{black} as estimated by (1) corresponds to the firewall system. Obviously, this timeout depends on the measured rate of attack and legitimate flows as well as on the maximum load of the IDS system. Particular interest is expressed in specific attack flows, or more precisely in the long-term behavior of such flows. Therefore, the load of attack flows is measured for each flow separately by λ_{bi}. We used this measure to punish flows that reappear multiple times in the blacklist. The constants C_1 and C_2 explicitly define the system behavior. Two different values are required because of the different scaling of t_1 and $t_2...t_4$.

$$TO_{black} = C_1 \underbrace{\frac{\lambda_{bi}}{\lambda}}_{t_1} + C_2 \left(\underbrace{\frac{\lambda_b}{\lambda}}_{t_2} + \underbrace{\frac{\lambda}{\lambda_w}}_{t_3} + \underbrace{\frac{\lambda}{\lambda_{IDS}}}_{t_4} \right) \quad (1)$$

On the other hand, TO_{white} as computed by (2) corresponds to the monitoring probe. Here, we only focus on the behavior of the firewall system and the capacity of the IDS represented by λ_b and λ_{IDS}, respectively. Again, C_3 is used to define the system behavior.

$$TO_{white} = C_3 \left(\underbrace{\frac{\lambda}{\lambda_{IDS}}}_{t_5} + \underbrace{\frac{\lambda}{\lambda_b}}_{t_6} \right) \quad (2)$$

The single terms ($t_1...t_6$) are discussed in the following. In principle, all terms belonging to one timeout are summed up and scaled by a constant ($C_1...C_3$). In our simulation experiments, we tried to find appropriate values for these constants. In a next step, the constants themselves can be adapted to the current scenario.

- t_1 – Ratio of the i-th attack flow to the overall attack rate. Used for penalizing previously discovered attack flows. This term must be scaled separately using C_1 because it is usually very small.

$$t_1 = \begin{cases} \frac{\lambda_{bi}}{\lambda} & \text{if } \lambda \neq 0 \\ 0 & \text{if } \lambda = 0 \end{cases} \quad (3)$$

- t_2 – Similar to t_1 but it defines the ratio of arriving attack traffic to the overall throughout. The larger t_2 is, the more aggressive the attack.

$$t_2 = \begin{cases} \frac{\lambda_b}{\lambda} & \text{if } \lambda \neq 0 \\ 0 & \text{if } \lambda = 0 \end{cases} \qquad (4)$$

- t_3 – This term describes the safety of the arriving traffic. The larger the amount of "white" packets, the smaller the requirement for large timeouts at the firewall.

$$t_3 = \begin{cases} \frac{\lambda}{\lambda_w} & \text{if } \lambda_w \neq 0 \\ 0 & \text{if } \lambda_w = 0 \end{cases} \qquad (5)$$

- t_4 – This term is a measure for the overload of the attack detection system.

$$t_4 = \begin{cases} \frac{\lambda}{\lambda_{IDS}} & \text{if } \lambda_{IDS} \neq 0 \\ 0 & \text{if } \lambda_{IDS} = 0 \end{cases} \qquad (6)$$

- t_5 – Similar to t_4 but used at the monitor instead of the firewall system.

$$t_5 = \begin{cases} \frac{\lambda}{\lambda_{IDS}} & \text{if } \lambda_{IDS} \neq 0 \\ 0 & \text{if } \lambda_{IDS} = 0 \end{cases} \qquad (7)$$

- t_6 – Similar to t_3 but defining the potential risk of arriving packets.

$$t_6 = \begin{cases} \frac{\lambda}{\lambda_b} & \text{if } \lambda_b \neq 0 \\ 0 & \text{if } \lambda_b = 0 \end{cases} \qquad (8)$$

5 Simulation Model and Evaluation

5.1 Simulation Model

In order to evaluate the potentials of the described feedback-based adaptation, we implemented a simulation model using the JAVA-based discrete simulation tool AnyLogic. The simulation model is depicted in figure 5. Several meters have been included to measure the performance of the overall system. In all setups, we executed a set of simulations showing the reduction of packet data that is to be received and processed by the attack detection systems. The simulation model has to be interpreted as follows. The observation domain 'creates' packet data (see below) that is inspected by the `observationDomainLinkMeter`. Afterwards, a firewall element is used to filter packets according to the current blacklist configuration. The packets arriving at the monitor are measured by the `firewallLinkMeter`. Depending on the simulation setup, the monitor is providing IPFIX or PSAMP data (after applying the whitelist) that is measured by the `exportLinkMeter`. Finally, the IDS selects packets to belong to attack or legitimate traffic according to a probabilistic scheme. The results are used to adapt the blacklist and whitelist entries. All measured information is transmitted to the `meterLogger` object for subsequent performance analysis.

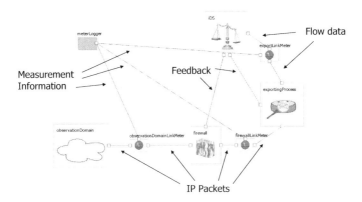

Fig. 5. Simulation model including message types on the communication channels: a) IP packets, b) flow data, c) feedback, d) measurement information

For reasonable comparability between the simulation environment and real communication behavior, we decided to use trace-based input modeling. We accumulated several traces in front of our workgroup server and directly at the Internet gateway of our university. As am example, figure 6 shows the throughput as observed by the monitor at the Internet gateway. The utilization is quite constant (58.6kpps ± 1.8kpps) as shown in figure 6 left. Due to few attacks, the firewall removes only few flows (around 0.5%). Therefore, the packet rate arriving at the monitor is quite the same but shaped (58.2kpps ± 1.8kpps, figure 6 right).

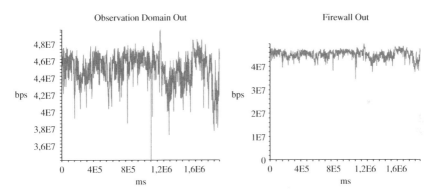

Fig. 6. Measurement at the Internet gateway: input ratio (left) and input ratio at the monitor (right)

5.2 Results using Flow Monitoring

In a first set of simulations we evaluated the capabilities of the proposed adaptive reconfiguration scheme for flow-based monitoring. Besides the main objective to assess the overall system behavior, the primary goal was to find adequate values for the scaling factors C_1, C_2, and C_3. In the following, selected simulation results are presented and discussed. In multiple experiments, we evaluated candidate values for the scaling factors. The parameters used for the presented simulation results were $C_1 = 9 * 10^8$, $C_2 = 236$, and $C_3 = 120$. The other simulation parameters are $\lambda_{IDS} = 0.6$, E[white]= 0.1, and E[black]= 0.01. As already mentioned, we monitored the traffic at the border gateway of our university. The utilization was quite constant (58.6kpps ± 1.8kpps). Due to few attacks, the firewall removes only few flows (around 0.5%). Therefore, the packet rate arriving at the monitor is similar but shaped (58.2kpps ± 1.8kpps).

Even though the number of discarded packets at the firewall is very small, this is primarily a result of the small timeout of each entry. The attack detection system is not overloading, therefore, it may analyze detected attacks over and over again. The behavior of the firewall is depicted in figure 7. In steady-state, around 6000 blacklist entries exist with an average timeout of 260s.

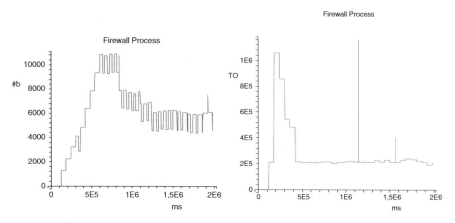

Fig. 7. Number of blacklist entries (left) and timeouts TO_{black} (right)

Finally, the numbers of whitelist entries and the corresponding timeouts have to be examined. As shown in figure 8, in steady-state, the number of whitelist entries is around 60000 and the average timeout is 320s. The number corresponds to the detection quality of legitimate traffic which is about ten times higher then for attacks. The timeout oscillates around the requested value.

In figure 9, the measured parameters λ_b (left) and λ_w (right) are depicted. These figures allow a more convenient analysis of the simulation results.

5.3 Results using Packet Sampling

For the packet sampling based measurements, we used the packet trace taken in front of out workgroup server. The monitor is executing a sampling algorithm that selects 50% of the packets (count-based). Obviously, the output packet rate is about one half of the input packet rate and the output byte rate represents the reduction due to keeping only parts of the packet (IP header including transport protocol information).

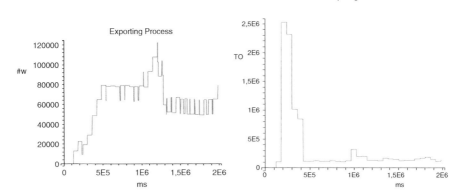

Fig. 8. Number of whitelist entries (left) and timeouts TO_{white} (right)

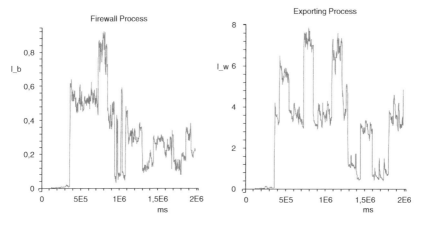

Fig. 9. Measured parameters λ_b (left) and λ_w (right)

In the following, we used the implemented blacklist and the whitelist representing the detected attack and legitimate flows. The most important parameters to analyze are: *Timeout* – A timeout value associated to each entry in the blacklist and whitelist. This defined how long an entry as determined by the attack detection system will be valid in the firewall system and the monitoring probe, respectively. *Detection ratio* – We assume a constant detection ratio resulting in new blacklist / whitelist entries. This is a presumption that does not correctly correspond with the behavior in a real network. Nevertheless, it reflects the behavior of the global system pretty well due to the proper configuration of the blacklist/whitelist.

In the following simulation results, we always display the output packet rate as issued by the monitoring probe and the input rate as a reference, because this reflects the corresponding amount of data that is to be processed by the detection system. In the first simulations, we examined the effect of the timeout value associated with the single entries in the blacklist and whitelist. For analyzing this behavior,

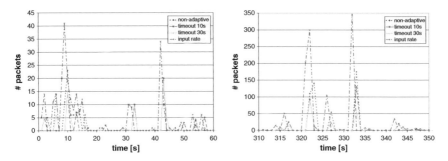

Fig. 10. Output rate for variable timeout and fixed detection ratio (10% whitelist, 1% blacklist)

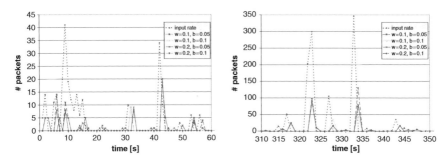

Fig. 11. Output rate for variable detection ratio and fixed timeout (10s)

we statically configured a detection ratio of 10% for new whitelist entries and 1% for new blacklist entries. These values seem to be adequate to reflect the behavior in real networks because about ten times as mush flows can be detected as legitimate traffic that as attack traffic and most of the network packets are not directly categorizable.

In figure 10, the simulation results are shown. Obviously, the adaptive configuration has a large impact on the amount of output packets. Additionally, it can be seen that a large timeout, which results also in large blacklist/whitelist tables and therefore, exhaustive search operations, does not lead to a massive reduction of the output packet rate. The reason for this behavior is the relatively short time a flow is lasting in the network. In figure 11, a second simulation run is shown. This time, the timeout was fixed at 10 seconds and the detection ration was modified. Interestingly, especially the whitelist has not that large impact on the amount of packets sent to the detection system as expected by the percentage of "white" packets (up to 20%). Additionally, it can be seen that the amount of data presented by the monitoring probe to the attack detection system can be adapted using an information exchange between all involved systems. An optimal adaptation to the behavior of the network seems to be very important and will lead to an optimized global system. This optimization step can be executed independently by each participating entity in the network by two meanings: local parameterization and communication/interoperation of neighboring entities. Therefore, we have shown that it is possible to self-organize the complete monitoring and analyzing environment for network security enhancement focusing on the monitoring part.

6 Conclusion

In this paper, we studied the adaptation of biologically inspired promoter / inhibitor schemes for adaptive parameter control in network security environments. Using an amplifying positive feedback loop and a suppressing negative feedback loop, we achieved a self-organizing autonomic behavior of the overall system. Especially the capabilities of the adaptation to reflect the local needs of an analyzing system in combination to the observation of the environment, i.e. the arrival rates at each subsystem make this approach useful for most monitoring scenarios. In a next step, we will evaluate the algorithm in an experimental setup to compare this evaluation to the simulation results.

References

1. H.-W. Braun, k. Claffy, and G. C. Polyzos, "A framework for flow-based accounting on the Internet," in *IEEE Singapore International Conference on Networks (SICON'93)*, Singapore, September 1993, pp. 847–851.
2. B. Caswell and J. Hewlett, "Snort Users Manual," The Snort Project, Manual, May 2004.
3. R. K. C. Chang, "Defending against Flooding-Based Distributed Denial-of-Service Attacks: A Tutorial," *IEEE Communications Magazine*, vol. 10, pp. 42–51, October 2002.
4. B. Claise, "Cisco Systems NetFlow Services Export Version 9," RFC 3954, October 2004.
5. ——, "IPFIX Protocol Specification," Internet-Draft (work in progress), draft-ietf-ipfix-protocol-22.txt, June 2006.
6. F. Dressler, "Adaptive network monitoring for self-organizing network security mechanisms," in *IFIP International Conference on Telecommunication Systems, Modeling and Analysis 2005 (ICTSM2005)*, Dallas, TX, USA, November 2005, pp. 67–75.
7. ——, "Efficient and Scalable Communication in Autonomous Networking using Bio-inspired Mechanisms - An Overview," *Informatica - An International Journal of Computing and Informatics*, vol. 29, no. 2, pp. 183–188, July 2005.
8. F. Dressler and I. Dietrich, "Simulative Analysis of Adaptive Network Monitoring Methodologies for Attack Detection," in *IEEE EUROCON 2005 - The International Conference on "Computer as a Tool"*, Belgrade, Serbia and Montenegro, November 2005, pp. 624–627.
9. F. Dressler and B. Krüger, "Cell biology as a key to computer networking," in *German Conference on Bioinformatics 2004 (GCB'04), Poster Session*, Bielefeld, Germany, October 2004.
10. F. Dressler and G. Münz, "Flexible Flow Aggregation for Adaptive Network Monitoring," in *31st IEEE Conference on Local Computer Networks (LCN): 1st IEEE LCN Workshop on Network Measurements (WNM 2006)*, Tampa, Florida, November 2006, pp. 702–709.
11. F. Dressler, G. Münz, and G. Carle, "CATS - Cooperating Autonomous Detection Systems," in *1st IFIP International Workshop on Autonomic Communication (WAC 2004), Poster Session*, Berlin, Germany, October 2004.

12. F. Dressler, C. Sommer, and G. Münz, "IPFIX Aggregation," Internet-Draft (work in progress), draft-dressler-ipfix-aggregation-03.txt, June 2006.
13. N. Duffield and M. Grossglauser, "Trajectory Sampling for Direct Traffic Observation," *IEEE/ACM Transactions on Networking (TON)*, vol. 9, no. 3, pp. 280–292, June 2001.
14. N. Duffield, "A Framework for Packet Selection and Reporting," Internet-Draft (work in progress), draft-ietf-psamp-framework-10.txt, January 2005.
15. A. Fessi, G. Carle, F. Dressler, J. Quittek, C. Kappler, and H. Tschofenig, "NSLP for Metering Configuration Signaling," Internet-Draft (work in progress), draft-dressler-nsis-metering-nslp-04.txt, June 2006.
16. Y. Hu, D.-M. Chiu, and J. C. Lui, "Adaptive Flow Aggregation - A New Solution for Robust Flow Monitoring under Security Attacks," in *IEEE/IFIP Network Operations and Management Symposium (IEEE/IFIP NOMS 2006)*, Vancouver, Canada, April 2006, pp. 424–435.
17. C. A. Janeway, M. Walport, and P. Travers, *Immunobiology: The Immune System in Health and Disease*, 5th ed. Garland Publishing, 2001.
18. B. Krüger and F. Dressler, "Molecular Processes as a Basis for Autonomous Networking," *IPSI Transactions on Advances Research: Issues in Computer Science and Engineering*, vol. 1, no. 1, pp. 43–50, January 2005.
19. T.-H. Lee, W.-K. Wu, and T.-Y. W. Huang, "Scalable Packet Digesting Schemes for IP Traceback," in *IEEE International Conference on Communications*, Paris, France, June 2004.
20. J. Mirkovic and P. Reiher, "A Taxonomy of DDoS Attack and DDoS Defense Mechanisms," *ACM SIGCOMM Computer Communication Review*, vol. 34, no. 2, pp. 39–53, April 2004.
21. M. Molina, "A scalable and efficient methodology for flow monitoring in the Internet," in *18th International Teletraffic Congress (ITC18)*, ser. Providing Quality of Service in Heterogeneous Environments, J. Charzinski, R. Lehnert, and P. Tran-Gia, Eds., vol. 5a. Berlin, Germany: Elsevier, August 2003, pp. 271–280.
22. V. Paxson, "Bro: A System for Detecting Network Intruders in Real-Time," *Computer Networks*, vol. 31, no. 23-24, pp. 2435–2463, December 1999.
23. J. Quittek, S. Bryant, B. Claise, and J. Meyer, "Information Model for IP Flow Information Export," Internet-Draft (work in progress), draft-ietf-ipfix-info-12.txt, June 2006.
24. M. Roesch, "Snort: Lightweight Intrusion Detection for Networks," in *13th USENIX Conference on System Administration*. USENIX Association, 1999, pp. 229–238.
25. R. F. Schmidt, F. Lang, and G. Thews, *Physiologie des Menschen*, 29th ed. Springer Verlag, 2005.
26. T. Zseby, M. Molina, N. Duffield, S. Niccolini, and F. Raspall, "Sampling and Filtering Techniques for IP Packet Selection," Internet-Draft (work in progress), draft-ietf-psamp-sample-tech-07.txt, July 2005.

Part II

System Design and Programming

Context Data Dissemination in the Bio-inspired Service Life Cycle

Carsten Jacob*, David Linner[†], Heiko Pfeffer[†], Ilja Radusch[†] and Stephan Steglich[†]

*Fraunhofer Institute for Open Communication Systems (FOKUS),
Kaiserin-Augusta-Allee 31, 10589 Berlin, Germany
carsten.jacob@fokus.fraunhofer.de
[†]TU Berlin, Sekr. FR 5-14, Franklinstraße 28/29, 10587 Berlin, Germany
{david.linner,heiko.pfeffer,ilja.radusch,stephan.steglich}@tu-berlin.de

Summary. Service provisioning in ad hoc and mobile environment introduced the need for new approaches to service adaptation and dissemination of appropriate context data for e.g. service management. Therefore, this paper introduces a novel bio-inspired service life cycle, leveraging enhanced service adaptation at the design time of services through guided service evolution, i.e. mutation and appropriate fitness selection of Service Individuals. Furthermore, we describe a context data dissemination algorithm to efficiently distribute the information needed in the bio-inspired service life cycle.

1 Introduction

Ever increasing processing power enabled the vision of smart and omnipresent devices available at the users demand. Additionally, connecting those heterogeneous devices seamlessly allow for new and unforeseen services, but also imposed new major research challenges as the maintenance needs of services and devices diverted subsequently. Putting the focus on the specific service to be provided to the user loosens the dependency on specific devices to execute this service on. This lead to more robust service provisioning as services are now provided in a collaborative fashion on one or several devices, which configuration can and will change throughout the service execution.

However, this introduced the problem of selecting the most appropriate set of devices for a given service to execute as well as maintaining this set of devices, since in ad hoc and mobile networks new and better suited devices can join or yet used devices can leave the network anytime. Current approaches mainly focus on more or less centralized traditional service orchestration with more or less capabilities to adapt to new and unforeseen changes in the environment or user needs. Furthermore, these approaches must rely on adaptation techniques already foreseen at the design time of services and, therefore, cannot cope with changes occurring during runtime. Furthermore, most traditional systems rely heavily on human intervention in the process of service creation, management, and deprecation.

Moreover, all distributed system building upon an ad hoc and mobile network possess the inherent need for efficient context data propagation through the network for various service management activities. Information of service performance, service quality, etc. needs to be transferred through the network to nodes where the management logic is situated. Obviously, centralized approaches introduce an often in mobile environment not feasible extensive load on the transmitting nodes. We therefore describe on the one hand a new approach to, distribute service management logic further into the network by rather describing and evaluating user driven Service Goals instead of explicit service calls. On the other hand we describe an approach to effectively disseminate context data used in various phases of our bio-inspired service life cycle throughout the network.

Section 2 describes our new bio-inspired service life cycle in comparison to traditional service life cycles. This bio-inspired service life cycle allows for holistic service autonomy through service self-management and enhanced service adaptation with bio-inspired evolution techniques. Section 3 continuous with describing the demand oriented dissemination algorithm for context data. Section 4 explains how both approaches are combined and Section 5 shortly describes the implementation of these approaches. Finally, Section 6 concludes this chapter with a summary.

2 Advances in Service Life Cycle Design

Services are subjected to a specific service life cycle, comprising their creation, lifetime and possible deprecation. Commonly, the creation of services is distinguished between addressing the explicit realization and implementation and the service execution on the other hand.

2.1 Traditional Service Life Cycles

The traditional service life cycle [1] is roughly dividable into two main phases as depicted in Figure 1: In a first step, the service is designed and modelled according to previously appointed requirements. It is then implemented, tested, simulated, and verified, where the order of these steps can vary and be interwoven in complex manner according to different software design principles. The first phase is referred to as the *Designtime* of a Service in the following. The complementary part is built by the so-called service *Runtime*, in which the service is first deployed (i.e. published or provisioned) and integrated. The running service can subsequently be adapted to the eventually occurring challenges or finally deprecated, enabled through monitoring, managing and evaluation mechanisms.

In terms of adaptability and self-management, traditional service life cycles have multiple shortcomings originated by a highly static service *Designtime*. First, all activities within the services' *Designtime* have to be performed manually, requiring

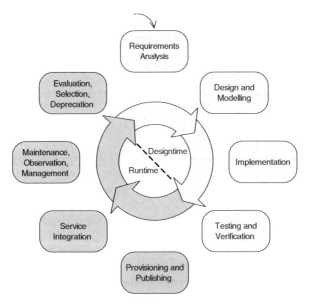

Fig. 1. Traditional Service Life Cycle [1]

noticeable system developer efforts for initial service creation. Furthermore, service adaptability is commonly restricted to predefined rule-based adaptation algorithms, changing the services behaviour triggered through context-awareness or user input [2].

In the following a bio-inspired service life cycle is conceptually introduced, addressing the two main weaknesses of traditional service life cycles in terms of autonomy. First, the service adaptability is upgraded to service evolution, i.e. services are not only able to adapt to current situations according to predefined transition table entries, but to evolve – i.e. changing and selecting service logic without prior planning – without with respect to their environment and assembled knowledge. Second, service autonomy principles are extended from *Runtime* to *Designtime* phase, paving the way for a holistic autonomous service life cycle enabled through biologically inspired concepts.

2.2 The Bio-inspired Service Life Cycle

The bio-inspired service life cycle is designed to allow services to emerge in an autonomic fashion in order to gain best support for user tasks while reducing the user's efforts in service creation [3]. This goal is regarded against the background of pervasive service environments, composed of services originating form the users' computing devices and those relevant in the users' context and their direct surrounding. In general, the service life cycle is supposed to address two facets of service emergence: The creation of service on user demand and the undirected service creation to achieve completely new functionalities not yet specifically requested by the user.

Thereby, we distinguish two basic structures, where life cycle activities are operating on. An atomic service, i.e., a service that cannot be broken up into more basic parts is referred to as a Service Cell. A composition of Service Cells is called Service Individual. Service Individuals may also be composed of other Service Individuals in a recursive way. However, decomposing Service Individuals completely results in a set of Service Cells, which cannot be broken into further parts. Service Individuals and Service Cells are supposed to have the same appearance, at least from an external point of view. So, when using the term Service in the following either a Service Cell or a Service Individual is meant. Service Cells underlie the traditional service life cycle described above, representing the basic building blocks the bio-inspired service life cycle builds upon.

The bio-inspired service life cycle, as illustrated in Figure 2, consists of four phases: Initiation, Integration, Evolution and Elimination.

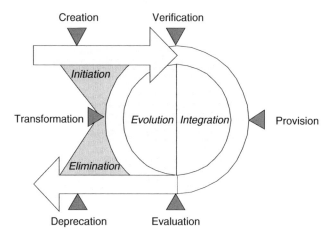

Fig. 2. : Bio-inspired Service Life Cycle

Services are created dynamically through service composition, i.e. Service Cells and Individuals are combined in order to either fulfil the user request or to come up with new service functionalities. Once a service is verified to be functional correct as well as to meet the requested non-functional requirements, it is provisioned to the service consumer – i.e. the user or other Service Individuals. The service integration phase is concluded with a consumer-side service evaluation, estimating the service's quality according to both functional and non-functional aspects. Depending on the evaluation results, a service either leaves the life cycle, i.e. enters the elimination phase leading to its deprecation, or re-enters it through a transformation procedure. Within the latter case, the transformation of services substitutes the initial service creation procedure, transforming the service according to biologically inspired concepts in order to come up with varied service functionalities and qualities. Transformed services therefore re-enter the service life cycle at the verification phase, proofing again functional correctness and ending the service's evolution phase in case it is considered as correct.

As mentioned above, the bio-inspired service life cycle aims at both extending service autonomy into the services' *Designtime* phases as well as to up value service adaptability to service evolution.

The Initiation and Evolution phases enable the extension of service autonomy through autonomous composition of services and their transformation, shifting the level of autonomy from *Designtime* to parts of the services' *Runtime* phases. The creation procedure bases on the composition of Service Cells and Service Individuals and is described by a set of Services and two directed graphs constituting the interrelation of the services, as illustrated in Figure 3. The first graph describes the flow of control data among the Services and therefore implicitly the execution sequence. Hence, it determines an initiation cascade along with an appropriate parameterization of the services. The second graph describes the flow of application data among the Services. It is distinguished between these two graphs since it is not expected that each Service in a composition participates in the creation or processing of application data. Service executions are regarded as state transitions, since the execution of a service either changes the state of the system providing and running the service or the state of the system consuming the service. The latter one may also mean the user, i.e., her informational state. However, this view also entails a certain schema for the flow of data exchanged among service consumer and service providers.

In order to upgrade service adaptation to service evolution, principles derived from evolutionary biology are overtaken, enabling the leaping of a complete service creation phases for each user request. Therefore, two genetic operations are defined, applicable to a previously introduced service composition graph. *Service Mutation* allows the modification of the service composition's execution flow, probably leading to a modified service behaviour or quality in terms of non-functional properties. Second, *Service Crossover* enables two services to exchange functional identical subgraphs, entailing a new service composition graph derived from the "genetic material" of the combined services.

Thus, whenever a service is built up through service composition, its composition plan is stored, where appropriate together with the related user request it was created for. These service composition plans can either be unintentionally modified by the genetic operations introduced above in order to aim at improving the service's quality or explicitly chosen for service transformation in case a similar user request is posed. By overtaking an already established service composition and varying it, the service creation phase can be skipped for a considerable amount of user requests. This advantage is of even greater significance when considering a collaborative processing of requests, which is introduced in the following section.

2.3 Collaborative Request Processing in State-Based User Environments

Users are supposed to be represented – together with their profiles and current context – as a state. Based on this abstraction, user requests can be regarded as state the user wishes to reach. In this context, a request formalizes the need for a

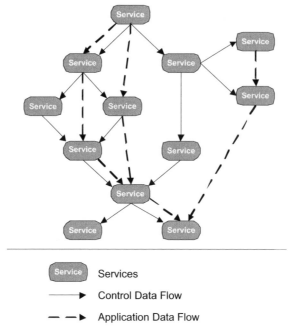

Fig. 3. Exemplary visualization of service graphs

service composition enabling the transformation of the current user state into the requested state.

User requests are treated in a collaborative manner from all nodes within the current network. Requests, i.e. states aimed at, are therefore placed on a shared interaction medium like a Semantic Data Space [4] where they can be accessed by all nodes. Each node capable (and willing) to contribute to a sub-problem implied by a request can apply an appropriate service in order to modify the current state (representing the request) into another state. As soon as a sequence of service appliances reaches the current state of the user that has initially posed the request to the medium, the request can be regarded as solved, since the reverse service appliance enables a transformation from the current user state to the state representing the user request. The concept of resolving a request in reverse order is known as backward chaining in terms of artificial intelligence graph algorithms [5].

Thus, by considering multiple nodes being able to exchange (probably partly) service composition plans through an interaction medium, service creation can increasingly be replaced by transformation procedures of already known service composition plans, reducing significantly the complexity of service creation, which is exponential in the number of services available within the current network.

3 Demand-oriented Dissemination of Context Data

The pervasive provision of context-aware services in mobile network environments assumes the instant availability of suitable context data and an easy way of accessing it. We identify challenges associated with these assumptions and propose an autonomous context data dissemination approach for peer-to-peer networks to address them. To intelligently control the propagation of context information a bio-inspired means for rating the relative importance of data is described.

3.1 Representation, Processing, and Provision of Context Data

Context data can be seen as one of the main enablers for sophisticated pervasive service provision in today's mobile and smart environments. Dey [6] describes context as being "any information that can be used to characterize the situation of an entity". To be able to reasonably use context data several complex activities have to be accomplished. For example, context data has to be gathered, represented, and processed in a suitable manner. Each of these issues can be further divided into distinct problem domains, whereas the optimal solution depends on multiple aspects such as the kind of the current network environment or the capabilities of available computing devices. Many approaches try to address those challenges by proposing so-called context-aware frameworks that define separate components for each task, e.g., for the inference of facts with the help of rules. Examples for such frameworks can be found in [7], [8], [9], each of which applying special components for context gathering and interpretation. Regarding the representation of such data, semantic description languages are used, which seem to be a good way for knowledge representation compared to key-value pairs or object oriented approaches [10].

Another framework approach incorporating semantic descriptions and inference is the Context-aware Service Adaptation Framework (CaSAF) [11], which focuses on the context-aware adaptation of legacy services. Here, components are defined being responsible for specific tasks, which are illustrated in Figure 4, to transform context data and service descriptions into adapted service calls.

Indeed, the depicted approaches have some shortcomings, which need to be addressed by an alternative and complementing approach. First, the reasonable provision of context-aware services is only possible provided that required context data with a satisfactory quality level is currently accessible. This is not assured due to mobile and disappearing nodes in distributed and unreliable peer-to-peer networks that we focus on. Second, extensive semantic descriptions, e.g., ontologies, are often presumed covering the preferences of the user, interfaces of services, or general knowledge. The complex creation of such descriptions has to be accomplished before being able to sensibly using context-aware services often preventing its reasonable application.

Main challenges in this area therefore pertain to the pervasive, robust, and decentralized provision of context data being addressed in Section 3.2, as well as to the reduction of extensive semantic descriptions that have to be defined by a particular user, which is further discussed in Section 4.2.

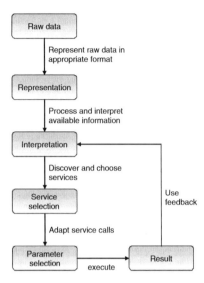

Fig. 4. CaSAF tasks and actions [11]

3.2 Loosely Coupled Data Dissemination

To address the pervasive access to context data identified as a challenge in the area of context-aware service provision in Section 3.1 we apply a data dissemination approach focusing on loosely coupled interactions. Here, data dissemination refers to the propagation of data in peer-to-peer networks in an autonomous manner.

For example, Directed Diffusion [12] considers a peer-to-peer network with nodes that are able to broadcast interests, being described by key-value pairs, in a controlled manner to its neighbors. In the same way, nodes can disseminate answers in the network. In contrast, gossiping algorithms [13] randomly propagate data in networks provided that peer nodes occasionally gossip with its neighbors or a number of them. For example, Construct [14], a middleware infrastructure, enables nodes to periodically send messages to neighbors selected at random, which are currently listed in a node specific contact list.

Extending the idea of gossiping with supportive components and loosely coupled interaction models, we introduce a flexible data dissemination mechanism focusing on dynamic and mobile network environments. Here, loose coupling refers to the common use of an interaction medium for communication, which can be either administrated by a particular node or a number of nodes. By using Semantic Data Spaces [15] as the interaction medium it is possible to decouple node communication with regard to the dimensions time, space, and representation. Therefore, the communication is temporally decoupled, nodes are addressed by their provided functionality instead of names, and semantic descriptions define a common vocabulary.

Based on this interaction model we define three types of components being important in the context data dissemination process. Apart from context providers, the Context Sources (CS), and context requesters, the Context-aware Applications

(CA), we consider hybrid Context Relays (CR), which are able to cache relevant information and provide it if needed. For example, due to the caching functionality sensor information can be provided by a Context Relay to Context-aware Applications even if the original sensor has disappeared. Furthermore, due to the assumed mobility of nodes, Context Relays are supposed to change its location very often causing the autonomous propagation of context data. This fact can be illustrated by means of defined Context Spaces, i.e., sets of mobile or immobile entities that commonly share information by means of the respective semantic data space they establish. Nodes, e.g., Context Relays, are able to switch between multiple Context Spaces by changing their location. Figure 5 illustrates an exemplary data dissemination process showing three distinct Context Spaces.

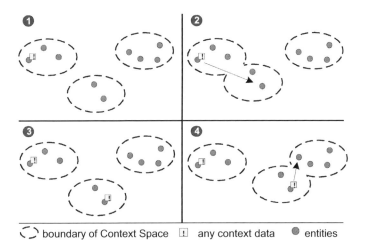

Fig. 5. Propagation of context information [16]

As nodes are highly mobile and Context Relays are able to cache interesting data, a particular piece of information is autonomously made available to remote Context Spaces.

Regarding the caching of data four distinct activities can be performed by a Context Relay as illustrated in Table 1: *Store*, *Update*, *Delete*, and *Spread*.

The Context Relay is able to perform each of the four activities provided that a particular condition is met. *Internal update* and *Spread* are active actions, i.e., requests or responses are published to the semantic data space. *Store* and *external update* are passive actions being performed provided that data is published by other nodes to the semantic data space.

3.3 Effective Data Dissemination through Bio-inspired Rating

The update and deletion activities of data stored by Context Relays depend on conditions that have to be met. To not store each kind of information published by means of

Activity	Type	Comment
Store	On response	Data is stored by the Context Relay if published by other sources
Update	Internal control	Stored data may be updated by actively requesting it if a particular condition is met.
	External control	Published up-to-date data may replace stored data
Delete	Deprecation	Data is deleted if a specific condition is met, e.g., it is not requested for a longer time
Spread	On request	Stored information is published if requested and a specific condition is met, e.g., no response is monitored in a fixed period of time

Table 1. Possible Context Relay activities

the interaction medium, a mechanism is required to evaluate information. For this, we apply a bio-inspired approach supporting the autonomous nature of the relaying of data. Here, the importance of data is determined by the number and type of applications requesting it following pheromone-based interaction approaches [17], [18]. The pheromone density of a request is used to compare the importance of two requests and to define conditions regarding the update or deletion. Three means of influencing a request's pheromone density can be distinguished: *pheromone accumulation*, *evaporation*, and *spreading*.

Pheromone accumulation refers to the increase of the pheromone density value belonging to a specific context data request if respective context data is requested by other nodes in the same Context Space. Pheromone evaporation describes the decrease in the pheromone density value due to suitable missing request in a fixed period of time. Spreading refers to the opportunity for Context Relays of publishing the current density value together with the suitable context data request in the internal update process. This causes other monitoring Context Relays to adapt its local value accordingly and is supposed to indirectly control the dissemination process as data with a higher density value is requested more often by Context Relays increasing the possibility of meeting a suitable context provider.

This evaluation approach aims at distinguishing important from unimportant data in a request-driven manner. Important information is supposed to be further propagated in the network, while unimportant information may be deleted or not even stored. For example, all requests with a pheromone density below a defined value are automatically deleted after a defined period of time.

Figure 6 illustrates an exemplary progress of the pheromone density value of a specific request cached by a specific Context Relay. Storing the request and increasing the pheromone density value at $t = 1$ is caused by the request publication

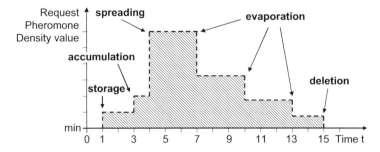

Fig. 6. Example of request pheromone density value progression

by another node via the current semantic data space. Another published request increases the current value at $t = 3$ referred to as pheromone accumulation. Another request is observed in $t = 4$ having a high pheromone density value attached. Therefore, the request's value is increased accordingly (pheromone spreading). Since no other request is published to the semantic data space for a longer time, the pheromone evaporation causes the decrease of the density value in $t = 7, 10, 13$, which finally leads to the deletion of the cached data in $t = 15$. In this example, the Context Relay may have switched the Context Space between $t = 4$ and $t = 7$, which would explain the sudden absence of context data requests. Note, that the pheromone density value is independent from the existence of context data fulfilling the respective request, but only actively influences the behavior of the Context Relay in terms of active updates. For example, due to the high pheromone density value between $t = 4$ and $t = 7$ the possibility is very high that the Context Relay published the respective request to retrieve and cache the potentially published response.

This approach allows us to effectively distribute context data through the service execution environment. This data is then used to evaluate and monitor service provisioning and performance in our Service Life further described in the following chapter.

4 Context Data in the Service Life Cycle

As outlined above context data disseminated through the network can be utilized at various stages in the bio-inspired service life cycle. The following sections describe the usage and benefits of context data in more detail.

4.1 Service Initiation

Above we already introduced Service Initiation as the first step in the bio-inspired service life cycle. Similar to traditional service life cycles, we distinguish two phases: *Service Goal Decomposition* and *Service Creation*. But, contrary to traditional service life cycles where services are designed, modeled and implemented in varying degrees by a human engineer, services in our bio-inspired eco-system are exposed to evolution resulting in optimizing and changing service functionality as well as service functions in ways not foreseen or requested by the engineer or user.

Service Goal Decomposition

Since the implementation of a service could change throughout its stay in the bio-inspired service life cycle and can therefore hardly be used as a description for the task of a given service, we use an additional meta description called the *Service Goal*. Thereby, the service goal does not convey how a certain service should perform a given task, but more generically what the goal of the user wants to achieve, thus the user request. Thereby, this goal can be either stated explicitly by the user or derived from the context data surrounding the user. Furthermore, non functional parameters such as priority, expected service quality level, etc. can be included in the Service Goal.

However, obviously, describing a user request formally in as a Service Goal is by itself a challenging task. Below, Figure 7 outlines the general dimensions of describing a user request, ranging from formal and imperative descriptions such as programming instructions to colloquial and rather declarative descriptions.

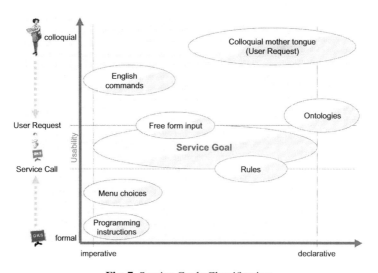

Fig. 7. Service Goals Classification

So far, traditional approaches to user request formalization rely on specific formal interface descriptions, whereas our approach tries to open this limitation by supporting a flexible mapping from abstract user requests to specific service calls utilizing the context data available in the network. Of course, the full potential but also the challenges of this approach rely in the quality and the possibilities of the mapping.

To allow for a generic mapping of Service Goals we propose a general architecture depicted in Figure 8. As outlined above, the objective is to provide a high colloquial abstraction most suitable for user interaction that is later grounded through various mappings to formal service calls on Service Individuals.

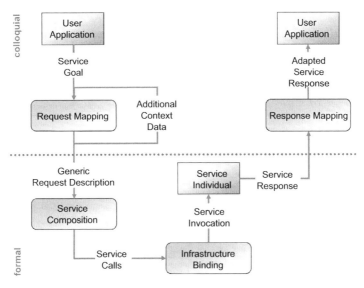

Fig. 8. Request Flow

Thereby, a rather generic *Service Goal* issued by the user is subsequently distributed and transformed into a *Generic Request Description*, which embeds all information needed for later request execution independent of the user environment but including appropriate context data. During this process the initial Service Goal might be extended with information by additional service objects such as personalization information, transformed into one or more other subsequent Service Goals, or accompanied with additional Service Goals, which then are reintroduced into the request mapping process and therefore also processed into Generic Request Descriptions. Please note, that these Generic Request Descriptions does not contain the names of Service Individuals to invoke, but rather information what kind of services are needed along with necessary input values, i.e. the inputs, outputs, preconditions, and effects of the services required. Furthermore, all kind of additional data, such as purpose descriptions, context information, or trust and privacy records, are added by appropriate Service Individuals. Once the Generic Request Description is complete, the Service Creation can start to create the appropriate Service Individual based on the available Service Cells. These Service Individuals are then fed into the service provisioning of bio-inspired service life cycle for binding to the infrastructure, i.e. the appropriate Service Individuals.

Service Creation

Traditional approaches to autonomous service composition rely mostly on searching more or less systematically through the full set of permutations of the available Service Individuals and Service Cells. Once again, in our bio-inspired service life cycle we try duplicate an approach from biology, where Service Cells are rather randomly composed into Service Individuals and later selected according to their distance to the targeted Service Goal. We visualized this process with our Service Tier Evolution Visualization Environment (stEVE) where Service Cells transform an initial state towards the requested Service Goal.

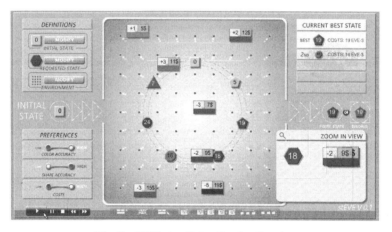

Fig. 9. stEVE visualizing Service Creation

For the sake of simplicity, we visualized the user state to abstract geometric figures with a specific color and a specific number of corners as shown in Figure 9. Furthermore, we reduced the number of possible state transitions to reducing or increasing the number of corners and changing the color of a given state.

During Service Creation Service Cells can change the initial state by influencing the number of corners and color of the state generating costs specific to each Service Cell. Thereby, the successive application of various Service Cells is recorded as a Service Individual. Each time a new state is reached, the distance to the Service Goal – i.e. regarding the number of corners and the color – is calculated. Based on that, Service Individuals are then either deprecated or selected for further mutation.

Using stEVE we can visualize how service creation can be achieved in our bio-inspired service life cycle. Once the Service Individuals are created or selected by traditional service composition approaches, they are verified to be functional correct and then fed to the next phase: integration. Once integrated Service Individuals enter the actual Service Provisioning, where they use context data to fulfill their Service Goal.

4.2 Service Provision

Section 3 describes the autonomous dissemination of context data to facilitate its omnipresent use, which is applicable to the notion of a bio-inspired service life cycle. Furthermore, the idea of Context Spaces introduced in Section 3.2 can be used to illustrate the way of providing context-aware services pervasively.

In a particular Context Space services are provided by respective Context-aware Applications (CAs), which can be divided into two types. Local CAs belong to the same Context Space the service is requested in and are therefore able to notice published requests. In contrast, global CAs are part of another Context Space or not considered as belonging to any Context Space. A global CA is represented by a CA mediator in a particular Context Space being responsible for forwarding request published to the respective interaction medium to the actual CA.

An example for a local CA is a jukebox that is located in a specific room and plays songs according to the current context or the preferences of the users in this room. CaSAF, which is described in Section 3.1, can be considered as a global CA being accessible with the help of a suitable interface that respective CA mediators are able to use. Indeed, both examples for Context-aware Applications do not only rely on context information, but also other data such as user preferences or semantic service descriptions.

To also take those data into account, the Context Relay can be extended by further caches, each being responsible for storing a particular type of request. For example, to also cover user preferences a rule cache can be applied in addition to the context data cache. This rule cache stores user specific rules, which can be requested by Context-aware Applications via the interaction medium. Furthermore, stored rules can influence the context data dissemination process, i.e., the pheromone-based evaluation procedure for context data is to be replaced or complemented by a rule-based method. For example, a rule pertaining to the current location of the user causes the Context Relay to actively request the user location.

Similar to context data, rules can then be evaluated and disseminated introducing the pheromone-based approach also for rules. Here, rules often requested by CAs, e.g., music rules requested by a juke box application, are rated more important than other rules being reflected in a higher pheromone density value. To achieve this, rule meta data tags have to be defined and attached to the rule, e.g., specifying the domains of a particular rule. This kind of data may either be set by the user at the creation time of the rule or be derived by parsing the rule for semantic concepts. For instance, a rule incorporating the concept "Song" may be classified as a music rule.

Context Relays also taking user specific data such as rules into account are defined as being Personalized Context Relays and supposed to reside on the user's mobile device. Figure 10 illustrates three exemplary Context Spaces with Context Sources (CS), Context-aware Applications (CA), and Personalized Context Relays (PCRs). Due to the assumed node mobility and reliability the topology of the depicted Context Spaces may change notably often. In this example, a particular PCR perambulates different Context Spaces and encounters a Context-aware Application acting as a mediator for another one, but also a local Context-aware Application.

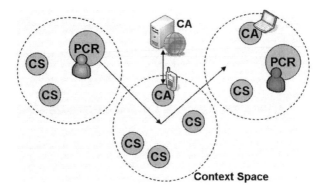

Fig. 10. Context Spaces example

The relaying of user specific data in addition to context data can be used to address the shortcomings identified in Section 3.1. The pervasive provision of requested context information is complemented by a means of propagating semantic descriptions such as rules. By this, the need of a user for defining extensive and complex descriptions is reduced since, similar to the idea of web communities, semantic descriptions are defined by multiple users in a distributed manner and conglomerated successively. For example, a user may mark particular rules as being public to be disseminated. Other users in the same Context Space may then decline or accept those rules and do not have the need for defining them on their own.

In this manner, the PCR approach and CaSAF can be utilized to adapt legacy services such as Internet search services to the user's context. The users PCR publishes a service request, user rules, and cached context data via the interaction medium, which then causes the CaSAF mediator to acknowledge the request, call CaSAF with the help of web interface, and publish returned results, e.g., links to web sites, accordingly.

4.3 Service Evolution

As introduced in Section 2, the bio-inspired service life cycle enables services to be transformed according to genetic procedures in order to enhance service performance in both functional and non-functional properties as well as to generate new service functionalities. Coevally, these service transformations render possible to skip verbose service creation phases by reusing already known service compositions.

As within nature, where the surrounding of individuals is the main driver for species evolution over time, context information is regarded as trigger for genetic transformations of services. The general influence of context information as well as personalized context information as introduced in sections 3.2 and 4.2, respectively, is illustrated in Figure 11. Every node builds up a virtual representation of the environment surrounding it by dint of context information. Thus, as depicted earlier, nodes exchange context information and personal information through CRs and PCRs, respectively, in order to obtain context data enabling them to create the most accurate view of their world possible. Changes in available context information leading to a varied perception of the environment can trigger genetic transformations unintentionally or based on personalized knowledge about the fitness of selected services within specific environments. For instance, a Crossover can be performed when the environment has changed in a way that parts of a service composition do not further work, in order to exchange these components with other (functionally equivalent) ones. Thereby, services evolve over a wide time span as aimed at in most evolutionary algorithms [19].

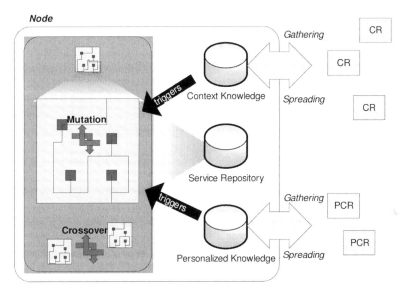

Fig. 11. Context Data triggered Service Evolution

Moreover, context data builds a basis for service fitness evaluation. Before a service is provisioned to a user, it is evaluated according to both its functional and nonfunctional properties in order to estimate its value for the user. A similar evaluation procedure is performed after service execution, including also possible feedbacks from the user or related services. In both evaluation phases, context information is relevant for service quality estimation as well as for knowledge aggregation. Latter aspect aims at deriving knowledge about a service's fitness within a specific environment, which can be spread through PCRs, possibly facilitating the fitness evaluation of services on remote nodes.

4.4 Service Elimination

Apart from Service Evolution which targets at optimizing and adapting Service Individuals in such way that they can be reintegrated into the service life cycle, Service Elimination affects Service Individuals no longer in use by the user or not fit enough compared to other Service Individuals. Thereby as in nature, Service Individuals can be superseded by new and more efficient instances providing similar functionality or suppressed by other Service Individuals providing different functionality currently more requested by the user. Depending on the resources available or type of Service Cells involved, Service Individuals are either deleted completely from the system or just suspended from actually using any resources in the system. This obviously ends the bio-inspired service life cycle of any given Service Individual.

5 Practical Perspective

To prove the feasibility and advantages of our model, the introduced concepts were implemented resulting in a Context Relay platform on top of a peer-to-peer framework. The realization was integrated in mobile smart spaces incorporating different devices, e.g., MICAz nodes [20] acting as Context Sources, and services. In this chapter we describe the way of realizing the Context Spaces approach including the notion of loosely coupled node interaction. Furthermore, Context Relays are described as well as the way these components are integrated into the Context Spaces realization. Finally, it is illustrated how Context-aware Applications, which were introduced in Section 4.2, can be realized.

5.1 Context Spaces Realization

Context Sources, Context-aware Applications, and Context Relays using the same interaction medium for communication form one Context Space. Semantic Data Spaces [15] are used as interaction medium, whereas our implementation is based on a REST [21] conform peer-to-peer framework. A shared naming service provided by one node in the respective Context Space enables peers to find its neighbours and is negotiated among all peers in case no naming service can be found.

The Semantic Data Space environment is realized in a similar manner with one or more nodes storing the respective data and being responsible for administration and classification purposes.

The Web Ontology Language (OWL) [22], a standardized language for describing ontologies and extending the Resource Description Framework (RDF) [23], is used for data representation. Particular benefits of OWL pertain to the possibility of distributing, merging, and classifying described knowledge. Furthermore, OWL files can be checked for integrity and validity. Regarding this, the Jena Semantic Web Framework [24] is utilized to process OWL. The Semantic Data Space additionally provides a means of classifying OWL data with the help of the Pellet reasoning engine [25], i.e., exploiting implicit knowledge and matching semantically described requests and responses.

The Semantic Data Space is composed of disjoint data spaces, each of them referring to a particular interaction subject, e.g., a context data request and respective response if available. A data space is further divided into multiple planes for each message to allow the tracking of changes and rollbacks.

Adding and retrieving data is geared to tuples space approaches such as Linda [26] and JavaSpaces [27], while data are added as entire RDF graphs. The data space environment allows for reading stored data immediately or subscribing for sub-graphs matching a predefined pattern, whereas these patterns are represented in the RDF query language SPARQL [28]. Since the peers in a Context Space are not necessarily supposed to process RDF or OWL, results are represented as a list of comma separated name-value pairs, which represent bindings for the free variable of a query.

5.2 Relaying Context Information

The Context Spaces implementation serves as a base for realizing Context Relays, which were introduced in Section 3.2. A Context Relay is considered as a normal peer in the Context Spaces realization and uses the Semantic Data Space for communication purposes. Additionally, a Context Relay owns local ontology caches to store context data requests and responses also applying OWL for this purpose. Therefore, request or response data can be cached as published except for the namespace of an interaction subject, which has to be adapted by the Context Relay accordingly. In case a particular Context Relay monitoring the Semantic Data Space defining the respective Context Space observes a suitable request for a stored response an internal timer is started that runs out shortly before the expiry date of the request. Observed responses are stored by the Context Relay updating old information. If not response is observed and the timer runs out the Context Relay publishes the locally cached data. Figure 12 illustrates the different layers of the Context Relay approach covering only one exemplary Context Space with a Context-aware Application (CA), Context Sources (CSs), and a Context Relay (CR). Note, that the data Space environment may also be distributed among multiple nodes.

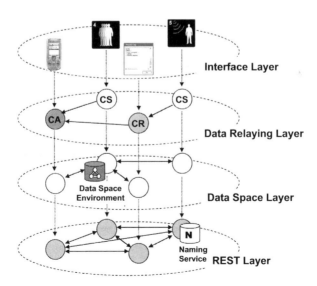

Fig. 12. Visualization of different layers in the Context Relay approach [16]

Pheromone density values as introduced in Section 3.3 are attached by a Context Relay to the respective request with the help of an appropriate OWL property. Pheromone accumulation is realized by simply increasing the stored value whenever a Context-aware Application publishes a suitable request. Whenever a Context Relay publishes a request to update a cached response it attaches the stored density value enabling other Context Relays to interpret this value, which represents the spreading of pheromones. Finally, pheromone evaporation denotes the decrease of the density value if a request is not observed for a longer time, i.e., the value is halved. The request and response data are deleted from the Context Relay's cache if the stored density value is below a particular value.

5.3 Personalized Service Provision in Context Spaces

In Section 4.2 the Personalized Context Relay (PCR) was introduced offering a possibility for the provision of personalized service provision in the scope of the Context Spaces approach. We realized the PCR approach as an alternative to Context Relays including a means for disseminating user preferences in terms of rules.

To achieve this, a new semantic concept, namely that of a rule, is introduced to be able to identify and classify published data. The CR implementation is extended by adding a second cache for rule objects being treated the same way as context data were before. Additionally, the Jena Semantic Web Framework is used to parse rules for semantic concepts influencing the dissemination process of context data. The rule cache of a PCR is initially filled by the respective user with the help of a suitable editor or, alternatively, by rule sensors offering domain specific rules that can be requested by PCRs in the same Context Space. We implemented two distinct types of Context-aware Application according to the classification described in Section 4.2.

A local CA is represented by a jukebox that has a predefined list of songs and selects the next song to play based on user specific rules, i.e., the jukebox requests all available music rules of PCRs in the current Context Space and infers a suitable band or genre. If multiple songs are possible, one entry is selected at random.

The Context-aware Service Adaptation Framework (CaSAF) is employed as a global CA and extended by an interface and a suitable mediator to provide rules and respective context data. In this manner, legacy services such as Internet search services are adapted. Returned results, such as web links, are again published to the Semantic Data Space by the mediator being then accessible by the originating PCR residing on the user's mobile device or even other suitable local CAs.

6 Conclusion

The bio-inspired service life cycle introduced above enables enhanced service adaptation at the design time of traditional service life cycles through service evolution. These services are thus prepared to cope with unforeseen changes in the environment as well as not specifically requested user demands.

Furthermore, we described a context data dissemination algorithm to efficiently distribute necessary information in ad hoc and mobile networks. Additionally, we outlined implementation details of the bio-inspired service life cycle, by introducing Context Spaces.

Continuous research based on this bio-inspired service life cycle will investigate more approaches for the initial service creation, i.e. state-based service composition, respectively even more advanced algorithms for Service Goal decomposition.

References

1. Q. Wall. Understanding the Service Lifecycle within a SOA: Run Time [Online]. Available: http://dev2dev.bea.com/pub/a/2006/11/soa-service-lifecycle-run.html
2. H. He, M. Potts, I. Sedukhin, Web Service Management: Service Life Cycle, February 2004, Available at http://www.w3.org/TR/wslc/
3. H. Pfeffer, D. Linner, I. Radusch, S. Steglich, "The Bio-inspired Service Life-Cycle: An Overview", In *Proceedings of the 3rd International Conference on Autonomic and Autonomous Systems* (ICAS 2007), Greece, 2007. *(to appear)*
4. D. Linner, I. Radusch, St. Steglich, St. Arbanowski: Request-driven Service Provisioning. Proceedings of the SICE-ICASE International Joint Conference 2006, Busan, Korea, October 18-21, 2006.
5. M.H. ter Beek, A. Bucchiarone and S. Gnesi, A Survey on Service Composition Approaches: From Industrial Standards to Formal Methods. Technical Report 2006-TR-15, Istituto di Scienza e Tecnologie dell'Informazione, Consiglio Nazionale delle Ricerche, 2006.
6. A. K. Dey, "Understanding and using Context," *Journal of Personal and Ubiquitous Computing*, Volume 5 (1), pp. 4-7, 2001.
7. Harry Chen, Tim Finin, and Anupam Joshi. Semantic Web in the Context Broker Architecture. In *Proceedings of PerCom 2004*, Orlando FL., March, 2004.
8. T. Gu, H. K. Pung, and D. Q. Zhang. A Middleware for Building Context-Aware Mobile Services. In *Proceedings of IEEE Vehicular Technology Conference (VTC2004)*, Milan, Italy, 2004.
9. X. Wang, J. S. Dong, C. Chin, S. Hettiarachchi, and D. Zhang Semantic Space: An Infrastructure for Smart Spaces. IEEE Pervasive Computing, Volume 3(3), pp. 32-39, 2004.
10. T. Strang and C. Linnhoff-Popien. A Context Modeling Survey. *Workshop on Advanced Context Modelling, Reasoning and Management*, UbiComp 2004, September, 2004.
11. C. Jacob, I. Radusch, and S. Steglich, "Enhancing Legacy Services through Context-enriched Sensor Data," in *Proceedings of the International Conference on Internet Computing (ICOMP'06)*, Las Vegas, Nevada, USA, pp. 278-287, June 26-29, 2006.
12. C. Intanagonwiwat, R. Govindan, D. Estrin, J. S. Heidemann, and F. Silva, "Directed diffusion for wireless sensor networking," *IEEE/ACM Transactions on Networking*, vol. 11, no. 1, pp. 2-16, February, 2003.
13. A. Datta, S. Quarteroni, and K. Aberer, "Autonomous Gossiping: A self-organizing epidemic algorithm for selective information dissemination in wireless mobile ad-hoc networks," in *Proceedings of the International Conference on Semantics of a Networked World (IC-SNW04)*, Paris, France, June, 2004.

14. G. Williamson, G. Stevenson, S. Neely, L. Coyle, and P. Nixon, "Scalable information dissemination for pervasive systems: implementation and evaluation," in *Proceedings of the 4th international workshop on Middleware for Pervasive and Ad-Hoc Computing (MPAC 2006)*, Melbourne, Australia, November 27-December 1, 2006, ISBN:1-59593-421-9.
15. David Linner, Ilja Radusch, Stephan Steglich, and Carsten Jacob, "Loosely Coupled Service Provisioning in Dynamic Computing Environments," in *Proceedings of ChinaCom 2006. First International Conference on Communications and Networking in China*, Beijing, China, October 25-27, 2006, IEEE Catalog Number: 06EX1414C, ISBN: 1-4244-0463-0, Library of Congress: 2006926274.
16. C. Jacob, D. Linner, S. Steglich, and I. Radusch, "Bio-inspired Context Gathering in Loosely Coupled Computing Environments," in *Proceedings of the 1st International Conference on Bio Inspired mOdels of NEtwork, Information and Computing Systems (BIONETICS) 2006*, CD-ROM, Cavalese, Italy, December 11-13, 2006, IEEE Catalog Number: 06EX1490C, ISBN: 1-4244-0539-4, Library of Congress: 2006930028.
17. M. Mamei, F. Zambonelli. Spreading Pheromones in Everyday Environments via RFID Technologies. *2nd IEEE Symposium on Swarm Intelligence*, June, 2005.
18. J. Wu and K. Aberer. Swarm Intelligent Surfing in the Web. *Third International Conference on Web Engineering (ICWE03)*, Oviedo, Asturias, Spain, July 14-18. 2003.
19. S. R. Hedberg, "Evolutionary Computing: The Rise of Electronic Breeding," *IEEE Intelligent Systems*, vol. 20, no. 6, pp. 12-15, Nov/Dec, 2005.
20. T. Luckenbach, P. Gober, S. Arbanowski, A. Kotsopoulos, and K. Kim, "TinyREST – a Protocol for Integrating Sensor Networks into the Internet," in *Proceedings of the Workshop on Real-World Wireless Sensor Networks (REALWSN'05)*, 2005.
21. R. T. Fielding, R. N. Taylor, "Principled design of the modern Web architecture," in *Proceedings of the 22nd International Conference on Software Engineering*, Limerick, Ireland, pp. 407-416, 2000.
22. OWL Web Ontology Language Reference. 2004. [Online]. Available: http://www.w3.org/TR/owl-ref/. [Accessed: March 20, 2007].
23. Resource Description Framework (RDF). [Online]. Available: http://www.w3.org/RDF/. [Accessed: March 20, 2007].
24. Jena - A Semantic Web Framework for Java. [Online]. Available: http://jena.sourceforge.net/. [Accessed: March 20, 2007].
25. Pellet OWL Reasoner. [Online]. Available: http://pellet.owldl.com/. [Accessed: March 20, 2007].
26. D. Gelernter, "Generative communication in Linda," *ACM Transactions on Programming Languages and Systems*, vol.7, pp.80-112. 1985.
27. Sun Microsystems, Inc. JavaSpace Specification, Revision 0.4, 1997.
28. E. Prud'hommeaux, A. Seaborne. (2006, February). SPARQL Query Language for RDF. [Online]. Available: http://www.w3.org/TR/rdf-sparql-query. [Accessed: March 20, 2007].

Eigenvector Centrality in Highly Partitioned Mobile Networks: Principles and Applications

Iacopo Carreras[1], Daniele Miorandi[1], Geoffrey S. Canright[2] and Kenth Engø-Monsen[2]

[1] CREATE-NET
Via Solteri 38
38100 – Trento (Italy)
name.surname@create-net.org

[2] Telenor R&I
Snaroyveien 30
N-1331 Fornebu (Norway)
name.surname@telenor.com

Summary. In this chapter we introduce a model for analyzing the spread of epidemics in a disconnected mobile network. The work is based on an extension, to a dynamic setting, of the eigenvector centrality principle introduced by two of the authors for the case of static networks. The extension builds on a new definition of *connectivity matrix* for a highly partitioned mobile system, where the connectivity between a pair of nodes is defined as the number of *contacts* taking place over a finite time window. The connectivity matrix is then used to evaluate the eigenvector centrality of the various nodes. Numerical results from real-world traces are presented and discussed. The applicability of the proposed approach to select on-line message forwarders is also addressed.

1 Introduction

In this chapter, we aim at characterizing the spreading power of nodes in a mobile disconnected network. In such a scenario, nodes exploit opportunistic contacts among themselves to diffuse information, mainly in the form of messages (i.e., chunks of packets). The problem is of interest for a number of reasons, the prominent one being the research efforts towards the definition of an architecture for Delay-Tolerant Networks (DTNs) [4, 5]. DTNs are networks in which the existence of a path between any pair of nodes is not taken for granted. Differently from conventional IP networking paradigm, DTNs are able to operate in the presence of frequent network partitions. In DTNs the conventional notion of "network" itself needs to be re-thought anew: information can diffuse in the system by means of (i) node mobility: a device conveys information while moving around (ii) opportunistic forwarding: messages are passed from a node to another one when they get in *contact*. Due to the high dynamism of such scenarios, some epidemic mechanism has to be used to spread information until it reaches the intended destination [6, 7, 8]. The problem we aim at tackling is to devise distributed mechanisms able to decide whether, upon meeting a given node, it should be used as a forwarder.

The starting point for our study is the work carried out by two of the authors on the spreading power of nodes in a static setting [9, 10]. The work therein focuses on the notion of *eigenvector centrality* (EVC), shown to be a meaningful measure of the ability of the nodes to spread an epidemic in the network. The EVC is computed as the eigenvector relative to the spectral radius (i.e., the largest eigenvalue) of the adjacency matrix of the network. Such a procedure is shown to produce a smooth measure, which can be used for studying the spread of epidemics [10]. Further, it implies a natural way of defining clusters in the network. The resulting network topography can be used to define regions, each region being characterized by the fact that epidemics spread extremely fast therein.

We would like therefore to extending the EVC concept to the case of highly partitioned mobile networks. The underlying idea is that EVC can be used as a metric for deciding whether a node encountered on the way should be used as a forwarder or not. This requires two main steps: (i) extending the definition of EVC to mobile disconnected scenarios, which requires to introduce an appropriate matrix able to describe well the "interaction pattern" among nodes (ii) introducing mechanisms and techniques for allowing a distributed on-line estimation of the EVC. In the chapter, we will presents techniques and results for issue (i), while some preliminary considerations will be presented for point (ii).

The main contribution of this chapter is the extension of the eigenvector centrality principle to more dynamic and disconnected scenarios. We will indeed focus on how to extend the topographic picture, with its obvious advantages for studying epidemic spreading, to a dynamic network — one in which the links are time-dependent. Our aim is to take a mobility pattern as input, to determine the time-dependent links, and finally to produce a topographic analysis analogous to the static case — with a smooth centrality function over the nodes, regions defined so as to correspond to well connected subgraphs, and a meaningful connection to the nodes' roles in spreading. Thus we want to define some kind of time-averaged or time-integrated *connectivity matrix*, with non-negative link weights. Given a suitable definition of connectivity matrix,[3] we can apply the EVC analysis to study its topographic properties. We introduce the notion of T-tolerant connectivity matrix, whose (i,j)-th entry equals the number of contacts between nodes i and j over a time window of length T.

The remainder of this chapter is organized as follows. In Sec. 2 we review the basic EVC concept and its application to static settings. In Sec. 3 we introduce the notion of T-tolerant connectivity matrix and show how to build it. In Sec. 4 we present some numerical results, obtained from real-world mobility traces, and discuss the properties of the correspondent systems. Sec. 5 concludes the chapter pointing out some directions for future work.

---

[3] The term "connectivity matrix" can be somehow misleading, in that we are mostly interested in disconnected scenarios, where a randomly taken snapshot of the network returns a disconnected graph.

2 Eigenvector Centrality, Topography, and Spreading

In this section we present a brief review of earlier work involving a "topographic" view of the structure of networks with undirected links. This work is presented in detail in [9] — which presents the basic structural analysis, based on eigenvector centrality or EVC [11] — and in [10] — which shows the utility of this analysis for understanding and predicting the progress of epidemic spreading over the network.

2.1 Previous Work

A fundamental problem in network analysis is the problem of *clustering*. That is: given a network topology (possibly with weights on the links), and assuming that the network is symmetric and connected, how do we then define and identify subgraphs of the whole network which, in some well defined sense, consist of nodes which "belong together"?

One meaningful and useful notion of "belonging together" is that of "being well connected". This point of view states that the clusters (subgraphs) of a network are better connected internally than they are to other clusters. Two of us [9] have constructed a precise version of this kind of clustering criterion. That is, one can define "well-connectedness" in a number of ways; but one definition — valid for a single node — is that the node be well connected to nodes that are well connected. This is of course a circular definition, which however is readily formulated mathematically [9, 11]. The resulting single-node measure of well-connectedness is termed "eigenvector centrality" or EVC [11]. Let us consider a graph model of the network topology and denote by A the corresponding adjacency matrix.

The EVC of a node i is defined being proportional to the sum of the the EVC of i's neighbors:

$$e_i = \frac{\sum_{j=nn(i)} e_j}{\lambda}, \quad (1)$$

where $nn(i)$ denotes the set of neighbors of node i. We can rewrite (1) in a compact matrix form:

$$A \cdot e = \lambda e, \quad (2)$$

where e represents the vector of nodes' centrality scores. Otherwise stated, e is the eigenvector of A relative to the eigenvalue λ. A number of reasons [9, 10] lead to the choice of the maximal eigenvalue λ_{max}. Due to the fact that A is nonnegative, $\lambda_{max} \geq 0$ and all the components of e are nonnegative [12].

EVC can be exploited to define clusters as follows. We note that, because of the dependence of a node's EVC on that of its neighbors, the EVC may be regarded as a "smooth" height function over the network. This smoothness motivates the appeal to topographic notions. That is: local maxima of the EVC are regarded as being the most well-connected node in a region of good connectivity. In topographic terms, these local maxima are mountain peaks. Each "mountain" (region of good connectivity) is then defined by its peak, plus a rule for membership of other (non-peak) nodes [9] in the well-connected cluster. A simple membership rule is that each node belongs to the same mountain as does its highest (EVC) neighbor. In other words: a node is connected to the cluster of its best connected neighbor. With this rule, essentially all nodes are assigned uniquely to one mountain; the mountains are then the well connected clusters of the graph. These are termed *regions* in the earlier work, and we will adhere to that usage here.

In more recent work [10] it has been shown that this smooth definition of well connectedness, and the resulting notion of regions, are very useful for describing and understanding the time and space (network) progression of an epidemic on an undirected network. The basic idea is that an infection front tends to move towards neighborhoods of high connectivity (EVC), because the spreading is fastest in such neighborhoods. This says in other words that the front naturally moves "uphill" over EVC contours—which in turn implies that, in general, spreading will be faster within regions (mountains) than between them. These ideas, and related ones, have been described in detail, and confirmed in simulations, in [10].

Some important points from this work which are relevant for the present chapter are:

- The importance of a node to the process of epidemic spreading may be roughly measured by the node's eigenvector centrality.
- *Regions*, as defined by the steepest-ascent rule, are clusters of the network in which spreading is expected to be relatively rapid and predictable. (Here "relatively" simply means compared to inter-region spreading.)
- Nodes whose links ("bridging links") connect distinct regions play an important role in the (less rapid, and less predictable) spreading from one region to another.

One can readily identify two important directions for extending such work, building on the work described in the previous subsection. First, the described topographic approach is, in its present form, only applicable to a *static* network. Secondly, the analysis is thus far only applicable when some entity (researcher, manager, engineer, machine) has access to knowledge of the full topology of the network. The latter constraint stems from the fact that the EVC of any node is in fact dependent on the entire topology of the network—because it is taken from the dominant eigenvector of the network's adjacency matrix, and because the network is (assumed) connected. Thus, an interesting question is how one can define local, distributed methods for finding the EVC distribution and the region structure of a given network. This question will not however be pursued in this chapter, whereas we will rather concentrate on the first issue, i.e., how to extend the EVC analysis to a mobile disconnected system.

3 Connecting Disconnected Networks

The results presented in the previous section justify the use of the EVC principle as a suitable metric for identifying the spreading power of nodes. The results therein clearly apply to a static network topology, where it is straightforward to define the term "connectivity matrix". Things change drastically when we look at mobile networks, where links come and go. One could consider taking a snapshot of the network at a randomly chosen time instant, but this clearly does not make sense for highly partitioned mobile networks as the ones subject of our study. It is therefore imperative, in order to extend the use of EVC to such scenario, to define a suitable matrix, whose entries reflect the actual level of "interaction" among nodes in the system. In this section, we will present a simple method for building such a matrix, that we call the T-tolerant connectivity matrix of the system.

3.1 Epidemic Spreading in Highly Partitioned Mobile Networks

Let us consider first the case of a wireless network in which all nodes are static. In this case, we can build a matrix A, that we call the *connectivity matrix*, by looking at the fact that two nodes are within mutual communication distance. In other words, the (i, j)-th entry is positive, $A(i, j) > 0$ if and only if nodes i and j are within mutual communication range. For example, $A(i, j)$ could be the rate at which two nodes are able to transmit data to each other. In this way we could account for SNR variations while still being able to define a matrix A representing the "topology" of the network. Things drastically change when we consider a mobile network. In this case, indeed, there is no standard notion of network topology we can rely on for building the connectivity matrix. This task gets even more challenging if we focus on networks operated in the *subconnectivity regime* [13], in which no giant component exists.

We are actually interested in something stronger, i.e., looking at networks in which, at any given time instant, each node is isolated with high probability. We have two reasons for looking at such scenarios. The first is technology-driven, in that there is a large number of application scenarios in which such assumptions hold. Most of them fall within the category of Delay Tolerant Networks (DTNs) [5]. DTNs are networks that are able to work in the absence of continuous connectivity. The connectivity in DTNs follows a random pattern, and is not taken for granted as in standard IP-based networks. DTNs include, e.g., Vehicular Ad hoc NETworks (VANETs), where the nodes are represented by cars, which may exchange data to run distributed services, including traffic monitoring, alert messages, personalized advertising etc.[14]. On the other hand, there is also a deep scientific reason for looking with interest at networks operated in this regime. Indeed, deep results in the field of network information theory state that connected wireless networks present poor scalability properties [15]. This is mainly rooted in the fact that network operations, in the wireless domain, are interference-limited. And the need for maintaining a connected network leads to the necessity of using a quite large communication range, at the expense of network capacity. On the other hand, nodes mobility can be exploited, for a network operating in the subconnectivity regime, to achieve a scalable network model [16].[4]

---

[4] In particular, for a connected network of n nodes, half of which act as sources and the other half as receivers, the per-connection throughput scales as $\Theta\left(\frac{1}{\sqrt{n}}\right)$, whereas in a mobile network with the two-hop relaying scheme in [16], the same quantity scales as $\Theta(1)$.

Given the aforementioned scenario, it remains to discuss how information can be successfully delivered in such systems. Standard routing algorithms were indeed built for wired networks, in which links are static. It is therefore sufficient to discover the route to the intended destination and use it for transmitting all data packets. In a highly mobile system, on the other hand, the notion of route itself looses its significant, since the destination becomes a "mobile target" and does not have a destination address any longer. Keeping track of the movements of all nodes is clearly not feasible, due to the obvious scalability problems related to the amount of control traffic. It is therefore imperative, for such scenarios, to resort to some form of epidemic forwarding [6]. In such protocols, data packets spread in the network like a virus or some other forms of epidemics. Nodes which carry the packets may infect nodes who had not received it yet (also called "susceptible" in standard epidemiology lexicon). Also in this case, in principle, scalability problems could arise, due to the fact that the number of copies of a message in the network grows exponentially over time. Mechanisms are needed to limit the spreading of epidemic data, in the form, e.g., of limiting the lifetime of a packet [1], the number of nodes a single user can infect (also called K-copy relaying protocols [20]) or by limiting the number of hops a packet can traverse. Also mechanisms based on the use of a "vaccine" or "immunization" in the form of so-called anti-packets were devised [2]. On the other hand, we envision to enhance these schemes by assigning to each node a value (a sort of fitness level) which describes their ability to spread an epidemic on the network. A node with a high fitness level has therefore a high level of interactions with many other nodes in the network, and configures itself as a good carrier of messages to the intended destination. Forwarding decisions (i.e., decisions whether to forward a message to an encounter or not) are taken on the basis of such fitness value. In this paper, we do not aim at specifying the exact mechanism based on which such system should work. Our target is to define a suitable method for *computing* such a fitness value.

In a disconnected scenario, where nodes are isolated most of the time, the only mean of transmitting information is given by *meetings* among nodes. Nodes i and j are said to meet at time t if, given an infinitesimal amount of time δt, at time $t - \delta t$ they were not able to transmit messages to each other, while they could do so at time $t + \delta t$. Meetings form a random pattern, which depends on (i) the mobility of nodes (ii) the random channel fluctuations. Meetings are characterized by (i) the IDs of nodes which get within mutual communication range (ii) the time at which the meeting takes place (iii) the duration of the meeting. The last parameter reflects the length of the time interval during which nodes are able to communicate. We actually overlook issue (iii), and consider that the duration of a meeting is sufficient for delivering all the data which needs to be transferred. This is reasonable since, with current wireless technologies (e.g., IEEE 802.11g), nodes moving even at vehicular speeds are able, in most situations, to transfer some MBytes of data [17].

In order for the EVC analysis to produce meaningful results, the matrix A should relate to a graph which is connected with high probability. On the other hand, given our assumptions, a randomly taken snapshot of the network status would report a graph which is almost surely disconnected. We need therefore to devise a mechanism able to enable us to pass from the *subcritical* to the *supercritical* regime.

3.2 The T-Tolerant Connectivity Matrix

The basic idea which we propose for building the matrix A is to consider an integrated version of the instantaneous network connectivity. Given a constant T, we construct the T-tolerant connectivity matrix A_T by considering all the meetings taking place over a time window of length T. The (i,j)-th entry $A_T(i,j)$ equals the number of meetings taking place in the time interval $[t_0, t_0 + T)$, where t_0 is a given initial time instant.

The underlying idea is the following. If the weight of a link increases with the strength or frequency of the connection — or, in other words, with the probability of spreading an "infection" over the link — then the analysis of this connectivity matrix may be expected to give useful information about how an infection (or message) is likely to be spread in a network of mobile nodes. Subsequent analysis is then the same as done in [9] and [10]; and because of this assumption about the link weights, the results should be useful for describing the process of diffusion of messages in the network, and for getting insight into the role played by the various nodes in the spreading process.

The definition of the T-tolerant connectivity matrix is worth some comments. In particular, the choice of an adequate time window T appears non-trivial. The value of T shall be consistent with the nodes' dynamics, so that the matrix A_T gives rise to non-trivial results. If T is too small, most entries of A_T would turn out to be zero, the network would show low connectivity and, hence, the EVC analysis would not carry any significant information. The parameter T shall be understood as a kind of time constant of the system. In particular, we are interested in working around the value of T for which the graph associated to the matrix A_T gets connected. Such value depends on the mobility and is therefore out of the control of system designers. On the other hand, such value determines the time horizon over which the system is able to deliver messages, placing therefore a constraint on the class of applications which can be supported by the system.

Recent results by the first two authors [18] show that, given a system with n nodes and where meetings take place at intensity λ[5], it can be shown under some independence assumptions that a *phase transition* takes place at T given by:

$$T = \Theta\left(\frac{n \log n}{\lambda}\right). \tag{3}$$

At such value, the graph corresponding to the skeleton of the matrix A_T passes from the subcritical to the supercritical result.

This time-integrated version of the connectivity matrix presents, however, also a shortcoming, and cannot be used — in general — to define the ability of the system to deliver data. Consider the example reported in Fig. 1, where we considered the graph $\mathcal{G}_T$ (called the T-tolerant connectivity graph) which we can naturally associate with the matrix A_T. The graph $\mathcal{G}_T$ has vertex set corresponding to the nodes in

---

[5] In this case, λ has to be understood as the intensity of the point process that can be easily constructed to reflect the dynamics of meetings taking place in the system.

the network. There is a link between i and j if and only if there is a non-zero value in the corresponding entry of the matrix A_T. The link is labelled (or weighted) with the value of the corresponding matrix entry. In the situation considered in Fig. 1 there are three nodes, A, B and C. At time t_1, nodes A and B get within mutual communication range. Thus, we set the corresponding entry in the connectivity matrix to 1, and add an edge (A, B) of unitary weight in the graph $\mathcal{G}$. At time t_2, node A meets with node C. Also in this case we add an edge of unitary weight. At time t_3, node A gets again in contact with node C. The weight of the corresponding link is thus increased to 2. In the resulting graph, there is a path connecting node B to node C. However, there is a timing (or ordering) issue that is not accounted for in this construction. Indeed, node B can successfully transmit information to node C (by using A as relay), but not vice versa.

Noting this potential problem, we proceed next to experiments. We find from our experiments that the EVC analysis, when applied to such a matrix A_T, returns useful results for understanding the spread of epidemics in disconnected mobile networks. We will give some discussion of why this is so in the final section.

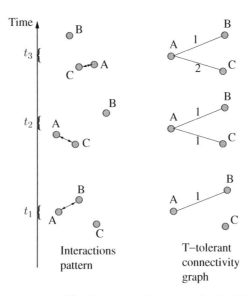

Fig. 1. Graphical representation of the timing issue in constructing the T-tolerant connectivity graph. In the resulting connectivity graph, there exists a path between B and C. Due to the dynamics, data originated from B can be delivered to C, but not vice versa

4 Experimental Results: Real-World Mobility Traces

In this section, we aim at verifying the properties, in terms of EVC distribution and spreading power of nodes, of various systems based on the deployment of disconnected mobile networks. While the analysis is carried out off-line, working on a log file describing all the contacts having taken place over the experimentation duration, we believe it to be a useful first step for understanding the ability of EVC, coupled with the T-tolerant connectivity matrix, to describe the spreading power of nodes in such framework. In particular, we considered four classes of traces, generated in different experiments and available on the CRAWDAD database at Dartmouth College [19].

The first set of experiments were performed by Intel Cambridge and reported in [20]. In this case, the devices were iMotes, equipped with a Bluetooth radio interface, and carried around by people in (i) a lab at Cambridge University (ii) people in the Intel lab at Cambridge (iii) attendees of IEEE INFOCOM05. Each iMote periodically scans the frequency range to check if other devices are present; each contact event is registered together with its time-stamp and duration.

The second set of experiments refers to the DieselNet project at University of Massachusetts at Amherst [21]. In this case, a number of buses has been equipped with IEEE 802.11b-compliant access points, used for bus-to-bus communications. Each device records the time and location of the contacts with other buses in the system. In our analysis, the trace 30122005 has been considered.

The third set of experiments comes from the Reality Mining project at MIT [22]. Data refers to a very long period (approximately one year), over which a set of students/researchers were tracked by means of mobile phones equipped with a Bluetooth interface. As for the Intel experiment, the Bluetooth device-discovery feature is exploited in order to detect the proximity of other Bluetooth-enabled devices. Each meeting is traced, and subsequently stored in a central repository for later processing.

The fourth set of experiments comes from the Student-Net project at University of Toronto [23]. In the experimentation, students were equipped with Bluetooth-enabled hand-held devices, capable of issuing inquiries and tracing any pairwise contact between users. The inquiry period was set so to preserve a 8–10 hours battery life-time. This resulted in a 16 s scan period. They performed two separate studies. The first one involved approximatively 20 graduate students only for a duration of two-and-a-half weeks, while the second one involved undergraduate students only, for a duration of 8 weeks.

All trace files have been preprocessed. We indeed observed in the traces that a single contact opportunity could lead (due to either SNR fluctuations or to the methods used to log contacts) to multiple entries in the trace file. This results in a series of "meetings" between a couple of nodes, with intermeeting times of the order of a few milliseconds, clearly not due to the devices mobility. Only the first entry in the trace file is kept for each meeting.

In Table 1, some details of the three experimentations are reported. It is worth noticing that the three experiments are referring to extremely different settings in terms of the underlying mobility pattern. As an example, Intel Exp2 refers present the results derived from nodes meeting in a relatively closed environment, such as the one performed at IEEE INFOCOM05, while the MIT Reality Mining project is considering the meetings of a nodes occurring at any time and place during the 1 year of experimentation. This results in extremely different meeting patterns, experimentation duration and number of nodes participating in the measurements.

Trace	Number of Nodes	Measurement Time (s)	Number of Records
Intel Exp1	128	359190	2766
Intel Exp2	223	522387	6732
Intel Exp3	264	255168	28216
UMass	22	83890	1504
Reality Mining	2135	3167388	25000
STUDNEWT Data 1	21	1371585	64160
STUDNEWT Data 2	23	3030076	14796

Table 1. Traces details for the four sets of data considered

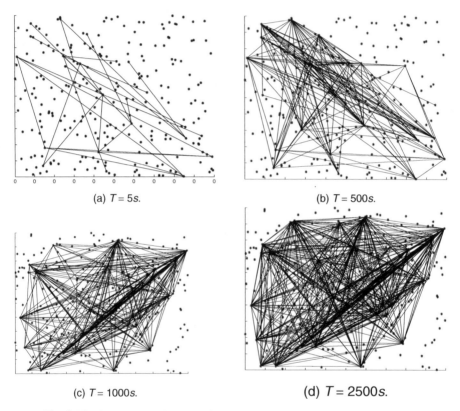

Fig. 2. T-tolerant connectivity graph for various values of T, INTEL Exp3 trace

As a first example, in Fig. 2, we plotted the resulting T-tolerant connectivity graph for the Intel Exp3 trace, with four different choices of the T value. Each node is assigned a random position in the unit square, chosen accordingly to a uniform distribution. A link between any two nodes is drawn if and only if a contact between them was observed within the specified time window T. The values considered for T are $5, 500, 1000, 2500$ seconds. Clearly, the specific connectivity graph evolution depends from the particular trace chosen. For the considered case, it can be observed that after 2500 seconds a significant number of nodes already experienced more than one meeting.

For all the datasets considered, we took T as the measurement times of the traces. It turned out, indeed, that taking long values of T does not impact the EVC analysis, at least for the traces we are considering.

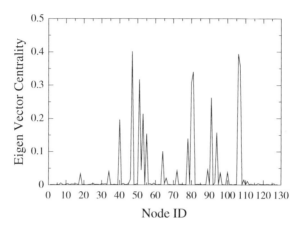

Fig. 3. Eigenvector centrality in the case of the INTEL Exp1 trace

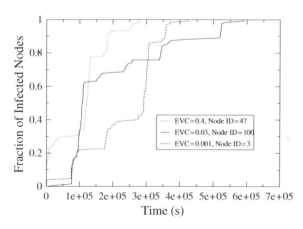

Fig. 4. Fraction of infected nodes over time, in the case of the INTEL Exp1 trace, and a variable *infecting* node

We started by considering both EVC and the spreading process as a function of the initiator (infecting) node. For all the experiments, we first computed the EVC, as detailed in Sec. 2. Then, we emulated the process of spreading (using the contacts in the corresponding trace file) starting from different "infecting" nodes. The infected process did not include any limitation in the number of copies each node could make or on the number of hops each message could traverse.

At time 0, one node is classified as "infected", and the remaining ones as "susceptible". Infected nodes spread the epidemics at any contact with susceptible ones. In order to avoid border effects due to the finiteness of the traces, traces were arranged in a cyclic fashion. The spreading process is considered concluded when all the nodes in the network are in the infected state.

In Fig. 3, we plotted the distribution of the EVC in the case of the Intel Exp1 trace, for the different nodes in the system. As it can be seen, the EVC is highly non-uniform, with nodes 1 and 9 presenting an EVC value far above the other nodes. In Fig. 4, we plotted the corresponding fraction of infected nodes vs. time for an epidemic starting from node 1 ($EVC = 0.65$), node 12 ($EVC = 0.03$) and node 7 ($EVC = 0.142$). As expected, the EVC turns out to give a significant measure of the ability of the node to initiate an epidemic. Please note, however, that the ability to spread an epidemics and the ability to initiate a spreading are, in general, different. Indeed, the choice of a given initiator node impacts mainly the first phase of the spreading process. For example, the curve corresponding to node 12 shows a sudden increase at about $0.85 \cdot 10^5$ seconds, and a detailed trace analysis showed that this corresponds to the time instant it meets with node 1, the one with the largest EVC value.

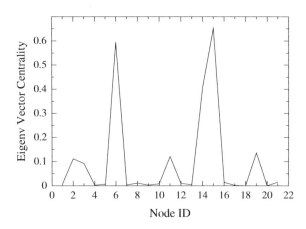

Fig. 5. Eigenvector centrality in the case of the STUDNETW Data1 trace

Similarly, Fig. 5 depicts the EVC distribution in the case of the STUDENT-Net data1 experimentation. Also in this second case, the EVC is not distributed uniformly, meaning that is reasonable to expect a different capacity of the nodes to diffuse information. This is confirmed in Fig. 6, where it is possible a observe a different infection depending from the initiator node. In this case, the difference does not seem to be as pronounced as in the INTEL Exp 1 case. This is mostly due to

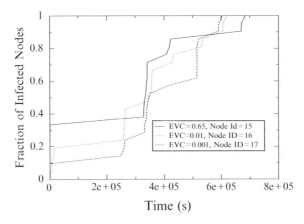

Fig. 6. Fraction of infected nodes over time, in the case of the STUDNETW Data 1 trace, and a variable *infecting* node

the limited number of nodes participating to the experimentation, and to the limited duration of the measurement.

The same analysis is applied to the Intel Exp2 trace, with the epidemics starting from nodes 3 ($EVC = 0.66$), node 19 ($EVC = 0.07$) and node 2 ($EVC = 0.002$), respectively. The results are depicted in Fig. 7 and Fig. 8. Also in this case, centrality scores are highly non-uniform, and the graph shows that the choice of the infecting node has a remarkable impact on the speed at which an epidemic spreads in the system. As for the Intel Exp3 trace, node 25, characterized by a low EVC value, at time $7 \cdot 10^4$ presents a sudden spreading increase derived from the meeting with a node with a high EVC value. It is also worth noticing that the time taken to infect the whole system seems to be not very sensitive to the EVC of the initiator node. We conjecture that such effect is due to the presence in the system of a low number of nodes which seldom meet other ones in the system, representing therefore a sort of "bottleneck" for the epidemics to spread the entire network.

We have then considered the traces from the UMASS DieselNet project experiment. Also for such case we evaluated the EVC on the whole trace duration (in this case, one day) and emulated the spreading process with different starting nodes. The results, in terms of eigenvector centrality and fraction of nodes infected with various initiator nodes, are reported in Fig. 9 and Fig. 10. It can be seen that the distribution is highly non-uniform, corresponding to a different ability of nodes to spread an epidemic in the system. Differently from other experiments, in the UMASS DieselNet trace the reported meetings occurred only among nodes participating to the experiment.[6] As a result, the meetings pattern shows a higher regularity, and the EVC distribution is more uniform, if compared with the Intel Exp1 and Intel Exp2 EVC analysis. Correspondingly, to higher values of EVC correspond a higher capability

---

[6] When leveraging on the Bluetooth device-discovery capabilities for reporting neighboring nodes, any Bluetooth-enabled device is reported.

of nodes both to initiate a spreading and to spread an epidemics. As an example, in Fig. 10 node 19 (EVC=0.5) not only presents the highest initial infection rate, but also infects the entire network in a short time, if compared with nodes with lower EVC values (i.e., node 2 and node 6).

In Fig. 11 and Fig. 12, the EVC and the spreading power analysis has been extended to the MIT Reality Mining experiment. Also in this case, the EVC distribution is highly non-uniform; this means that there is a large variation in spreading power among the nodes of the network. Node 3 presents the highest EVC value, and, correspondingly, a regular spreading pattern can be observed from Fig. 12. Conversely, node 22 presents an extremely low EVC score. This is reflected in an extremely low infection for the first $5 \cdot 10^5$ seconds, and a sudden and extremely fast diffusion afterwards. This behavior can be easily explained with node 22 meeting a node characterized by an extremely large value of EVC. Also in this case the fact that the various curves tend to converge can be due to a small set of student in the experiment which tend to remain isolated from the rest of the network.

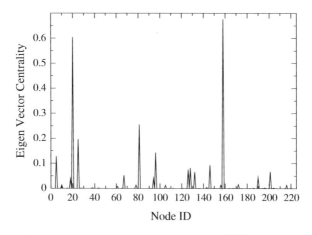

Fig. 7. Eigenvector centrality in the case of the INTEL Exp2 trace

The outcomes are different for the Intel Exp3 trace. Indeed, as it can be seen from the EVC, plotted in Fig. 13, just a fraction of the nodes in the system shows a non-negligible level of interactions. This can be due to the fact that all contacts with Bluetooth-enabled devices are recorded, not just with the other people in the experiment. A large fraction of the nodes appear for a limited number of times in the traces (in most cases, actually just once). The result is that the dependence on the EVC of the initiating node is only partial, in that it is valuable only for the nodes "truly" in the system. The other nodes appear mostly just once in the traces, so they cannot be infected until the very only meeting happens. For this reason, we decided not to report the graph of the fraction of infected nodes.

From all the considered real world deployments, we can reasonably conclude that the EVC analysis, applied to the T-tolerant connectivity matrix, returns meaningful

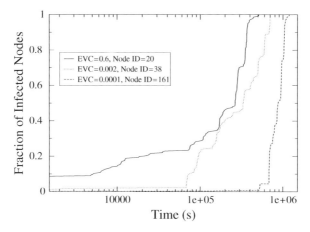

Fig. 8. Fraction of infected nodes over time, in the case of the INTEL Exp2 trace, and a variable *infecting* node

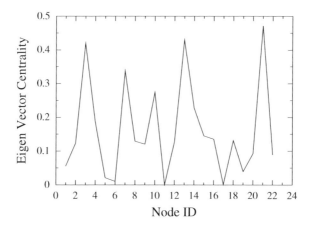

Fig. 9. EVCentrality in the case of the UMASS trace

performance metrics for assessing the ability of nodes to spread epidemics within a mobile disconnected network. The impact of such metric on the actual epidemics spreading pattern depends on the particular scenario considered. There might be cases in which nodes are able to infect the entire system in very short time, while in other circumstances a node with a large EVC may initiate a spreading process which, after a fast initial ramp-up, takes very long times to reach all nodes. The rationale underpinning this behavior resides in the level of partition of the network. Nodes meeting on a more regular basis contribute to the initial boost of the spreading process, while nodes meeting only rarely (or only once) are responsible for the heavy tail of the epidemic spreading process. The EVC value of the initiator node impacts

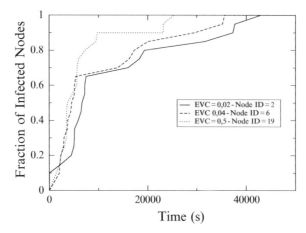

Fig. 10. Percentage of infected nodes over time, in the case of the UMASS trace, and a variable *infecting* node

mostly the initial part of the spreading, and only partially the subsequent spreading of messages in the system.

Region Analysis

In the following, a simplified region analysis is performed for the data set corresponding to the case of the UMass DieselNet project trace, with a time window of 8000 seconds. By following the steepest ascent rule [9] an ancestor is iteratively assigned to each node, until a local maximum is reached. The local maximum corresponds to a *centre* of the region [9]. From the conducted analysis, the network results to be partitioned in 2 distinct regions, node 1 being the centre of the first region with a EVC of 0.3822, and node 10 of the second one with an EVC of 0.5537. The first region consists of a total of two nodes, whereas the latter region consists of 13 nodes.

In Fig. 14, the resulting graph is depicted, with the corresponding EVC values. Region centres are depicted with larger circles; the nodes belonging to the region of node 10 are gridded, while the ones belonging to the region of node 1 are plotted in black. As can be seen from the figure, centre nodes are characterized by a high connectivity degree. Further, note that the two centres are not directly connected. (They cannot be, since each one is a local maximum of the EVC.) Thus, any path between the two centres has to go through another node which bridges the two regions.

Finally, in Fig. 15 and Fig. 16, the effect of the regions centre is highlighted in the case the UMass DieselNet project trace, with a time window of 8000 seconds. In Fig. 15, node 3 is starting the epidemics spreading. When node 10 becomes infected, it is possible to observe a significant boost in the spreading process. This effect is even more evident when the epidemics spreading reaches nodes 1: in less than 5000 seconds the remaining nodes of the network are infected. This is due to the well-connectedness of the region centre nodes. It is possible to observe a similar behavior

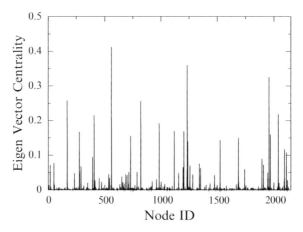

Fig. 11. Eigenvector centrality in the case of the MIT trace

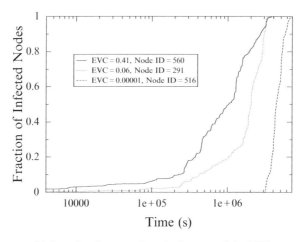

Fig. 12. Percentage of infected nodes over time, in the case of the MIT trace, and a variable *infecting* node

in Fig. 16, where node 2 is the node initiating the epidemics spreading. Node 2 belongs to the region of node 1, and, therefore, node 1 is infected before node 10 (as opposed to the previous case).

The region analysis has been applied to a variety of traces. We took a set of 6 traces, and considered for each of them two cases. In the first one, we used the whole trace, whereas in the second one we limited the analysis to only the first half of the records. This corresponds to choosing two different values of T. The results are summarized in Table 2. Some interesting remarks can be inferred from the results therein. First, all traces whose analysis was limited to the first half of the records returned a disconnected (highly partitioned) graph. On the other hand, all full trace files lead to a connected system. This emphasizes the importance of a correct choice

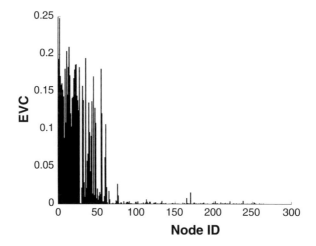

Fig. 13. Eigenvector Centrality in the case of the INTEL Exp3 trace

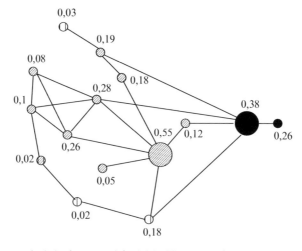

Fig. 14. Region analysis in the case of the UMASS trace, and time window of 8000 seconds

of the parameter T, which plays a key role in ensuring that the EVC analysis returns meaningful results. Second, all full traces returned a single region. We conjecture that such result is due to the fact that all experiments targeted rather homogeneous scenarios (people in a lab, attendees at a conference, student in one class, buses in a town etc.), where nodes tend naturally to form one single cluster. It remains therefore still an open issue to understand what the results of such analysis could be when applied to a real-world deployment, not limited to the simple experiments we considered in this chapter.

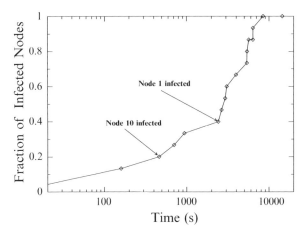

Fig. 15. Fraction of infected nodes, with a time window of 8000 seconds, and node 3 starting the epidemics spreading

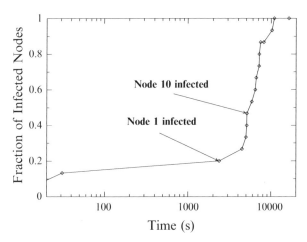

Fig. 16. Fraction of infected nodes, with a time window of 8000 seconds, and node 2 starting the epidemics spreading

5 Conclusions & Open Issues

In this chapter we have addressed the problem of enhancing forwarding mechanisms in disconnected wireless networks with information concerning the "fitness" of the nodes, understood as a measure of their ability to spread epidemics. The reason for such a study comes from the observation that, in the case of highly partitioned mobile networks, information can be diffused system-wide only by opportunistically exploiting the contacts among nodes, and some form of epidemic forwarding is necessary to achieve such purpose.

Trace	Number of Regions	Connected	Connected Components
INTEL Exp1 (full)	1	1	1
INTEL Exp1 (half)	25	0	25
INTEL Exp2 (full)	1	1	1
INTEL Exp2 (half)	117	0	117
INTEL Exp3 (full)	1	1	1
INTEL Exp3 (half)	92	0	92
MIT (full)	1	1	1
MIT (half)	389	0	389
STUDNETW Data1 (full)	1	1	1
STUDNETW Data1 (half)	1	1	1
STUDNETW Data2 (full)	1	1	1
STUDNETW Data2 (half)	1	1	1
UMASS 2072005 (full)	1	1	1
UMASS 2072005 (half)	3	0	3

Table 2. Results of the region analysis for various trace files

Based on earlier work addressing the problem of understanding the spread of epidemics in static wired networks, we identified the EVC as a suitable metric. The T-tolerant connectivity matrix concept was introduced as a mean to enable the application of the EVC analysis to the disconnected wireless scenario under study.

The resulting method has been applied to a wide range of trace files, coming from real deployments of delay-tolerant networks. The results reported show that the EVC, coupled with the T-tolerant connectivity matrix, is able to provide a good understanding of the ability of nodes in such systems to spread information on a network-wide scale.

It is not obvious that the EVC approach—which works well for networks with a fixed topology—should also work well when applied, via use of the T-tolerant connectivity matrix, to a dynamic network. We recall the discussion around Figure 1, where we showed that the use of the connectivity matrix can imply the existence of transmission possibilities which in fact do not exist in the true, dynamic network. So we ask, why do we find such a good correspondence between the predictions of the static EVC picture and the true epidemic simulations?

We believe that the answer lies in the fact that, for these traces, many pairs of nodes meet repeatedly. We note, for example, that if A and B meet repeatedly in the course of the experiment depicted in Figure 1, then the communication paths shown on the right hand side (from the connectivity matrix) will also in fact exist in the true, dynamic, network. That is, repeated meetings eliminate most or all of the artifactual errors introduced by forming the connectivity matrix.

Furthermore, we note that we give most weight in the connectivity matrix to links which involve repeated meetings. Conversely, links involving only a single meeting—which is thus never repeated in the course of the trace—receive the lowest possible nonzero weight in the connectivity matrix. Hence, our approach gives most weight to node pairs which meet repeatedly; and our EVC analysis works best when such pairs dominate the analysis. Thus we get a consistent picture of why the EVC analysis can work well for such dynamic networks.

While the results presented in this chapter show that EVC has the potential for enhancing existing forwarding protocols proposed for disconnected wireless networks, many open issues are still present. The first one is concerned with the possibility of computing on-line the EVC in a distributed fashion. While this problem has been tackled in [3], the application to the present scenario is not straightforward. Each node could indeed well count the meetings having taken place (over a time interval T) with other nodes; this would give each node its own row (= column) of the T-tolerant connectivity matrix. With this information, each node could also exchange its current EVC value with those nodes it meets. We recall from [3] that a distributed EVC calculation needs two things: (i) iterated weight passing (taking also into account link weights), and (ii) periodic normalization of the weight vector (which otherwise grows to infinity). Now we see that operation (i) is possible for such a dynamic net. The problem is then (ii).

Normally, one wishes to rescale the weight vector by dividing by a local estimate of the eigenvalue. In [3], a method is given for finding this estimate. This method assumes however constant connectivity, so that it can converge much faster than the slower, weight passing operation. Here we study networks which do not have such constant connectivity. Hence we consider other approaches.

- Each node can simply use the *sum* of its link weights as an estimate of its centrality. This is equivalent to a single iteration of the EVC power method (starting from a uniform start vector). For cases where the power method converges rapidly, this may be a good estimate of the true EVC; and in any case, it represents useful and easily accessible information.
- Two iterations are also possible. That is, each node gets (after enough meetings) the one-iteration estimates (column sums) for all of its neighbors, multiplies these by the corresponding link weights, and takes the sum of these, over all neighbors, as its two-iteration estimate.
- We see from considering one and two iterations that, in principle, any finite number n of iterations may be managed (with a corresponding cost in storage, signalling, and clock time needed). Each estimate, when finally obtained by a node, must simply be stored, labelled (by its iteration number), and passed on (when the opportunity arises). Thus, in real-world networks, we can imagine that $n > 2$ can be practical, giving a good, or even very good, approximation to the true EVC—*without* the need to ever estimate the global eigenvalue λ_{max}.

We see from the above that a reasonable estimate of the EVC may be obtained by dynamic networks of the type we study here. An implicit precondition for this idea to work is that the nodes' pattern of movements and meetings must be reasonably stable and repetitive. The reason for this is that (a) the nodes need a 'startup' time T to obtain estimates of the link weights; (b) the nodes then need some time (depending on the desired iteration number n) to compute the centrality estimates based on the stored connectivity matrix; and (c) if the pattern of meetings has *changed* significantly since the startup time, then the answer will cease to be valid. In other words, the mobility pattern must be, in some statistical sense, stable over a time scale which is considerably longer than the time needed to compute the EVC estimates—which is, roughly, $O(n \times T)$.

If this precondition is not even met for $n = 1$, we see no way for such a network to compute, using its own meetings, any reasonable EVC estimate. However, any network with this degree of dynamism will in fact render invalid the use of the T-tolerant connectivity matrix. We note that we get some indication of repetitive movement from the good results we have found in this paper; but of course this question must be studied more carefully, by examining the traces in detail. We also leave for future work the problem of how the network itself can discover which n (if any) is appropriate to its own mobility patterns.

Finally, it remains also to devise how such a metric — whatever way it gets computed — can be effectively used for enhancing the performance of epidemics-style forwarding mechanisms for disconnected wireless networks. The knowledge of the fitness value of the encounters represents indeed additional information which could be used to optimize the performance of the system. While it is easy to introduce simple techniques for making use of such knowledge, it is nonetheless unclear at the moment what could be the optimal way of exploiting it for optimizing the system performance.

Acknowledgments

The work of I. Carreras and D. Miorandi has been partially supported by the European Commission within the framework of the BIONETS project IST-FET-SAC-FP6-027748, www.bionets.eu. The work of G. Canright and K. Engø-Monsen was partially supported by the Future and Emerging Technologies unit of the European Commission through Project DELIS (IST-2002-001907).

References

1. A. A. Hanbali, P. Nain, and E. Altman, "Performance of two-hop relay routing protocol with limited packet lifetime," in *Proc. of ValueTools*, Pisa, Italy, 2006.
2. X. Zhang, G. Neglia, J. Kurose, and D. Towsley, "Performance modeling of epidemic routing," in *Proc. of Networking*, 2006.
3. G. Canright, K. Engø-Monsen, , and M. Jelasity, "Efficient and robust fully distributed power method with an application to link analysis," Department of Computer Science, University of Bologna, Tech. Rep. UBLCS-2005-17, 2005. [Online]. Available: http://www.cs.unibo.it/bison/publications/2005-17.pdf
4. V. Cerf, S. Burleigh, A. Hooke, L. Torgerson, R. Durst, K. Scott, K. Fall, and H. Weiss, "Delay-tolerant network architecture," 2005, iETF Internet Draft. [Online]. Available: http://www.dtnrg.org/wiki
5. K. Fall, "A delay-tolerant network architecture for challenged Internets," in *Proc. of ACM SIGCOMM*, Karlsruhe, DE, 2003.
6. A. Khelil, C. Becker, J. Tian, and K. Rothermel, "An epidemic model for information diffusion in manets," in *Proc. of ACM MSWiM*, 2002.

7. I. Carreras, I. Chlamtac, F. De Pellegrini, and D. Miorandi, "Bionets: Bio-inspired networking for pervasive communication environments," *IEEE Trans. Veh. Tech.*, 2006, in press. [Online]. Available: http://www.create-net.org/$\sim$dmiorandi
8. T. Small and Z. Haas, "The shared wireless infostation model Ű a new ad hoc networking paradigm (or where there is a whale, there is a way)," in *Proc. of ACM MobiHoc*, 2003, pp. 233–244.
9. G. Canright and K. Engø-Monsen, "Roles in networks," *Science of Computer Programming*, vol. 53, pp. 195–214, 2004.
10. G. Canright and K. Engo-Monsen, "Spreading on networks: a topographic view," in *Proc. of ECCS*, Paris, 2005.
11. P. Bonacich, "Factoring and weighting approaches to status scores and clique identification," *Journal of Mathematical Sociology*, vol. 2, pp. 113–120, 1972.
12. H. Minc, *Nonnegative matrices*. New York: J. Wiley and Sons, 1988.
13. R. Meester and R. Roy, *Continuum Percolation*. New York: Cambridge Univ. Press, 1996.
14. U. Lee, E. Magistretti, B. Zhou, M. Gerla, P. Bellavista, and A. Corradi, "MobEyes: smart mobs for urban monitoring with vehicular sensor networks," UCLA CSD, Tech. Rep. 060015, 2006. [Online]. Available: http://netlab.cs.ucla.edu/wiki/files/mobeyestr06.pdf
15. P. Gupta and P. R. Kumar, "The capacity of wireless networks," *IEEE Trans. on Inf. Th.*, vol. 46, no. 2, pp. 388–404, Mar. 2000.
16. M. Grossglauser and D. Tse, "Mobility increases the capacity of ad hoc wireless networks," *IEEE/ACM Trans. on Netw.*, vol. 10, no. 4, pp. 477–486, Aug. 2002.
17. J. Burgess, B. Gallagher, D. Jensen, and B. N. Levine, "MaxProp: routing for vehicle-based disruption-tolerant networks," in *Proc. of IEEE INFOCOM*, Barcelona, ES, 2006.
18. F. D. Pellegrini, D. Miorandi, I. Carreras, and I. Chlamtac, "A graph-based model for disconnected ad hoc networks," in *Proc. of IEEE INFOCOM*, 2007.
19. CRAWDAD, the community resource for archiving wireless data at Dartmouth. [Online]. Available: http://crawdad.cs.dartmouth.edu/
20. A. Chaintreau, P. Jui, J. Crowcroft, C. Diot, R. Gass, and J. Scott, "Impact of human mobility on the design of opportunistic forwarding algorithms," in *Proc. of IEEE INFOCOM*, Barcelona, ES, 2006.
21. The disruption tolerant networking project at UMass. [Online]. Available: http://prisms.cs.umass.edu/diesel/
22. Machine perception and learning of complex social systems. [Online]. Available: reality.media.mit.edu/
23. J. Su, A. Goel, and E. de Lara, "An empirical evaluation of the student-net delay tolerant network," in *Proc. of MOBIQUITOUS*, San Jose, US, July 2006.

Toward Organization-Oriented Chemical Programming: A Case Study with the Maximal Independent Set Problem

Naoki Matsumaru, Thorsten Lenser, Thomas Hinze and Peter Dittrich

Bio Systems Analysis Group
Jena Centre for Bioinformatics (JCB) and
Department of Mathematics and Computer Science
Friedrich-Schiller-University Jena
{naoki, thlenser, hinze, dittrich}@minet.uni-jena.de

Summary. Biological systems are considered as a source of inspiration to make a system robust, self-organizing, adaptive, fault-tolerant, and scalable. These features are achieved by an orchestrated, decentralized interplay of many relatively simple asynchronous components. Since information is processed in living organisms using interconnected chemical reactions, the chemical reaction metaphor has been proposed as a novel computation paradigm. A couple of approaches are already using the chemical metaphor, such as, Gamma, MGS, amorphous computing, membrane computing, and reaction-diffusion processors. When employing a large number of components into a system, however, it becomes hard to control and program the system behavior. Therefore, new programming techniques are required. Here we describe how chemical organization theory can serve as a tool for chemical programming. The theory allows to predict the potential behavior of a chemical program and thus supports a programmer in the design of a chemical-like control system. The approach is demonstrated by applying it to the maximal independent set problem. We show that the desired solutions are predicted by the theory as chemical organizations. Furthermore the theory uncovers "undesirable" organizations, representing uncompleted halting computations due to insufficient amount of molecules.

1 Introduction

In order to implement a system to be more reliable, adaptable, and fault tolerant, inspirations have been sought from biological systems [1]. For instance, inter- and intracellular signal transduction processes are compared with a computer network system to address network security problem [2]. Immune system mechanisms have been employed to implement autonomous intrusion detection systems [3]. Self-organizing collective behavior observed in ant colonies has led to a new method [4, 5] to balance the network load in telecommunication networks [6] or in a distributed peer-to-peer system [7]. Other biologically inspired techniques for distributed computing can be found in [8]. These applications of biologically inspired methods are rather focused on decentralized, asynchronous distributed computing systems though they are not restricted to that. This application emphasis on distributed systems can be rationalized by general principles of biological systems. Out of an orchestrated, decentralized interplay of many relatively simple components, biological system behavior emerges in the global level. Moreover, conventional engineering methods have faced difficulties in this field. A fundamental of this situation is the immaturity of theoretical developments of the emergent behavior resulting from extensive local interactions [9]. The global system behavior cannot be understood by looking at isolated local parts of the system.

The chemical reaction metaphor has been proposed as a source of inspiration for a novel computation paradigm [10, 11] since all known life forms process information using chemical processes [12]. In chemical computing, the solution appears as an emergent global behavior based on a manifold of local interactions [13]. Analysis of such behavior is hard because of its nonlinear characteristics. In a non-trivial situation, it is impossible to predict the behavior by methods that are more efficient than simulations (proof by reduction to the halting problem). Since a prerequisite for programming by construction is the ability to predict how a chemical program defined by, for example, a list of reaction rules behaves [14], a theoretical analysis of emergent behavior in chemical computing is necessary. We have suggested chemical organization theory [15] as a tool to help chemical programming, constructing a reaction network, and analysis, describing and understanding the reaction system behavior. A contribution towards establishing a theory of (chemical) emergence is also conceived [16].

Applying chemical organization theory, the constructed reaction networks are decomposed into overlapping sub-networks called "(chemical) organizations" according to qualitative steadiness estimated from the network structure. Although individual molecules may appear or disappear, qualitative states of the system characterized by molecular types (species) present are less likely to change if the qualitative state equals to an organization. It has been proven that only organizational set of species can constitute a stationary state, given an ordinary differential equation system describing the dynamics of the constructed reaction network system [15]. Employing this analysis interactively, a chemical reaction network is programmed to behave in the system level as desired. This "organization-oriented chemical programming" is envisioned as a design methodology for programming chemical-like system.

In this article, we show how chemical organization theory helps programming distributed processes of chemical computing. The maximal independent set (MIS) problem serves as an example. In the next section (Section 1.1), we briefly review chemical organization theory including necessary definitions. Section 2 describes a chemical programming to solve the MIS problem, converted and adapted from the algorithm proposed in [17]. Specific problem instances and the corresponding chemical programs are followed. For each instance, the chemical program is analyzed using chemical organization theory, investigating the organizational structures embedded in the constructed reaction network. From these examples, the association between the MIS and the organizations with the most species is observed. In Section 2.2, a proof of this association is given. Finally, we discuss progressing work to evaluate our approach quantitatively.

We note that our focus here is where chemistry stimulates the development of new computational paradigms. These approaches can be distinguished by whether real or artificial chemistries are used. *Real chemical computing* employs real molecules and chemical processes to compute. For example, the simplest nonlinear function XOR can be performed with reaction-diffusion behavior of palladium chloride [18] or with the context sensitive enzyme malate dehydrogenase [19]. Flip-flops, basic components of memory devices with bistable states, can be implemented by a gene regulatory network *in vivo* [20, 21]. On the other hand, *artificial chemical computing* is where the chemical metaphor is utilized to program or to build electronic computational systems. This takes the chemical metaphor as a design principle for new software or hardware architectures built on conventional silicon devices. Artificial chemical computing, thus, includes constructing chemical-like formal systems in order to model and master concurrent processes [11, 22, 23].

1.1 Chemical Organization Theory

The target of chemical organization theory are reaction networks. A reaction network consists of a set of molecules $\mathcal{M}$ and a set of reaction rules $\mathcal{R}$. Therefore, we define a reaction network formally as a tuple $\langle \mathcal{M}, \mathcal{R} \rangle$ and call this tuple an algebraic chemistry in order to avoid conflicts with other formalizations of reaction networks.

Definition 1 (algebraic chemistry [15]). *Given a set $\mathcal{M}$ of molecular species and a set of reaction rules given by a relation $\mathcal{R} \subseteq \mathcal{P}_M(\mathcal{M}) \times \mathcal{P}_M(\mathcal{M})$. We call the pair $\langle \mathcal{M}, \mathcal{R} \rangle$ an* algebraic chemistry, *where $\mathcal{P}_M(\mathcal{M})$ denotes the set of all multisets with elements from $\mathcal{M}$.*

A multiset differs from an ordinary set in that it can contain multiple copies of the same element. A reaction rule is similar to a rewriting operation [22, 24] on a multiset. Adopting the notion from chemistry, a reaction rule is written as $A \to B$ where both A and B are multisets of molecular species. The elements of each multiset are listed with "+" symbols between them. Instead of writing $\{s_1, s_2, \ldots, s_n\}$, the set is written as $s_1 + s_2 + \cdots + s_n$ in the context of reaction rules. We also rewrite $a + a \to b$ to $2a \to b$ for simplicity. Note that "+" is not an operator but a separator of elements.

A set of molecular species is called an organization if the following two properties are satisfied: closure and self-maintenance. A set of molecular species is closed when all reaction rules applicable to the set cannot produce a molecular species that is not in the set. This is similar to the algebraic closure of an operator in set theory.

Definition 2 (closure [25]). *Given an algebraic chemistry $\langle \mathcal{M}, \mathcal{R} \rangle$, a set of molecular species $C \subseteq \mathcal{M}$ is closed, if for every reaction $(A \to B) \in \mathcal{R}$ with $A \in \mathcal{P}_M(C)$, also $B \in \mathcal{P}_M(C)$ holds.*

The second important property, self-maintenance, assures, roughly speaking, that all molecules that are consumed within a self-maintaining set can also be produced by some reaction pathways within the self-maintaining set. The general definition of self-maintenance is more complicated than the definition of closure because the production and consumption of a molecular species can depend on many molecular species operating as a whole in a complex pathway.

Definition 3 (self-maintenance [15]). *Given an algebraic chemistry $\langle \mathcal{M}, \mathcal{R} \rangle$, let i denote the i-th molecular species of $\mathcal{M}$ and the j-th reaction rules is $(A_j \to B_j) \in \mathcal{R}$. Given the stoichiometric matrix $\mathbf{M} = (m_{i,j})$ that corresponds to $\langle \mathcal{M}, \mathcal{R} \rangle$ where $m_{i,j}$ denotes the number of molecules of species i produced[1] in reaction j, a set of molecular species $S \subseteq \mathcal{M}$ is self-maintaining, if it there exists a flux vector $\mathbf{v} = (v_{A_1 \to B_1}, \ldots, v_{A_j \to B_j}, \ldots, v_{A_{|\mathcal{R}|} \to B_{|\mathcal{R}|}})^T$ satisfying the following three conditions:*

(i.) $v_{A_j \to B_j} > 0$ if $A_j \in \mathcal{P}_M(S)$
(ii.) $v_{A_j \to B_j} = 0$ if $A_j \notin \mathcal{P}_M(S)$
(iii.) $f_i \geq 0$ if $s_i \in S$ where $(f_1, \ldots, f_i, \ldots, f_{|\mathcal{M}|})^T = \mathbf{Mv}$.

These three conditions can be read as follows: When the j-th reaction is applicable to the set S, the flux $v_{A_j \to B_j}$ must be positive (Condition 1). All other fluxes are set to zero (Condition 2). Finally, the production rate f_i for all the molecular species $s_i \in S$ must be nonnegative (Condition 3). Note that we have to find only one such flux vector in order to show that a set is self-maintaining.

Taking closure and self-maintenance together, we arrive at an organization:

Definition 4 (organization [15, 25]). *A set of molecular species $O \subseteq \mathcal{M}$ that is closed and self-maintaining is called an organization.*

We visualize the set of all organizations by a Hasse diagram, in which organizations are arranged vertically according to their size in terms of the number of their members (e.g. Fig. 1). Two organizations are connected by a line if the lower organization is contained in the organization above and there is no other organization in between.

Finally, a relevant theorem from Ref. [15] states that given a differential equation describing the dynamics of a chemical reaction system and the algebraic chemistry corresponding to that system, then the set of molecular species with positive concentrations in a fixed point (*i.e.*, stationary state), if it exists, is an organization. In other words, we can only obtain a stationary behavior with a set of molecular species that are both closed and self-maintaining.

2 Chemical Programming for the MIS Problem

In this section we present a procedure for designing chemical reaction networks solving the maximal independent set (MIS) problem (see Table 1 for a short recipe). An application of this problem lies in wireless sensor networks to determine clusterheads, which manage logical clustering structures in the networks. Sensor nodes that

---

[1] Formally, this can be defined as $m_{i,j} = \#(i \in B_j) - \#(i \in A_j)$, where $\#(i \in A_j)$ denotes the number of occurrences of species i on the left-hand side of reaction j and $\#(i \in B_j)$ the number of occurrences of species i on the right-hand side of reaction j.

can communicate directly with the cluster-head form a cluster, and no cluster-heads are neighbors. Assigning a role to each sensor node benefits to increase lifetime of the whole network [26]. Conventional algorithms to solve the MIS problem are theoretically studied (*e.g.*, [27]). In the context of distributed computing, algorithms are investigated for self-stabilization [17, 28, 29].

Let an undirected graph $G = \langle V, E \rangle$ be defined by a set of N vertexes $V = \{v_1, \ldots, v_N\}$ and a set of edges E. When two vertexes v_p and v_q are connected, the pair of the vertexes is in the set of edges: $(v_p, v_q) \in E$. Note that the order of the pair is insignificant, that is, $(v_p, v_q) = (v_q, v_p)$. A set of vertexes $I \subset V$ is independent if no two vertexes in the set are adjacent: $(\forall v_p, v_q \in I : (v_p, v_q) \notin E)$. An independent set is maximal if no vertex can be added to the set while keeping the property of independence. Including another vertex in the MIS should violate the independence property.

Given the undirected graph G, an algebraic chemistry $\langle \mathcal{M}, \mathcal{R} \rangle$ is designed as follows: For each vertex v_j, we assign two molecular species s_j^0 and s_j^1 representing the membership of the vertex in the MIS. The subscript of the species name corresponds to the index number of the vertex. High concentration of species s_j^1, higher than a threshold chosen to be smaller than any positive coordinate of any fixed point, means that the vertex v_j is included in the MIS. High concentration of species s_j^0 expresses that the vertex v_j is *not* included in the MIS. Thus the set of molecular species $\mathcal{M}$ contains $2N$ molecular species:

$$\mathcal{M} = \{s_j^0, s_j^1 \mid j = 1, \ldots, N\} \quad (1)$$

The set of reaction rules $\mathcal{R}$ is constructed by assembling reactions for each vertex:

$$\mathcal{R} = \bigcup_{i=1}^{N} \mathcal{R}^i = \bigcup_{i=1}^{N} (\mathcal{V}^i \cup \mathcal{N}^i \cup \mathcal{D}^i). \quad (2)$$

For each reaction set $\mathcal{R}^i$, there are three sorts of reactions. The first two sorts are adapted from two predicates constituting a program for any distributed processor to solve the MIS problem under a central scheduler [17]. A reaction rule to produce species s_i^1 is the first:

$$\mathcal{V}^i = (\overbrace{s_j^0 + s_k^0 + \cdots + s_l^0}^{n_i} \to n_i s_i^1) \quad (3)$$

where n_i is the number of vertexes connected to vertex v_i and $v_j, v_k, \ldots, v_l$ are its neighboring vertexes, that is, $(v_i, v_j), (v_i, v_k), \ldots, (v_i, v_l) \in E$. The left hand side of the reaction contains n_i terms, and this reaction is interpreted as follows: When no neighboring vertex is included in the MIS, the target vertex v_i should be included in the set.

The negation of this predicate is considered by a set of n_i reactions:

$$\mathcal{N}^i = \{s_j^1 \to s_i^0 | (v_i, v_j) \in E\}. \quad (4)$$

Table 1. Recipe for mapping an undirected graph to a chemical reaction network.

Input: Undirected graph $G = \langle V, E \rangle$ where V is a set of N vertexes $V = \{v_1, \ldots, v_N\}$ and E is a set of edges. When two vertexes v_p and v_q are connected, $(v_p, v_q) \in E$.
Output: Algebraic chemistry $\langle \mathcal{M}, \mathcal{R} \rangle$ (a set of molecular species $\mathcal{M}$ and a set of reaction rules $\mathcal{R}$) representing the chemical program to solve the maximal independent set problem.
Algorithm:
1. For each vertex v_j:
(a) Add two molecular species, s_j^0 and s_j^1, to $\mathcal{M}$;[2]
(b) Add one *destructive reaction* of the form $s_j^0 + s_j^1 \to \emptyset$ to $\mathcal{R}$;
(c) Add one reaction to $\mathcal{R}$ of the form:

$$(\cdots + s_i^0 + \ldots \to n_j s_j^1)$$

where n_j is the number of edges connected to vertex v_j and $(v_j, v_i) \in E$.
(d) Add a set of n_j reactions to $\mathcal{R}$:

$$\{s_i^1 \to s_j^0 | (v_i, v_j) \in E\}.$$

---
[2] As a naming convention of molecular species in this paper, the superscript indicates the membership for the maximal independent set.

This is the second type of reactions, which produce species s_i^0 from any species corresponding to the neighboring vertexes with superscript 1. This rule can be interpreted as follows: If there exists at least one neighboring vertex included in the MIS, then the target vertex v_i should be excluded from the maximal independent set (otherwise the definition of the MIS would be violated). Generating species s_i^0 forces vertex v_i not to be included in the set.

The last component of set $\mathcal{R}^i$ is a *destructive reaction*. Since the membership of the MIS is a binary state, the state becomes undefined when neither or both of the species are present. In order to avoid the latter case, the two opposite molecular species are defined to vanish upon collision:

$$\mathcal{D}^i = s_i^0 + s_i^1 \to \emptyset. \tag{5}$$

Note that the algebraic chemistry is defined such that molecules react only if they are located on the same vertex or are neighbors. Thus, the resulting (artificial) chemical system can be interpreted as a spatially distributed compartmentalized reaction system, where a compartment j holds only the two chemical species representing a vertex v_j, namely s_j^0 and s_j^1 and where the topological structure of the compartments is equivalent to the undirected graph.

2.1 Example of Chemical Programming to Solve the MIS Problem

Linear Graph with Three Nodes

Provided that an undirected graph $G = \langle V, E \rangle$ consists of three vertexes and those vertexes are connected linearly as shown in Fig. 1 (A):

$$G = \langle \{v_1, v_2, v_3\}, \{(v_1, v_2), (v_2, v_3)\} \rangle. \tag{6}$$

Following the recipe, an algebraic chemistry $\langle \mathcal{M}, \mathcal{R} \rangle$ is constructed. The set of molecular species $\mathcal{M}$ consists of six species because the graph contains $N = 3$ vertexes:

$$\mathcal{M} = \{s_1^0, s_1^1, s_2^0, s_2^1, s_3^0, s_3^1\}. \tag{7}$$

Our naming convention for the species is that the subscript of the species name is associated with the index of the graph vertex and that the superscript stands for the membership of the MIS. For example, species s_2^1 stands for vertex v_2 being included in the MIS, and s_2^0 represents the opposite for the same vertex.

For each vertex v_1, v_2, and v_3, reaction rules are constructed. The destructive reactions are:

$$\mathcal{D} = \bigcup_{i=1}^{3} \mathcal{D}^i = \{s_1^0 + s_1^1 \to \emptyset, s_2^0 + s_2^1 \to \emptyset, s_3^0 + s_3^1 \to \emptyset\}.$$

The reaction rules to produce positive membership species are composed of three reactions:

$$\mathcal{V} = \bigcup_{i=1}^{3} \mathcal{V}^i = \{s_2^0 \to s_1^1, s_1^0 + s_3^0 \to 2s_2^1, s_2^0 \to s_3^1\}$$

Finally, the non-membership species are also produced:

$$\mathcal{N} = \bigcup_{i=1}^{3} \mathcal{N}^i = \{s_2^1 \to s_1^0, s_1^1 \to s_2^0, s_3^1 \to s_2^0, s_2^1 \to s_3^0\}$$

The whole set of reactions $\mathcal{R}$ results in:

$$\begin{aligned} \mathcal{R} &= \mathcal{V} \cup \mathcal{N} \cup \mathcal{D} \\ &= \{s_2^0 \to s_1^1, s_2^1 \to s_1^0, s_1^0 + s_3^0 \to 2s_2^1, \\ &\quad s_1^1 \to s_2^0, s_3^1 \to s_2^0, s_2^0 \to s_3^1, s_2^1 \to s_3^0, \\ &\quad s_1^0 + s_1^1 \to \emptyset, s_2^0 + s_2^1 \to \emptyset, s_3^0 + s_3^1 \to \emptyset\} \end{aligned} \tag{8}$$

The algebraic chemistry is analyzed for its hierarchical organizational structure within the reaction network. When applying chemical organization theory (Section 1.1), the chemical reaction network is decomposed into a hierarchy of overlapping sub-networks, called organizations. These organizations provide an overview of the potential (emergent) behavior of the system because only a set of molecular species forming an organization can be stable [15]. Furthermore, the dynamics of the system can be explained as a transition between organizations instead of a movement in the potentially more complex state space.

In our example, the reaction network $\langle \mathcal{M}, \mathcal{R} \rangle$ possesses five organizations:

$$O = \{\emptyset, \{s_1^0\}, \{s_3^0\}, \{s_1^0, s_2^1, s_3^0\}, \{s_1^1, s_2^0, s_3^1\}\} \tag{9}$$

Figure 1 (B) visualizes these organizations as a Hasse diagram. In passing we note that the organizations do not form a lattice, because there is not a unique largest organizations. The two largest organizations represent the two desired solutions to the MIS problem, namely "010" and "101". This explains that in a dynamical reaction system implementing the designed algebraic chemistry, the species combinations representing desired solutions are more likely to stay in the dynamical system and the other solutions consisting of species that are not an organization cannot stably exist (cf. [15]). Interestingly the analysis has also uncovered three smaller organizations. These organizations represent uncompleted computations due to a lack of molecules. For example, the empty organization trivially implies: if there are no molecules in the system, no molecule will enter the system and there will be no computation. If we setup our chemical computing system such that these small organizations are avoided, the system must produce a solution.

We can now ask whether these solutions, i.e. organizations $\{s_1^0, s_2^1, s_3^0\}$ and $\{s_1^1, s_2^0, s_3^1\}$, are stable or whether the system, once they have been found, might move spontaneously down to a smaller organization below them. In general, this type of question requires to investigate the dynamics, such as, rate constants, in detail. Here, however, we can see already by looking at the reaction rules that organization $\{s_1^1, s_2^0, s_3^1\}$ must be stable, because all reactions are mass-conserving so that the empty organization (the only organization below) can never be reached. The deficiency value [30] for the organization is calculated to be zero so that the asymptotically stable state is contained, assuming mass action kinetics. The situation with organization $\{s_1^0, s_2^1, s_3^0\}$ is more complicated. It contains also the small organizations $\{s_1^0\}$ and $\{s_3^0\}$, and the deficiency value for the organization is one. Hence, we cannot use the same argument as before. The stability of that organization depends on the kinetics applied (not shown here).

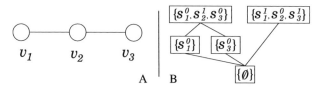

Fig. 1. Analysis of a chemical program with organization theory. The figure is adopted from [31]. (A) Graph structure and (B) hierarchy of organizations within the chemical reaction network for the maximal independent set problem for the linear 3-vertex graph

Graph with Six Vertexes

Two circular graphs with three vertexes are connected as shown in Fig. 2 (A) so that both circles and lines are contained. Since the graph consists of six vertexes, the algebraic chemistry holds 12 molecular species. Twenty-six reactions among those species constitute the reaction network. Within that reaction network, there are 49 organizations in total. In Fig. 3, a whole hierarchy of the organizations is shown, and only the largest organizations with six species are listed in Fig. 2 (B). Focusing on the largest organizations within the reaction network, only the sets of species representing solutions to the maximal independent set problem are found to be the organization.

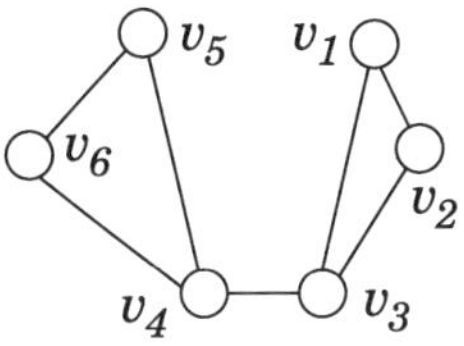

$$\{s_1^1, s_2^0, s_3^0, s_4^0, s_5^0, s_6^1\} \quad \{s_1^0, s_2^1, s_3^0, s_4^1, s_5^0, s_6^0\}$$
$$\{s_1^1, s_2^0, s_3^0, s_4^1, s_5^0, s_6^0\} \quad \{s_1^0, s_2^1, s_3^0, s_4^0, s_5^1, s_6^0\}$$
$$\{s_1^1, s_2^0, s_3^0, s_4^0, s_5^1, s_6^0\} \quad \{s_1^0, s_2^0, s_3^1, s_4^0, s_5^0, s_6^1\}$$
$$\{s_1^0, s_2^1, s_3^0, s_4^0, s_5^0, s_6^1\} \quad \{s_1^0, s_2^0, s_3^1, s_4^0, s_5^1, s_6^0\}$$

Fig. 2. Analysis of a chemical program with organization theory. The figure is adopted from [31]. (A) Graph structure with six vertexes and seven edges. (B) The largest organizations within the chemical reaction network for the maximal independent set problem with respect to the graph. Each organization with the size of six corresponds to a solution to the maximal independent set problem

2.2 Proof of Exact Correspondence between MISs and Organizations of Size N

Given an undirected graph $G = \langle V, E \rangle$ where $V = \{v_1, \ldots, v_N\}$ is a set of N vertexes and E is a set of edges, an algebraic chemistry $\langle \mathcal{M}, \mathcal{R} \rangle$ can be constructed as described in Section 2, where $\mathcal{M} = \{a_1, \ldots, a_{2N}\} = \{s_i^0, s_i^1 | i = 1, \ldots, N\}$ is a set of $2N$ molecular species and $\mathcal{R} = \{(A_j \to B_j)\}$ is a set of reaction rules. Here, we show with a proof that the constructed reaction network contains organizations with N species forming the largest and those organizations correspond to MIS in the given graph. To prove this, we first introduce the following lemma stating the maximum number of species each organization in the constructed reaction network can contain.

Lemma 1. *In the reaction network $\langle \mathcal{M}, \mathcal{R} \rangle$ constructed as described in Section 2, no organization can contain species s_k^0 and s_k^1 together. Therefore, no organization with a size (number of species) greater than N can exist.*

Proof. Let $O \subset \mathcal{M}$ be an organization, and suppose the organization contains s_k^0 and s_k^1 simultaneously ($s_k^0, s_k^1 \in O$) regarding vertex v_k. From the definition of the organization to be self-maintaining, there exists a flux vector $\mathbf{v} = (v_1, \ldots, v_{|\mathcal{R}|})^T$ satisfying the three conditions listed in Definition 3. Due to the third condition, the production rates f_i with respect to species belonging to the organization is greater than or equal to zero. Sum of those production rates should also be greater than or equal to zero.

$$\sum_{\{i | a_i \in O\}} f_i = \sum_{\{i | a_i \in O\}} \sum_{j=1}^{|\mathcal{R}|} m_{ij} v_{r_j} \geq 0.$$

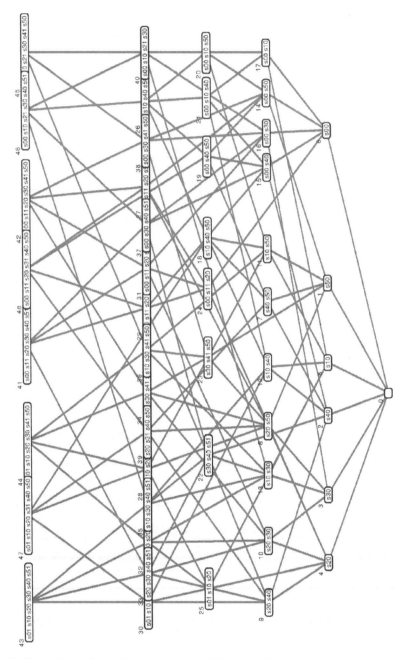

Fig. 3. Hierarchy of chemical organizations within the reaction network programmed to solve the maximal independent set problem in the graph structure depicted in Fig. 2 (A). There are 49 organizations in total, and eight organizations with six species are the largest on top of the diagram. The potential dynamical behaviors of the reaction network to solve the maximal independent set problem appear as the largest organizations. The figure is adopted from [31]

Since organization O is also closed, the production rates for the species excluded from O should be zero: $f_i = 0$ if $a_i \notin O$. Therefore, this equation can be extended to:

$$\sum_{i=1}^{|\mathcal{M}|} f_i = \sum_{\{i|a_i \in O\}} f_i + \sum_{\{i|a_i \notin O\}} f_i = \sum_{i=1}^{|\mathcal{M}|}\sum_{j=1}^{|\mathcal{R}|} m_{ij} v_{r_j} \geq 0 \quad (10)$$

Here, j-th reaction $A_j \to B_j$ is denoted as r_j.

The set $\mathcal{R}$ of reaction rules can be divided into the three sets $\mathcal{V}, \mathcal{N}, \mathcal{D}$ (Section 2):

$$\sum_{i=1}^{|\mathcal{M}|} f_i$$
$$= \sum_{i=1}^{|\mathcal{M}|} \left[\sum_{\{j|r_j \in \mathcal{V}\}} m_{ij} v_{r_j} + \sum_{\{j|r_j \in \mathcal{N}\}} m_{ij} v_{r_j} + \sum_{\{j|r_j \in \mathcal{D}\}} m_{ij} v_{r_j} \right]$$
$$= \sum_{\{j|r_j \in \mathcal{V}\}} \sum_{i=1}^{|\mathcal{M}|} m_{ij} v_{r_j} + \sum_{\{j|r_j \in \mathcal{N}\}} \sum_{i=1}^{|\mathcal{M}|} m_{ij} v_{r_j} \quad (11)$$
$$+ \sum_{i=1}^{|\mathcal{M}|} \sum_{\{j|r_j \in \mathcal{D}\}} m_{ij} v_{r_j}$$
$$= \sum_{\{j|r_j \in (\mathcal{V} \cup \mathcal{N})\}} v_{r_j} \left(\sum_{i=1}^{|\mathcal{M}|} m_{ij} \right) + \sum_{i=1}^{|\mathcal{M}|} \sum_{\{j|r_j \in \mathcal{D}\}} m_{ij} v_{r_j}$$

For the reactions of type $\mathcal{V}$ and $\mathcal{N}$, sum of the stoichiometric coefficients is arranged to be zero:

$$\forall j | r_j \in (\mathcal{V} \cup \mathcal{N}) : \sum_{i=1}^{|\mathcal{M}|} m_{ij} v_{r_j} = 0 \quad (12)$$

Thus, the last term of Equation (12) must be non-negative:

$$\sum_{i=1}^{|\mathcal{M}|} \sum_{\{j|r_j \in \mathcal{D}\}} m_{ij} v_{r_j} \geq 0 \quad (13)$$

To keep this inequality, fluxes for the reactions of type $\mathcal{D}$ should be zero because all of the stoichiometric coefficients for the type $\mathcal{D}$ reactions are negative. The flux for reaction $(s_k^0 + s_k^1 \to \emptyset) \in \mathcal{D}$ must be set to a positive value, however, if both s_k^0 and s_k^1 are contained in organization O at the same time (Condition 1). The sum of the production rates cannot be positive, and at least one production rate has to be negative. This contradicts the definition of the organization, and hence, two species s_k^0, s_k^1 cannot coexist in the organization. □

Next we define the set of species induced by a set of vertexes:

Definition 5. *Let $I \subset V$ be a set of vertexes. We call $S_I = \{s_1^{b_1}, \ldots, s_{|V|}^{b_{|V|}}\}$ the set of species that is induced by I if $b_i = 1$ when $v_i \in I$ and $b_i = 0$ when $v_i \notin I$.*

Given a subset of vertexes $I \subset V$ in an undirected graph $G = \langle V, E \rangle$, a set of species S_I can be "induced" such that the subscript of the species name specifies the vertex

158   N. Matsumaru et al.

identification number and the superscript is the binary state whether the vertex v_i is a member of set I ($b_i = 1$) or not ($b_i = 0$). The constructed set of species consists of $|V|$ species so that all of the vertexes in the graph are considered.

Using this notation, the exact correspondence between maximal independent sets and organizations of size $N = |V|$ can be stated as follows:

Theorem 1. *Given an undirected graph $G = \langle V, E \rangle$, a set $I \subset V$ of vertexes is a maximal independent set iff the induced set S_I of species is an organization.*

Proof. The first part of the proof is to show the necessary condition, namely a set of species induced by a MIS is an organization. Let $I \subset V$ be a MIS and S_I be the set of species induced by I. To show that S_I is an organization, two criteria will be tested: closure and self-maintenance.

Closure: Assume that S_I is not closed, i.e. there exists a reaction $(A_j \rightarrow B_j) \in \mathcal{R}$ that is applicable to S_I and that produces a species that is not in the set S_I. If such a reaction has the form of $(s_j^1 \rightarrow s_k^0) \in \mathcal{N}$ for an edge $(v_j, v_k) \in E$, then we know $s_j^1 \in S_I$ and thus $v_j \in I$. Since that reaction is assumed to violate the closure property, $s_k^0 \notin S_I$. The induced set of species S_I is defined to include either s_k^0 or s_k^1, so $s_k^1 \in S_I$. As a result, set S_I contains s_j^1 and s_k^1, meaning that set I includes both v_j and v_k. This leads to a contradiction that set I containing v_j and v_k is a MIS even though there is an edge $(v_j, v_k) \in E$.

The reaction to violate the closure property can have the form $(s_h^0 + s_l^0 + \cdots + s_m^0 \rightarrow n_k s_k^1) \in \mathcal{V}$. In that case no neighboring vertexes v_p with respect to v_k are included in set I because $s_p^0 \in S_I$ for any p such that $(v_p, v_k) \in E$. To violate the closure, s_k^1 should be excluded from S_I so that vertex v_k is not in I. Since none of neighboring vertexes are in I, however, v_k can be added to I with keeping the independence. This contradicts the fact that the independent set I is maximal.

The third type of reactions ($\mathcal{D}$) can be neglected because there are no products. From these arguments, no reaction can produce new species that is not in S_I. It follows that S_I is closed.

Self-maintenance: To satisfy the conditions of self-maintenance, the flux vector $\mathbf{v}$ is set as: $v_{r_j} = 1$ if $A_j \in \mathcal{P}_M(S_I)$ and $v_{r_j} = 0$ if $A_j \notin \mathcal{P}_M(S_I)$. Given this $\mathbf{v}$, we show that production rates for all species in S_I are non-negative. From the definition of set S_I induced by a set of vertexes, either s_m^0 or s_m^1, not both, is included in S_I. Therefore, The fluxes for reaction type $\mathcal{D}$ are set to zero.

Assume a species with the superscript of one, s_m^1, is in S_I. This implies $s_p^0 \in S_I$ for any neighboring vertexes v_p of vertex v_m since v_m is in MIS but none of neighboring vertexes v_p are included. There is only one reaction in $\mathcal{V}$ producing s_m^1 with the stoichiometry of n_m, where n_m is the number of vertexes connected to vertex v_m, and containing reactants $A_j \in \mathcal{P}_M(S_I)$. Production rate of species s_m^1 caused by reaction type $\mathcal{V}$ is calculated to n_m because flux for the reaction is set to 1. In the type of $\mathcal{N}$, there are n_m reactions with s_m^1 as the reactant. Setting the fluxes to 1 for these reactions, the production rate caused by this type of reactions is $-n_m$. Combining those, the production rate of species s_m^1 is 0.

Next, a species with the superscript of zero, s_m^0, is assumed to be in S_I. Since $v_m \notin I$ from the definition, at least one or maximum n_m neighboring vertexes are in the MIS I. Let $1 \leq g \leq n_m$ be the number of neighboring vertexes in the MIS I. There are g reactions of type $\mathcal{N}$ applicable to S_I depending on I. Stoichiometric coefficients of these reactions for s_m^0 are all 1. In the reactions of type $\mathcal{V}$, the same number of reactions as type $\mathcal{N}$ are applicable. For a vertex in the MIS, no neighbors are included in the MIS according to the definition. It follows that a reaction in $\mathcal{V}$ should be applicable for each vertex included in the MIS. If v_m has g neighboring vertexes included in the MIS I, there have to be g reactions of type $\mathcal{V}$ with the coefficients -1. Hence, production rate of the species s_m^0 is equal to zero.

The second part shows the sufficient condition. Namely, if a set of species induced by a set of vertexes is an organization, then this set of vertexes is a MIS. Given a set of vertexes I and its induced set of species S_I, which is an organization, we need to show that I is a MIS. Taken any of two vertexes v_p and v_q from the set I, we know that $s_p^1 \in S_I$ and $s_q^1 \in S_I$ from the definition of the induced set of species. Suppose there exists an edge $(v_p, v_q) \in E$ between those two vertexes. The reaction $s_p^1 \to s_q^0$ defined for that edge would produce s_q^0 inside the organization S_I. In order to keep S_I as an organization, s_q^1 and s_q^0 should coexist, which is impossible according to Lemma 1 given above. Therefore, we conclude that $(v_p, v_q) \notin E$ and I is an independent set.

If I does not represent a "maximal" independent set, we could add a vertex $v_{p'} \in V \setminus I$ to I, and $I \cup \{v_{p'}\}$ remains an independent set. To be more general, there exists a non-empty set of vertexes $I' \subseteq V \setminus I$ such that the union of these sets $I \cup I'$ becomes the MIS. Since we showed in the first part of this proof that a MIS induces an organization, $S_{I \cup I'}$ induced by set $I \cup I'$ is an organization whereas S_I is also an organization from the assumption. Set $S_{I \cup I'}$ differs from S_I with respect to any vertex $v_{p'}$ in I': $s_{p'}^0 \in S_I$, $s_{p'}^1 \in S_{I \cup I'}$. Those indexes p' have to be chosen such that vertexes $v_{p'}$ are not in I and no neighboring vertexes $v_{q'}$ are in set I. If vertex $v_{q'}$ is already contained in I and vertex $v_{p'}$ is added, then set $I \cup I'$ cannot be an independent set. However, the absence of those neighboring vertexes $v_{q'}$ in S_I would produce $s_{p'}^1$ by the reactions in $\mathcal{V}$, which violates the closure property of the organization S_I because the set only contains $s_{p'}^0$. Thus, there are no such indexes p', and set I' is an empty set. □

3 Discussion and Outlook

In chemical computing, the result emerges as a macroscopic phenomenon resulting from many microscopic reaction events. It is, in general, very difficult to anticipate the macroscopic behavior from the microscopic interactions. Since programming chemical reaction systems is to manipulate the reaction rules in the microscopic level, the ability to anticipate the behavior of a program in the macroscopic level is required, however. The micro-macro gap has to be bridged, at least partially, to allow programming the reaction systems.

We have shown that chemical organization theory can serve as a tool to predict the potential behavior of a chemical program given its "microscopic" reaction rules, without need to know the kinetics in detail. The desired solutions to the MIS problem

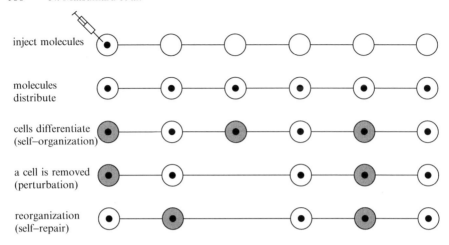

Fig. 4. Scenario of a benchmark problem, where a set of sensor nodes have to differentiate such that pairwise neighbors are in different states. The figure is adopted from [31]

appeared as organizations. Furthermore, the organizational analysis uncovers organizations representing incomplete computations due to a lack of molecules. Chemical organization theory can now guide further improvements of the chemical program, which aim at reducing or even removing completely these "undesired" organizations.

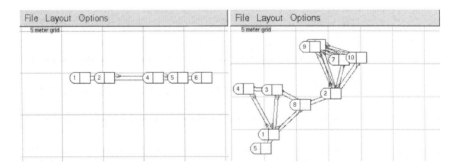

Fig. 5. Screen shots of a simulation on EmStar. The simulation is being developed for a quantitative performance analysis of chemical programs on ad-hoc networks, namely wireless sensor networks

To allow not only qualitative but also quantitative evaluation of our approach, a benchmark problem is desirable. We envision a variant of the maximal independent set problem in a sensor network scenario as sketched in Fig. 4: In this scenario, we assume that sensor nodes are arranged linearly. Specific molecules are distributed over the network. Then the network should self-organize such that pairwise neighboring nodes are in different states, for example, one class should perform a measurement at night the other at daytime. When nodes are removed or added dynamically, spontaneous reconfiguration should occur (self-repair). The recovery time or number of acceptable perturbations can serve as a quantitative measure of the systems performance. Implementation of this benchmark problem is being developed on top of a software for wireless sensor networks called EmStar [32, 33]. Sample screen shots of the simulation output are given in Fig. 5. For each node, the dynamics of the constructed chemical program is specified as ordinary differential equations (ODEs), and Systems Biology Markup Language (SBML) ODE solver [34] is integrated into the simulation. This work is still in progress.

Acknowledgment

The authors would like to thank our colleagues of our group for their support and useful advice. We are very grateful to Gerd Grünert and Christoph Kaleta for their help with the software development. Financial supports by the German Research Foundation (DFG) Grant DI 852/4-1, by the European Union, NEST-project ESIGNET no. 12789, and by the Federal Ministry of Education and Research (BMBF) Grant 0312704A are acknowledged.

References

1. Lodding, K. N. (2004) The hitchhiker's guide to biomorphic software. *Queue*, **2**, 66–75.
2. Krüeger, B. and Dressler, F. (2004) Molecular processes as a basis for autonomous networking. *International IPSI-2004 Stockholm Conference: Symposium on Challenges in the Internet and Interdisciplinary Research (IPSI-2004 Stockholm)*.
3. D'haeseleer, P., Forrest, S., and Helman, H. (1996) An immunological approach to change detection: Algorithms, analysis and implications. *Proceedings of the 1996 IEEE Symposium on Security and Privacy*, pp. 110–119, IEEE Computer Society Press.
4. Dorigo, M., Caro, G. D., and Gambardella, L. (1999) Ant algorithms for discrete optimization. *Artif. Life*, **5**, 137–172.
5. Bonabeau, E., Dorigo, M., and Theraulaz, G. (2000) Inspiration for optimization from social insect behaviour. *Nature*, **406**, 39–42.
6. Schoonderwoerd, R., Bruten, J. L., Holland, O. E., and Rothkrantz, L. J. M. (1996) Ant-based load balancing in telecommunications networks. *Adapt. Behav.*, **5**, 169–207.
7. Montresor, A. and Babaoglu, O. (2003) Biology-inspired approaches to peer-to-peer computing in bison. *The Third International Conference on Intelligent System Design and Applications*.
8. Babaoglu, O., et al. (2006) Design patterns from biology for distributed computing. *ACM Transactions on Autonomous and Adaptive Systems*, **1**, 26 – 66.
9. Müller-Schloer, C. (2004) Organic computing: On the feasibility of controlled emergence. *Proceedings of the 2nd IEEE/ACM/IFIP International Conference on Hardware/Software Codesign and System Synthesis, CODES+ISSS2004*, pp. 2–5, ACM Press, New York.
10. Banâtre, J.-P. and Métayer, D. L. (1986) A new computational model and its discipline of programming. Tech. Rep. RR-0566, INRIA.

11. Dittrich, P. (2005) The bio-chemical information processing metaphor as a programming paradigm for organic computing. Brinkschulte, U., Becker, J., Hochberger, C., Martinetz, T., Müller-Schloer, C., Schmeck, H., Ungerer, T., and Würtz, R. (eds.), *ARCS '05 - 18th International Conference on Architecture of Computing Systems 2005*, pp. 95–99, VDE Verlag, Berlin.
12. Küppers, B.-O. (1990) *Information and the Origin of Life*. MIT Press.
13. Banzhaf, W., Dittrich, P., and Rauhe, H. (1996) Emergent computation by catalytic reactions. *Nanotechnology*, 7, 307–314.
14. Zauner, K.-P. (2005) From prescriptive programming of solid-state devices to orchestrated self-organisation of informed matter. Banâtre, J.-P., Giavitto, J.-L., Fradet, P., and Michel, O. (eds.), *Unconventional Programming Paradigms: International Workshop UPP 2004*, vol. 3566 of *LNCS*, pp. 47–55, Springer, Berlin.
15. Dittrich, P. and Speroni di Fenizio, P. (2006) Chemical organization theory. *Bull. Math. Biol.*, (accepted).
16. Matsumaru, N., Centler, F., Speroni di Fenizio, P., and Dittrich, P. (2007) Chemical organization theory as a theoretical base for chemical computing. *International Journal of Unconventional Computing*, (in print).
17. Shukla, S. K., Rosenkrantz, D. J., and Ravi, S. S. (1995) Observations on self-stabilizing graph algorithms for anonymous networks. *Proceedings of the Second Workshop on Self-Stabilizing Systems*, pp. 7.1–7.15.
18. Adamatzky, A. and De Lacy Costello, B. (2002) Experimental logical gates in a reaction-diffusion medium: The XOR gate and beyond. *Phys. Rev. E*, 66, 046112.
19. Zauner, K.-P. and Conrad, M. (2001) Enzymatic computing. *Biotechnology Progress*, 17, 553–559.
20. Gardner, T. S., Cantor, C. R., and Collins, J. J. (1999) Construction of a genetic toggle switch in escherichia coli. *Nature*, 403, 339–342.
21. Weiss, R., Homsy, G., and Knight, T. (1999) Toward in vivo digital circuits. *Proceedings of the Dimacs Workshop on Evolution as Computation*.
22. Banâtre, J.-P., Fradet, P., and Radenac, Y. (2004) Principles of chemical programming. Abdennadher, S. and Ringeissen, C. (eds.), *RULE'04 Fifth International Workshop on Rule-Based Programming*, pp. 98–108, Tech. Rep. AIB-2004-04, Dept. of Comp. Sci., RWTH Aachen, Germany.
23. Păun, G. (2002) *Membrane Computing: An Introduction*. Nat. Comput. Ser., Springer.
24. Suzuki, Y. and Tanaka, H. (1997) Symbolic chemical system based on abstract rewriting system and its behavior pattern. *Artificial Life and Robotics*, 1, 211–219.
25. Fontana, W. and Buss, L. W. (1994) 'The arrival of the fittest': Toward a theory of biological organization. *Bulletin of Mathematical Biology*, 56, 1–64.
26. Reichenbach, F., Bobek, A., Hagen, P., and Timmermann, D. (2006) Increasing lifetime of wireless sensor networks with energy-aware role-changing. *Proceedings of the 2nd IEEE International Workshop on Self-Managed Networks, Systems & Services (SelfMan 2006)*, Dublin, Ireland, pp. 157–170.
27. Luby, M. (1986) A simple parallel algorithm for the maximal independent set problem. *SIAM Journal on Computing*, 15, 1036–1055.
28. Herman, T. (2003) Models of self-stabilization and sensor networks. Das, S. R. and Das, S. K. (eds.), *IWDC*, vol. 2918 of *LNCS*, pp. 205–214, Springer, Berlin.
29. Ikeda, M., Kamei, S., and Kakugawa, H. (2002) A space-optimal self-stabilizing algorithm for the maximal independent set problem. *Proceedings of the Third International Conference on Parallel and Distributed Computing, Applications and Technologies (PDCAT)*, pp. 70–74.

30. Feinberg, M. and Horn, F. J. M. (1974) Dynamics of open chemical systems and the algebraic structure of the underlying reaction network. *Chem. Eng. Sci.*, **29**, 775–787.
31. Matsumaru, N. and Dittrich, P. (2006) Organization-oriented chemical programming for the organic design of distributed computing systems. *Proc. of Bionetics*, December 11-13, pp. Cavalese, Italy, IEEE, available at `http://www.x-cd.com/bionetics06cd/`.
32. Elson, J., et al. (2003) EmStar: An Environment for Developing Wireless Embedded Systems Software. Tech. Rep. CENS Technical Report 0009, Center for Embedded Networked Sensing, University of California, Los Angeles.
33. Girod, L., Elson, J., Cerpa, A., Stathopoulos, T., Ramanathan, N., and Estrin, D. (2004) Emstar: a software environment for developing and deploying wireless sensor networks. *Proceedings of the 2004 USENIX Technical Conference*, Boston, MA, USENIX.
34. Machné, R., Finney, A., Müller, S., Lu, J., Widder, S., and Flamm, C. (2006) The sbml ode solver library: a native api for symbolic and fast numerical analysis of reaction networks. *Bioinformatics*, **22**, 1406–1407.

Evolving Artificial Cell Signaling Networks: Perspectives and Methods

James Decraene, George G. Mitchell and Barry McMullin

Artificial Life Laboratory
Research Institute for Networks and Communications Engineering
School of Electronic Engineering
Dublin City University, Glasnevin, Dublin, Ireland
james.decraene@eeng.dcu.ie

Summary. Nature is a source of inspiration for computational techniques which have been successfully applied to a wide variety of complex application domains. In keeping with this we examine Cell Signaling Networks (CSN) which are chemical networks responsible for coordinating cell activities within their environment. Through evolution they have become highly efficient for governing critical control processes such as immunological responses, cell cycle control or homeostasis. Realising (and evolving) Artificial Cell Signaling Networks (ACSNs) may provide new computational paradigms for a variety of application areas. In this paper we introduce an abstraction of Cell Signaling Networks focusing on four characteristic properties distinguished as follows: Computation, Evolution, Crosstalk and Robustness. These properties are also desirable for potential applications in the control systems, computation and signal processing field. These characteristics are used as a guide for the development of an ACSN evolutionary simulation platform. Following this we describe a novel class of Artificial Chemistry named Molecular Classifier Systems (MCS) to simulate ACSNs. The MCS can be regarded as a special purpose derivation of Hollands Learning Classifier System (LCS). We propose an instance of the MCS called the MCS.b that extends the precursor of the LCS: the broadcast language. We believe the MCS.b can offer a general purpose tool that can assist in the study of real CSNs in Silico The research we are currently involved in is part of the multi disciplinary European funded project, ESIGNET, with the central question of the study of the computational properties of CSNs by evolving them using methods from evolutionary computation, and to re-apply this understanding in developing new ways to model and predict real CSNs.

1 Introduction

Cell Signaling networks (CSNs) are bio-chemical systems of interacting molecules in cells [12, 17]. Typically, these systems take as inputs chemical signals generated within the cell or communicated from outside. These trigger a cascade of chemical reactions that result in changes of the state of the cell and (or) generate some (chemical) output, such as prokaryotic chemotaxis, coordination of cellular division, or even to order the death of a cell (in the context of multi-cellular organisms).

As signal processing systems, CSNs can be regarded as special purpose computers [4]. In contrast to conventional silicon-based computers, the computation in CSNs is not realized by electronic circuits, but by chemically reacting molecules in the cell. The most important molecular components of CSNs are proteins and nucleic acids (DNA, RNA). There is an almost infinite variety of potential molecular species, each of which would have distinct chemical functionality and could engage in interactions with other molecules with varying degrees of specificity.

We distinguish CSNs as being networks made up of more than one distinct cell signaling *pathway*, which interact with each other.

An example of a simple chemotaxis signaling pathway is shown in Figure 1.A. Chemotaxis is a phenomenon where simple organisms such as bacteria move toward higher concentrations of specific chemicals in their surroundings. In this diagram, we distinguish six intracellular proteins (denoted as A, B, R, W, Y and Z) and the membrane receptors which can bind to the corresponding stimulatory element. The input level is determined by the concentration of bound molecules. This affects the

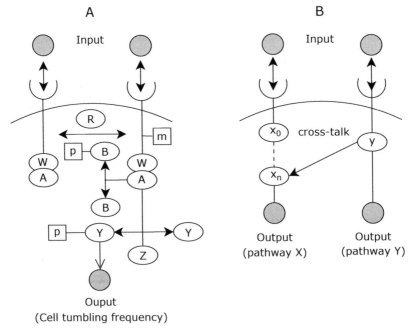

Fig. 1. A: Schematic representation of bacterial chemotaxis signaling pathway, adapted from [1]. The output is designated by the tumbling frequency which is determined from the input, the concentration level of ligand bound to the membrane receptors. This signal transduction is carried out by the reaction cascade depicted by the proteins A, B, R, W, X and Z. Details on chemical reactions can be found in [23]. B: A CSN composed from two distinct cell signaling pathways with unique input and output, an interaction between pathways occurs as molecule y interacts with x_n, this modulates the output of pathway X

output represented by the tumbling frequency which governs the bacteria direction. Figure 1.B shows, in schematic form, a simple Cell Signaling *Network* made up of two such interacting signal *pathways*. We distinguish this work from previous work on real CSNs [12, 17] by focusing purely on Artificial Cell Signaling Networks (ACSNs). Through the use of evolutionary computing techniques we allow ACSNs to spontaneously emerge and adapt to the environment. Potentially of interest for the Biologist may be the insight that ACSNs gives as to how real CSNs evolved and how they operate. This synthetic biology approach allows us to incorporate the present knowledge of real CSNs into ACSNs. This biological understanding provided guiding points that directed the design of the MCS, these points also guide the evolution of ACSNs in silico. This may, for example, facilitate the prediction of missing signaling pathway information in real CSNs [15].

Given our motivation to maintain the biological plausibility of ACSNs, we are interested in investigating the use of ACSNs to implement computation, signal processing and (or) control functionality. This is motivated by preliminary studies which demonstrated that real CSNs could be considered for computational and engineering purposes:

- In [20], Lauffenburger presents his approach to cell signaling pathways which could be thought of and modelled as control modules in living systems.
- Yi et al. [26] demonstrated that CSNs may have some of the essential properties of an integral feedback control. This is a basic engineering strategy to ensure that a system outputs desired values independent of internal and external perturbations.
- Deckard and Saura [6] used and evolved artificial biochemical networks capable of certain simple forms of mathematical computation such as a square root function.

One way to design ACNs to carry out such complex operations is to use artificial evolutionary techniques. A significant insight related to the evolution of signaling networks in silico, was suggested by Holland [15]. Holland proposed examining a simple agent-based model where the agents' behavior and adaptation was determined by the use of Learning Classifier System [13, 5]. Based on this machine learning approach Holland suggested that signaling networks could be modeled with LCS in a *top-down* fashion. We show how this work does not lend itself as a method for addressing our project.

We propose a variation of the LCS called the Molecular Classifier System which allows the emergence and evolution of signaling networks. This approach may be considered as viewing the evolution and design of signaling networks from a *bottom-up* manner complementing Holland's approach.

In section 2 we examine the nature of ACSNs by examining the critical issues that these raise. In Section 3 we first present and discuss Holland's approach and following this we propose the structure of our ACSN evolutionary simulation platform, the Molecular Classifier System.

2 Artificial Cell Signaling Networks

As an abstraction of real CSNs, ACSNs are differentiated and simplified by some key properties. The selection of these particular characteristics is motivated by the will to employ Artificial Cell Signaling Networks for computational and control engineering purposes. Four issues are distinguished and presented: Computation, Evolution, Crosstalk and Robustness.

2.1 Computation

In the simplest cases, CSNs can be approximately modelled by systems of continuous differential equations, where the state variables are the concentrations of the distinct species of interacting molecules. As a "computational" device, this is most naturally compared to a traditional *analog* computer. Analog computers are precisely designed to model the operation of a target dynamical system, by creating an "analogous" system which shares (approximately) the same dynamics. Electronic analog computers (based on the "operational amplifier" as the core computational device) have long been displaced by digital computers, programmed to numerically solve the relevant dynamical equations, due to their much greater ease of programming and stability.

Nonetheless, there may be applications where a molecular level analog computer, in the form of a CSN, may have distinct advantages. Specifically, CSNs may offer capabilities of high speed and small size that cannot be realised with solid state electronic technology. More critically, where it is required to interface computation with chemical interaction, a CSN may bypass difficult stages of signal transduction that would otherwise be required. This could have direct application in so-called "smart drugs" and other bio-medical interventions.

While CSNs are typically treated in this "aggregate" manner, where the signal or information is carried by molecular concentration, one can also consider the finer grained behaviours of individual molecules are computational in nature. Thus a single enzyme molecule can be regarded as carrying out pattern matching to identify and bind target substrates, and then executing a discrete computational operation in transforming these into the product molecule(s). This has clear parallels with a wide variety of so-called "rewriting systems" in computational theory.

However, it also clearly differs in important ways, such as:

- Operation is stochastic rather than deterministic.
- Operation is intrinsically *reflexive* in that all molecules can, in principle, function as both "rules" (enzymes) and "strings" (substrates/products).

Dittrich [9] provides a more extended discussion of the potential of such "chemical computing".

2.2 Evolution

Evolutionary Algorithms (EAs) are non-deterministic search and optimisation algorithms inspired by the principles of neo-Darwinism. They have been applied successfully in a variety of fields [14, 11, 16]. Generally based on genetic operations such as crossover and mutation, EAs initially generate a wide range of candidate solutions. Over time, through selection, this can be reduced to an optimized set. Evolutionary computation can therefore deliver useful results without requiring a priori knowledge of the entire search space [11, 16].

Such techniques are relevant to the study of ACSNs because:

- The complex, and unpredictable, interactions between different components of CSNs, make it very difficult to design them "by hand" to meet specific performance objectives.
- However, natural evolution shows that in suitable circumstances, effective CSNs functionality can be achieved through evolutionary processes.

For example, Deckard and Saura [6] used such evolutionary techniques to construct (simulated) biochemical networks capable of certain simple forms of signal-processing. In this model (called Lakhesis), computational "nodes" represent molecule species with an attribute for concentration. Connections between nodes designate reactions defined by the type and rate of the reaction.

Another ESIGNET project contribution is given by Lenser et al. [24], in this System Biology approach, a multi-level EA is proposed to evolve biochemical networks (represented in SBML) for performing pre-specified tasks such as reconstructing real CSNs. The base level EA searches for optimal network topologies and the second level explores the kinetic parameters.

2.3 Crosstalk

"Crosstalk" phenomena happen when signals from different pathways become mixed together. This arises very naturally in CSNs due to the fact that the molecules from all pathways may share the same physical reaction space (the cell). Depending on the relative specificities of the reactions there is then an automatic potential for any given molecular species to contribute to signal levels in multiple pathways. An example is shown in Fig. 1.

In traditional communications and signal processing engineering, crosstalk is regarded as a defect—an *unintended* or *undesigned* interaction between signals, that therefore has the potential to cause system malfunction. This can also clearly be the case of crosstalk in CSNs. However, in the specific case of CSN's, crosstalk also has additional potential functionality, which may actually be constructive:

- Even where an interfering signal is, in effect, adding uncorrelated "noise" to a functional signal, this may sometimes improve overall system behaviour. This is well known in conventional control systems engineering in the form of so-called "dither". Compare also, [2, 25] on constructive biological roles of noise.

- The crosstalk mechanism provides a very generic way of creating a large space of possible modifications or interactions between signaling pathways. Thus, although many cases of crosstalk may be immediately negative in their impact, crosstalk may still be a key mechanism in enabling incremental evolutionary search for more elaborate or complex cell signaling networks.

2.4 Robustness

It is argued that key properties in biochemical networks are to be robust, this is so as to ensure their correct functioning [3]. Similar works include research carried out at the Santa Fe institute in studying Cytokine signaling networks to design distributed autonomous networks, that are robust to small perturbations and responsive to larger ones [18]. Potential applications are distributed intelligent systems such as large fleets of robots working together, for automated response in computer security, for mobile computing networks, etc.

Alon et al. have demonstrated from studying *Escherichia coli* chemotaxis that molecular interactions can exhibit robustness [1, 20]. In this case it means that after a change in the stimulus concentration (input), the tumbling frequency (output) managed to reach a steady state that is equivalent to the pre-stimulus level. This is illustrated in Figure 2.

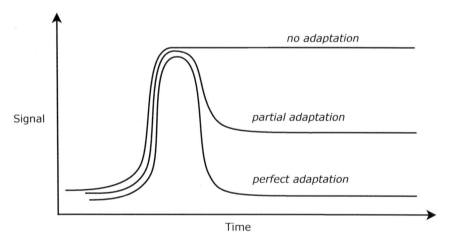

Fig. 2. Representation of dynamic responses of a system to a stimuli adapted from [20]. No adaptation is observed when the system response attains a new steady state following the change in input. Partial adaptation describe a partial recovery, the difference between the initial state and the new state is lower than the one observed in the previous case. Perfect adaptation is met when the system is able to come back to its initial state

Such properties are highly desirable in dynamic engineered systems when subjected to internal and external uncertainty and perturbation.

3 An Evolutionary Approach to Implement ACSNs

In the following, we first examine a specific class of Evolutionary Algorithm called Learning Classifier System (LCS) devised by Holland in 1976 [13]. In 2001 Holland [15] identified the possibility of using LCS to implement signaling networks (biochemical circuits). However Holland's work was never actually implemented. We use Holland's proposition as the seminal point for the development of the first Classifier System based ACSN implementation, our Molecular Classifier System (MCS).

3.1 Learning Classifier Systems

Learning Classifier Systems are systems constructed from condition-action rules called *classifiers*. The classifiers can be viewed as IF/THEN statements in the form IF "rule" THEN "action". The condition section of the classifier examines all of the messages in the system and identifies those that satisfy the *rules* conditions. Once this is accomplished the action part instructs that a message is to be sent. Holland's initial work was modified a number of times and at present many different varieties of learning classifier systems are available [19].

In Holland's LCS the system receives an input from its environment as a binary encoded data. This is then stored in an internal data store termed the *message list*, see Figure 3. The LCS then evaluates the input and determines an appropriate response, indicated by the action. This action typically alters the current state of the environment. Any desired behaviour that is exhibited is then rewarded through a scalar reinforcement. The system iterates the cycle of response, reinforcement and discovery for each discrete time-step.

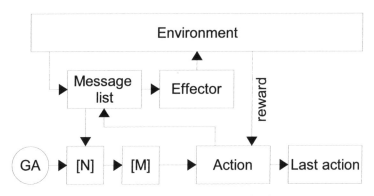

Fig. 3. Schematic of Holland's Learning Classifier System

The rule-base consists of a population of N classifiers. Both parts of the classifier are randomly initialized. The rule conditions and actions (the classifiers) can be characterized by strings formed from a ternary alphabet 0,1,#. The use of the # provides a single character wildcard which allows for the potential matching of a greater number of strings e.g. 10# would match two potential inputs 100 or 101. The use of the wildcard character also provides for string processing at the action stage, for example: in responding to the input 110, the rule IF 1#0 THEN 0#1 would produce the action 011. Each classifier also has an associated fitness measure, quantifying the *usefulness* of a rule in attracting external reward.

On receiving an input message, a typical LCS processes as follows: initially the input message rule-base is scanned and all rules whose condition matches the external message are added to the "match set" denoted as [M], see Figure 3. Secondly any other rules matching messages in the message list are also added to [M]. Rules that contribute significantly to the targeted learning task may then be reinforced through the use of *bidding techniques*. For a comprehensive introduction to Learning Classifier System, see [5].

In [15] Holland proposed an agent-based model where the agents' behavior and adaptation are determined by the use of Learning Classifier System. This work provided an existence proof that LCS could be used to evolve a simple repertoire of condition-action rules to a more complex goal directed set of rules. In typical biochemical networks, interactions between molecules follow the same condition-action mechanisms. Thus Holland suggested that this approach could be used to simulate and evolve signaling networks. His proposition to design signaling networks was to start with a LCS-based "over-general" model of a biological phenomenon (e.g. transformation of a healthy cell to a cancer cell, see Table 1). Then a general phe-

Table 1. Over-general model of the transformation of healthy cell to cancer cell

Rule Condition	Action
(1) If healthy cell and DNA damage	Then apoptosis or immortality
(2) If immortality	Then stable existence or genetic instability
(3) If genetic instability	Then ephemeral clonal expansion or robust clonal expansion

nomenon can be refined through several iterations. At each iteration, the details of the occurring interactions are detailed, see Table 2. These iterations were continued until the desired CSN level was reached, where the biomolecular elements are specified

Table 2. Refinement of rule 1

Rule Condition	Action
(1.1) IF healthy cell and DNA damage	Then apoptosis or mutation for resistance to apoptosis
(1.2) IF resistance to apoptosis	Then susceptibility to growth inhibitory signals or mutation for loss of susceptibility to growth inhibitory
(1.3) IF loss of susceptibility to immortality	THEN selective growth advantage and growth inhibitory signals

(e.g. protein ligand, receptor, ions etc.), see Table 3 for an example of such a rule. This refining process clearly shows the top-down methodology to design signaling networks.

Table 3. A biomolecular level rule

Rule	Condition	Outcome
(x.x.x.x)	If apropos growth factor	Then gf receptor activated

Despite this, the LCS-based approach to specify CSNs sounded promising, actual implementation was never performed. Importantly, this approach does not meet the requirements of our project. First, we do not distinguish a demarcation between rules and messages, in our context, the chemical operations are reflexives. Secondly, Holland's suggestion was to initially model known real CSNs, however from our bottom-up perspective, we require the ACSNs to evolve from very simple networks to more complex networks that exhibit the known real CSNs properties. As a consequence we propose a variation of Holland's LCS to fulfill the requirements of our project.

3.2 The Molecular Classifier System

We define the Molecular Classifier System (MCS) as a class of string-rewriting based Artificial Chemistries. This approach is inspired by Hollands Learning Classifier Systems (LCS). In Hollands LCS, a demarcation is distinguished between *rules* and *messages*, however as mentioned earlier operations in a biochemical networks are intrinsically *reflexive* in the sense that all molecules can function as both rules (enzymes) and messages (substrates/products). The MCS addresses this issues by removing this rules/messages demarcation found in the LCS.

The behavior of the condition/binding properties and action/enzymatic functions is specified by a "chemical" language defined in the MCS. The chemical language defines and constrains the complexity of the chemical reactions that may be represented and simulated with the MCS. For example, a MCS model using a limited number of computational functions may only fatefully represent very simplistic chemical reactions.

Before describing the nature of the enzymatic functions (action part of a molecule), the binding properties of the molecules must be identified. We have thus far identified the following potential properties: In the MCS approach, a reaction between molecules may only occur if the informational string of a first molecule satisfies/binds with the conditional part of a second molecule. The second molecule may be the same as the first molecule leading to self-binding. The condition part refers to the binding properties of a molecule whereas action refers to the computational ("enzymatic") function. This pattern matching occurring implies a notion of *specificity* or "binding strength". A molecule having a high specificity would have less chance to react with another one. Whereas a molecule having a low specificity is likely to bind to another more often. Therefore we could translate this into an effect on reaction rate/kinetics.

When two molecules can bind and consequently react to each other, the action part of one of the molecules is used to carry out the enzymatic operations upon the binding molecule (substrate). This operation results in producing another offspring (product), see Fig. 4. This is analogous to the action part of a LCS rule used by Holland [15]. When a reaction occurs, the symbols contained in the MCS action part are processed in a sequential order (parsed from left to right). The outcome (product) of the reaction depends on the nature of the symbols' functionality.

In [21], a minimalist approach to the MCS was proposed to investigate protocell computation. In this study, a protocell is modelled as a container for artificial molecules (Molecular Classifier Systems). The latter may interact with each other to generate new molecular offspring. The chemical language used in this instance of a MCS for protocells employs a minimal set of computational components to reduce the conceptual gap between artificial and real chemistry. To represent, simulate and evolve ACSNs, more computational functions are necessary, nevertheless

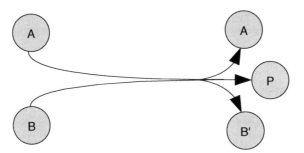

Fig. 4. Schematic of a reaction in the MCS: When a molecule A can react with a molecule B, the action statement of molecule A is "executed" upon the informational string of the binding molecule B. A is viewed as an enzyme and B as a substrate, thus A's structure is not affected by the reaction whereas B's structure is degraded and a product P is generated. A's action statement operators take as inputs the symbols of B's string. An offspring molecule P is generated as a result of these operations

the definitive set of operations is still under investigation as we are trying to understand what are the minimal operational requirements to allow a primitive ACSN to spontaneously emerge. In the remainder of this section, we present a candidate solution based on a variant of the Holland Broadcast Language.

3.3 The Holland Broadcast Language

The Broadcast Language is a programming formalism introduced by Holland in 1975 [14, 7], which can be thought of as the precursor for the LCS, the latter being a simplification of the Broadcast Language. The Broadcast Language was originally intended to solve some undesirable issues arising out of the use of Genetic Algorithms (GAs) [14, 11]. Holland argued that GAs provide an efficient method of adaptation, however in the case of long-term adaptation, the efficiency of GAs could be limited by the representation used to encode the problem. This problem representation is usually fixed and influences the complexity of the fitness function. During long-term evolution, this may limit the performances of the GA. To overcome this limitation, Holland proposed to adapt the problem representation used by the fitness function. Adapting the representation may then generate correlations between the problem representation and the GA performance.

A key property shared between the MCS and the Broadcast Language is the removal of any demarcation between messages and rules. A second beneficial property is the ability of the Broadcast Language to provide a straightforward representation to a variety of natural models such as the operon-operator model (a Genetic Regulatory Network model).

The Broadcast Language basic components are called *broadcast units* which can be viewed as condition/action rules. Whenever a broadcast unit conditional statement is satisfied, the action statement is executed. This means that whenever a broadcast unit detects in the environment the presence of (a) specific signal(s), including themselves, then the broadcast unit would broadcast an output signal.

Some broadcast units may broadcast a signal that may constitute a new broadcast unit. Similarly, a broadcast unit can be interpreted as a signal detected by another broadcast unit. Broadcast units may also process a given signal, in the sense that, a broadcast unit may output a signal that is some modification of the detected/input signal. As a result, a broadcast unit may create new broadcast units or detect and modify an existing broadcast unit. A set of broadcast units, combined as a string, designates a *broadcast device*.

Table 4. Comparison of biological and broadcast language terminology

Biology	Broadcast Language
sequence of amino acids from $\{A, R, N, D, C, E, \ldots\}$	string of symbols from $\Lambda = \{0, 1, *, :, \Diamond, \nabla, \blacktriangledown, \triangle, p, '\}$
substrate	input signal
product	output signal
protein with no enzymatic function	null unit
enzyme	broadcast unit
protein complex	broadcast device
cellular milieu	list of strings from Λ

As a summary, Table 4 presents a comparison between the biological and the broadcast language terminology.

The Broadcast Language alphabet Λ is finite and contains ten *symbols*, Λ^* is the set of strings over Λ. The symbols constitute the atomic elements of the language.

$$\Lambda = \{0, 1, *, :, \Diamond, \nabla, \blacktriangledown, \triangle, p, '\}$$

Let I be an arbitrary string from Λ^*, in I, a symbol is said to be quoted if it is preceded by a symbol $'$. A broadcast unit I_n is an arbitrary string from Λ^* which contains neither unquoted $*$ or unquoted $:$.

The finite collection of broadcast devices can be described by its *state* S at each timestep t. For example $S(0) = \{011 : *\triangle 011 : 11, 101, 100, 0111 : 01 :\}$ describes the set of broadcast devices at timestep $t = 0$, which corresponds to the initial state of the collection. Four types of broadcast unit can be distinguished, any broadcast units that do not follow one of the four schemes (see below) are null units. Broadcast units may engage in the following interactions based on discrete timesteps:

(i.) $*I_1 : I_2$
 If a signal of type I_1 is detected at time t then the signal I_2 is broadcast at time $t + 1$.

(ii.) $* : I_1 : I_2$
 If there is no signal of type I_1 present at time t then the signal I_2 is broadcast at time $t + 1$.

(iii.) $*I_1 :: I_2$
 If a signal of type I_1 is detected at time t then a *persistent* string of type I_2 (if any) is removed from the environment at the end of time t.

(iv.) $*I_1 : I_2 : I_3$
 If a signal of type I_1 and a signal of type I_2 are both present at time t then the signal S_3 is broadcast *at the same time* t unless the string I_3 contains unquoted symbols $\{\triangledown, \blacktriangledown, \triangle\}$ or singly quoted occurrence of $*$, in which case the string I_3 is broadcast a time $t + 1$.

The Symbols

The interpretation of each symbol in $\Lambda = \{0, 1, *, :, \Diamond, \triangledown, \blacktriangledown, \triangle, p, '\}$ is now presented. In particular cases, some symbols may not be interpreted by a given broadcast unit, these *ignored* symbols are simply overlooked. However, they remain important as they may get activated at a later stage where broadcast units undergo recombination.

$\{0, 1\}$ 0 and 1 are the basic elements to specify a signal. A string such as 010110 can be regarded as the signature of a particular signal. This signature can be employed by a broadcast unit to detect and identify a signal. For example: let $I_1 = *10111 : 00$ be a broadcast unit and $I_2 = 10111$ a signal, both strings are present at time t in the environment: $S(t) = \{*10111 : 00, 10111\}$. At time t, I_2 is detected by I_1, this triggers the activation of broadcast unit I_1, as a result: $S(t+1) = \{*10111 : 00, 10111, 00\}$.

$*$ This symbol indicates that the subsequent *symbols* until the next unquoted $*$ (if any) are to be interpreted as a broadcast unit. If a broadcast device I does not contain any unquoted $*$ then I is a null unit.

: This symbol is used as a punctuation mark to differentiate the arguments of a broadcast unit. The symbol : (position and frequency) determines the type of the broadcast unit as presented earlier. If more than two unquoted : are found in a broadcast unit then the third : and anything to the right of it are ignored.

$\Diamond$ When this symbol is met in the argument of a broadcast unit, it indicates that a signal detected by the broadcast unit may present any symbol at this position. This specific symbol occurring in the detected signal does not affect its acceptation or rejection by the broadcast unit.

For example, with $S(t) = \{*10\Diamond 11 : 00, 10011\}$ we obtain at time $t + 1$ $S(t+1) = \{*10\Diamond 11 : 00, 10011, 00\}$, $*10\Diamond 11$ broadcasts 00 if a signal containing $10\ldots 11$ is present ($\ldots$ indicates any arbitrary symbol from Λ).

Also if $\Diamond$ occurs at the rightmost position in the argument of a broadcast unit, then it indicates that a signal detected by the broadcast unit may present any suffix without affecting acceptance or rejection.

For example, with $S(t) = \{*1011\Diamond : 00, 101101101\}$ we obtain $S(t+1) = \{*1011\Diamond : 00, 101101101, 00\}$, $*1011\Diamond : 00$ would broadcast 00 if any signal containing the prefix 1011 is detected.

$\triangledown$ When this symbol occurs in the arguments of a broadcast unit, it designates any arbitrary initial (prefix) or terminal (suffix) strings of symbols. This allows one to pass a string of symbols from the input signal to the broadcast signal ($\approx$ unit processing).

For example, with $S(t) = \{*10\triangledown : \triangledown, 10011\}$ we obtain at $t + 1$: $S(t+1) = \{*10\triangledown : \triangledown, 10011, 011\}$. In this case $\triangledown$ designates the suffix 011 occurring in the input signal 10011. whereas if $S(t) = \{*10\triangledown : \triangledown, 100100101\}$ then we obtain at $t + 1$: $S(t+1) = \{*10\triangledown : \triangledown, 100100101, 0100101\}$. If several occurrences of $\triangledown$ are found in the output argument of a given broadcast unit, then they all designate the same substring.

▼ This symbol is similar to $\triangledown$ but can also concatenate different inputs signals.

For example, with $S(t) = \{*10\triangledown : 11▼ : 000\triangledown▼, 10111, 1100\}$ we obtain at $t + 1$: $S(t+1) = \{*10\triangledown : 11▼ : 000\triangledown▼, 10111, 1100, 00011100\}$. In this case $\triangledown$ designates the suffix 111 occurring in the input signal 10111 and ▼ designates the suffix 00 found in the detected signal 1100. The format of the broadcast signal is $000\triangledown▼$, therefore we replace and concatenate $\triangledown$ and ▼ accordingly and we obtain the output signal 00011100.

$\triangle$ This symbol is employed in the same manner as $\triangledown$ and ▼ but designates an arbitrary *single* symbol whose position can be anywhere in the argument of a given broadcast unit.

For example, with $S(t) = \{*11\triangle 0 : 1\triangle, 1100\}$ the output signal 10 is broadcast and this produces at $S(t+1) = \{*11\triangle 0 : 1\triangle, 1100, 10\}$. Whereas if $S(t) = \{*11\triangle 0 : 1\triangle, 1110\}$ the output signal 11 is broadcast producing $S(t+1) = \{*11\triangle 0 : 1\triangle, 1100, 11\}$.

p When this symbol occurs at the first position of a string, it designates a *persistent* string. This string would then persist over time until it is deleted, even if the string is not an active broadcast unit. A null device occurring at time t which is not persistent exists only for one timestep and is removed at the end of time t.

$'$ This symbol is used to *quote* a symbol in the arguments of a broadcast unit. When a symbol is said to be quoted, it acts as a simple literal, i.e. a $'\triangle$ would only match $\triangle$.

For example, with $S(t) = \{*11'\triangle 0 : 11, 11\triangle\}$, the input signal $11\triangle$ is detected by the broadcast unit and thus the output string 11 is broadcast producing at $t+1$: $S(t+1) = \{*11'\triangle 0 : 11, 11\triangle, 11\}$.

The broadcast language presented by Holland in the original text omitted a number of interactions between broadcast devices, which could in certain cases present us with ambiguities regarding the expected action to be performed. Holland discussed some of these semantical conflicts [14], while the remaining ambiguities were addressed in [7, 8]. However in the context of this paper, this is not an important consideration.

Example: Building a NAND Gate

In this section we describe the construction of a NAND gate using the broadcast language. This is intended to demonstrate how the broadcast language can be considered as a logical universal computational formalism. Moreover using the Boolean abstraction, it is also possible to build qualitative models of natural networks such as Genetic Regulatory Networks. With the Boolean abstraction, a molecule is considered as a logical expression having two different possible states. One possible state is the ON state meaning that the molecule is present in the environment. When a molecule state is OFF, this indicates that the particular molecule is not present in the environment (cell).

In the remainder of this section we first present a simple example in which a NAND gate is constructed within a static environment (the inputs values do not change over time), then a second example follows in which the same gate is adapted to be used with a dynamic system:

(i.) We consider a NAND gate having for inputs signals A and B and for output signal C. To construct this logical gate with the broadcast language, we first represent each signal A, B, C as null broadcast devices (substrates): $A = p001$, $B = p010$ and $C = p000$.

We then declare the following active broadcast devices (enzymes): $I_1 = *p001 : 011$ and $I_2 = *p010 : 100$, these devices emit signaling molecules $S_1 = 011$ and $S_2 = 100$ upon detecting A and B respectively. Similarly, we define $I_3 = * : p001 : 101$ and $I_4 = * : p010 : 110$ which would emit $S_3 = 101$ and $S_4 = 110$ if A or B are not detected.

Finally the following broadcast devices are employed to output C according to the intermediary states of signaling molecules S_1, S_2, S_3 and S_4: $I_5 = *011 : 110 : p000$, $I_6 = *100 : 101 : p000$ and $I_7 = *101 : 110 : p000$. Using these broadcast devices, it is possible to obtain the state of C according to the states of input signals A and B, 2 time steps are necessary to propagate and process the signals A and B.

(ii.) Within a dynamic system, some modifications are necessary to maintain our NAND gate. These modifications are intended so as to allow the output signal C to degrade over time. First, the broadcast devices I_1, I_2 and I_3 are modified as follows: As currently defined, those broadcast devices output the signal $p000$ (which designates the persistent signal C) when satisfied, these output signals are replaced with an additional signaling molecule $S_5 = 111$. As a result, we obtain the broadcast devices: $I'_5 = *011 : 110 : 111$, $I'_6 = *100 : 101 : 111$ and $I'_7 = *101 : 110 : 111$.

We then declare a broadcast device $I_8 = *111 : p000$ that upon detecting S_5 would emit the output signal C. Finally we declare a broadcast device $I_9 = p000 :: p000$ that deletes (degrades) C upon detecting C. We note that as soon as a signal C appears at time t, it would be removed by I_9 at the end of time t. To counter balance that effect, we double the concentration of broadcast devices I_8 so that the production rate of C is higher than its degradation rate.

In Fig. 5 we present a simulation using such a NAND gate specified with the broadcast language. In this simulation, the inputs A and B are manually switched ON/OFF at different timesteps. We detail the states of the system at timestep 0,1 and 2:

$S(0) = \{A = p001,\ I_1 = *p001 : 011,\ I_2 = *p010 : 100,\ I_3 = * : p001 : 101,$
$I_4 = * : p010 : 110,\ I'_5 = *011 : 110 : 111,\ I'_6 = *100 : 101 : 111,$
$I'_7 = *101 : 110 : 111,\ I_8 = *111 : p000,\ I_8 = *111 : p000,\ I_9 = p000 :: p000\}$

At $t = 0$, the system is initialized with above broadcast devices, A is ON, and both B and C are OFF. We note that both broadcast devices I_1 and I_4 are satisfied leading to the production of S_1 and S_4 at $t = 1$:

$S(1) = \{A = p001,\ I_1 = *p001 : 011,\ I_2 = *p010 : 100,\ I_3 = * : p001 : 101,$
$I_4 = * : p010 : 110,\ I'_5 = *011 : 110 : 111,\ I'_6 = *100 : 101 : 111,$
$I'_7 = *101 : 110 : 111,\ I_8 = *111 : p000,\ I_8 = *111 : p000,\ I_9 = p000 :: p000,$
$S_1 = 011, S_4 = 110\}$

At $t = 1$, I'_5 is activated due to the presence of S_1 and S_4. I'_5 is a type 4 broadcast device that is able to output S_5 signal during same timestep. As a result, both I_8 broadcast devices are now activated and produce two instances of C molecule at $t = 2$. As S_1, S_4 and S_5 are not persistent, these signals are removed at the end of $t = 1$. However, as A is still ON and B is OFF, S_1 and S_4 are again produced at $t = 2$.

At the beginning of $t = 2$, two instances of C are contained in the system. However the I_9 broadcast device is now satisfied by instances of the C molecule resulting in the removal of one instance of C.

$S(2) = \{A = p001, I_1 = *p001 : 011, I_2 = *p010 : 100, I_3 = * : p001 : 101,$
$I_4 = * : p010 : 110, I'_5 = *011 : 110 : 111, I'_6 = *100 : 101 : 111,$
$I'_7 = *101 : 110 : 111, I_8 = *111 : p000, I_8 = *111 : p000, I_9 = p000 :: p000,$
$C = p0000, S_1 = 011, S_4 = 110\}$

3.4 Fusing MCS and the Broadcast Language

In [8], it was demonstrated that the Broadcast Language can model Genetic Regulatory Networks (GRNs). This was due to the ability of the Broadcast Language to mirror Boolean networks which illustrates the wide ranging processing power that Broadcast Systems are capable of. Nevertheless, it was also highlighted that the Broadcast Language is limited regarding the representation and simulation of CSNs. To address this issue, we propose to combine the MCS concept with the Broadcast Language in a new system termed "MCS.b". The MCS.b complements the broadcast language (syntax and semantics) and extends it by including the following refinements:

- Instead of processing all broadcast devices sequentially and deterministically during a time step, the MCS.b processes as follows: at each time step t, we pick n

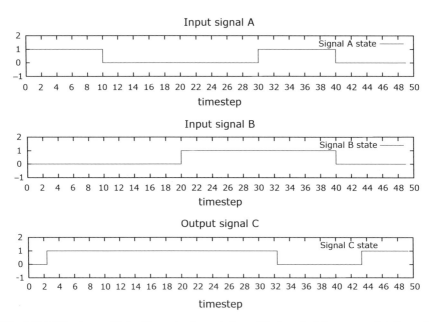

Fig. 5. NAND gate specified with our implementation of the broadcast language, details can be found in [7]. The output signal C state is initialized as OFF (0), at timestep 10, 20, 30 and 40, inputs A/B are manually switched ON/OFF (1/0), We note that the propagation time needed to process the switching of inputs A or B differs according to the nature of the switching involved and present states of A, B and C

pairs of broadcast devices at random. For each pair of devices, one of the broadcast devices is designated (at random) as the *catalyst device* and the second one as the *substrate device*. If the conditional statement of the catalyst device is satisfied by the signal of the substrate device, then the action statement of the catalyst device is executed upon the substrate device.
- n is a constant and designates the number of pairs of broadcast devices that will interact during a timestep. It is also plausible to consider n as the temperature in real chemistry. Temperature has an important role in chemical reactions, indeed molecules at higher temperature have a greater probability to collide with one another. In the broadcast language "universe", in order to increase the "temperature", one may increment the integer number n.
- In the broadcast language specification given by Holland, additional rules were required to resolve some ambiguities raised by the interpretation of broadcast devices. To facilitate this, the MCS.b simplifies the interpretation of broadcast units by preserving broadcast units of type 1 only.
- Similarly the notion of non-persistent devices is removed: by default all devices are considered as persistent molecules.
- As type 3 broadcast units and non-persistent devices no longer exist in this proposal, no molecule can be deleted from the population. However the deletion of molecules is needed to obtain evolutionary pressure. Our suggestion is as follows: each time two molecules react together, we pick a molecule at random and delete it from the population.

By combining the strength of both the MCS and Broadcast Language, we expect the MCS.b to be capable of modeling, simulating and evolving ACSNs in a more fateful manner. At present, we have conducted a number of preliminary experiments examining the spontaneous emergence of collective autocatalytic sets among others. This was expected to be trivial as this phenomenon was already demonstrated with other Artificial Chemistry Systems (such as Tierra, Alchemy, etc.). Initial results suggest that the MCS.b performs as expected, however before these results can be presented to the research community, validation against empirical biological data is required.

4 Future Work

In keeping with the four presented ACSN characteristic properties, we will focus on the following areas:

4.1 Computation

As part of the ESIGNET project, we will investigate the computational power of the MCS.b. This will include an examination of the MCS.b for Turing Completeness. As one of our project goals is to evolve ACSNs for computational purposes, incorporating completeness may be regarded as a crucial issue.

4.2 Evolution

An ACSN implies several cell signaling pathways interacting with each other. In order to evolve such a system of signaling networks *controlling* each other, it will be necessary to evaluate different Evolutionary Computational (EC) techniques. Because from biology it is natural to have a *hierarchical* system it may prove beneficial to investigate multi-level EC systems e.g. Hierarchical Genetic Algorithms [10].

4.3 Crosstalk

To obtain a better understanding of the crosstalk phenomenon and more specifically about the positive and negative effects of crosstalk. We will would like to see if it is possible to specify a network topology that allows optimal control of crosstalk effects.

A small world topology [22] may be of interest, as we may observe an analogy between CSNs and small world networks. This class of network, and more specifically scale-free networks are characterized by possessing nodes acting as "highly connected hubs". Although most nodes in these networks are of low degree. For example, a highly connected node could be referring to an ATP molecule that shares the same high degree of connectivity in real biochemical networks.

4.4 Robustness

We will investigate the ability of ACSNs to create and sustain specific internal conditions such as homeostasis. We would like to exhibit such robust behavior in simulated ACSNs, and how through evolutionary changes, robustness can be refined. Another consequent issue is to quantify the robustness of such systems to external shocks and changes of conditions.

5 Conclusion

In this paper we introduced an abstraction of Cell Signaling Networks focusing on four characteristic properties distinguished as follows: Computation, Evolution, Crosstalk and Robustness. We indicated how these attributes can be highly desirable properties for potential applications in the control systems, computation and signal processing field. Following this we described a novel class of Artificial Chemistry named Molecular Classifier Systems (MCS) which was inspired by Hollands Learning Classifier System (LCS). To simulate and evolve ACNS, we proposed an instance of the MCS called the MCS.b that extends the precursor of the LCS: the broadcast language. We completed the paper by examining further work that is required to conclusively validate our approach.

Acknowledgement

This work was funded by ESIGNET (Evolving Cell Signaling Networks in Silico), an European Integrated Project in the EU FP6 NEST Initiative (contract no. 12789).

References

1. U. Alon, M. G. Surette, N. Barkai, and S. Leibler. Robustness in bacterial chemotaxis. *Nature*, 397(6715):168–171, January 1999.
2. A. M. Arias and P. Hayward. Filtering transcriptional noise during development: concepts and mechanisms. *Nature Reviews Genetics*, 7(1):34–44.
3. N. Barkai and S. Leibler. Robustness in simple biochemical networks. *Nature*, 387(6636): 913–917, June 1997.
4. D Bray. Protein molecules as computational elements in living cells. *Nature*, 376(6538): 307–312, Jul 1995.
5. L. Bull and T. Kovacs. Foundations of Learning Classifier Systems: An Introduction. *Foundations of Learning Classifier Systems*, 2005.
6. A. Deckard and H. M. Sauro. Preliminary studies on the in silico evolution of biochemical networks. *Chembiochem*, 5(10):1423–1431, October 2004.
7. J. Decraene. The Holland Broadcast Language. Technical Report ALL-06-01, Artificial Life Lab, RINCE, School of Electronic Engineering, Dublin City University, 2006.
8. J. Decraene, G. G. Mitchell, B. McMullin, and C. Kelly. The holland broadcast language and the modeling of biochemical networks. In Marc Ebner, Michael O'Neill, Anikó Ekárt, Leonardo Vanneschi, and Anna Isabel Esparcia-Alcázar, editors, *Proceedings of the 10th European Conference on Genetic Programming*, volume 4445 of *Lecture Notes in Computer Science*, Valencia, Spain, 11–13 April 2007. Springer.
9. P. Dittrich. Chemical computing. In Jean-Pierre Banâtre, Pascal Fradet, Jean-Louis Giavitto, and Olivier Michel, editors, *UPP*, volume 3566 of *Lecture Notes in Computer Science*, pages 19–32. Springer, 2004.
10. B. Freisleben. Metaevolutionary approaches. In Thomas Bäck, David B. Fogel, and Zbigniew Michalewicz, editors, *Handbook of Evolutionary Computation*, pages C7.2: 1–8. Institute of Physics Publishing and Oxford University Press, Bristol, New York, 1997.
11. D. E. Goldberg. *Genetic Algorithms in Search, Optimization, and Machine Learning*. Addison-Wesley Professional, January 1989.
12. Ernst J. M. Helmreich. *The Biochemistry of Cell Signalling*. Oxford University Press, USA, 2001.
13. J.H. Holland. Adaptation. *Progress in theoretical biology*, 4:263–293, 1976.
14. J.H. Holland. *Adaptation in natural and artificial systems*. MIT Press, Cambridge, MA, USA, 1992.
15. J.H. Holland. Exploring the evolution of complexity in signaling networks. *Complexity*, 7(2):34–45, 2001.
16. J. R. Koza. *Genetic Programming: On the Programming of Computers by Means of Natural Selection (Complex Adaptive Systems)*. The MIT Press, December 1992.
17. G. Krauss. *Biochemistry of Signal Transduction and Regulation*. John Wiley & Sons, 2003.
18. S. Forrest L. Segel. Robustness of cytokine signalling networks. http://www.santafe.edu/research/signallingnetworks.php.

19. P. L. Lanzi, W. Stolzmann, and S. W. Wilson, editors. Springer-Verlag, April 2001.
20. D.A. Lauffenburger. Cell signaling pathways as control modules: complexity for simplicity? *Proc. Natl. Acad. Sci. USA*, 97(10):5031–3, 2000.
21. B. McMullin, C. Kelly, D. OŠBrien, G. G. Mitchell, and J. Decraene. Preliminary Steps toward Artificial Protocell Computation. In *Proceedings of the 2007 International Conference on Morphological Computation*, 2007. To appear.
22. M. E. J. Newman. Models of the small world: A review, May 2000.
23. R. C. Stewart and F. W. Dahlquist. Molecular components of bacterial chemotaxis. *Chem. Rev.*, 87:997–1025, 1987.
24. T. Lenser, T. Hinze, B. Ibrahim, and P. Dittrich. Towards Evolutionary Network Reconstruction Tools for Systems Biology. In *Fifth European Conference on Evolutionary Computation, Machine Learning and Data Mining in Bioinformatics*, 2007. To appear.
25. D. Volfson, J. Marciniak, W. J. Blake, N. Ostroff, L. S. Tsimring, and J. Hasty. Origins of extrinsic variability in eukaryotic gene expression. *Nature*, December 2005.
26. T. M. Yi, Y. Huang, M. I. Simon, and J. Doyle. Robust perfect adaptation in bacterial chemotaxis through integral feedback control. *Proc Natl Acad Sci USA*, 97(9): 4649–4653, April 2000.

Part III

Sensor and Actor Networks

Immune System-based Energy Efficient and Reliable Communication in Wireless Sensor Networks

Barış Atakan and Özgür B. Akan

Next Generation Wireless Communications Laboratory
Department of Electrical and Electronics Engineering
Middle East Technical University, 06531, Ankara, Turkey
atakan,akan@eee.metu.edu.tr

Summary. Wireless sensor networks (WSNs) are event-based systems that rely on the collective effort of densely deployed sensor nodes. Due to the dense deployment, since sensor observations are spatially correlated with respect to location of sensor nodes, it may not be necessary for every sensor node to transmit its data. Therefore, due to the resource constraints of sensor nodes, it is imperative to select the minimum number of sensor nodes to transmit the data to the sink. Furthermore, to achieve the application-specific distortion bound at the sink, it is also of great significance to determine the appropriate sampling frequency of sensor nodes to minimize energy consumption. In order to address these needs, the Distributed Node and Rate Selection (DNRS) algorithm which is based on the principles of natural immune system is developed. Based on the B-cell stimulation in immune system, DNRS selects the most appropriate sensor nodes that send samples of the observed event, i.e., *designated nodes*. The aim of the designated node selection is to meet the event signal reconstruction distortion constraint at the sink node with the minimum number of sensor nodes. DNRS enables each sensor node to distributively decide whether it is a designated node or not. In addition, to exploit the temporal correlation in the event data DNRS regulates the sampling frequency rate of each sensor node while meeting the application-specific delay bound at the sink. Based on the immune network principles, DNRS distributively selects the appropriate sampling frequencies of sensor nodes according to the congestion in the forward path and the event signal reconstruction distortion periodically calculated at the sink by Adaptive LMS Filter. Performance evaluation shows that DNRS provides the minimum number of designated nodes to reliably reconstruct the event signal and it regulates the sampling frequency of designated nodes to exploit the temporal correlation in the event signal with significant energy saving.

1 Introduction

Wireless Sensor Networks (WSN) are generally comprised of densely deployed sensor nodes collaboratively observing and communicating extracted information about physical phenomenon [1]. Due to the dense deployment of sensor nodes, sensor observations are spatially correlated according to the location of sensor nodes. This results in transmission of highly redundant sensor data which is not necessary to reconstruct the event signal at the sink. Therefore, due to the energy constraint of sensor nodes, it is necessary to select the minimum number of sensor nodes that send samples of the observed event, which are referred to as *designated nodes*. Furthermore, the nature of the energy-radiating physical phenomenon constitutes the temporal correlation between each consecutive observation of a sensor node [2]. Therefore, it is necessary to regulate the sampling frequency of sensor nodes to achieve minimum energy consumption while achieving a certain reconstruction distortion.

There has been some research efforts to study about correlated data gathering in WSN according to the different methods [3], [4], [5], [6], [7]. These works provide the great deal of capability to exploit the correlation at the sensor data. In [10], the model for point and field sources are introduced and their spatio-temporal characteristics are derived along with the distortion functions. In [9], exploiting spatial correlation at the MAC layer is achieved by collaboratively regulating medium access so that redundant transmissions from correlation neighbors are suppressed. In [9], the Iterative Node Selection (INS) algorithm is given to select the representative nodes which represent the set of nodes generating spatially correlated data. Since INS selects the representative nodes with centralized manner at the sink and it computes the distortion according to the Wide Sense Stationary (WSS) assumption on the sink which is neither scalable nor realistic. These works consider and capture the spatio-temporal model of physical phenomenon observed by sensor nodes for the realization of advanced efficient communication protocols. However, in these works since the event data is assumed to be Gaussian and wide-sense stationary (WSS) and the most process encountered in practice are non-stationary, these works are not realistic. Moreover, in the literature, there exist some distributed solutions [12], [13] to exploit the spatio-temporal correlation in WSN. However, these studies neither provide distributed selection of the sensor nodes which communicate with the sink node nor regulate the data rate to the sink node according to the estimation distortion at the sink node.

On the other hand, the natural Immune System has given the great inspiration for several research efforts from robotic to network security [14], [15], [16]. Because of its ability to self and non-self discrimination, it has been used for the data clustering and the computer security as an inspiration. In [14], an effective artificial immune system is presented, which is used as a simple classification tool. Using the clonal selection mechanism in Immune System, it attempts to group similar data items according to relationship between them. In [17], the problem of protecting computer systems is addressed as the problem of learning to distinguish self from other. It proposes a method for change detection which is based on the generation of T-cells in the immune system.

In this chapter, the Distributed Node and Rate Selection (DNRS) method [23], which is based on the principle of natural immune system, is introduced. Based on the B-cell stimulation model in Immune System, DNRS selects the most appropriate sensor nodes that send samples of observed event, are referred to as designated nodes. The aim of the designated node selection is to meet the distortion constraint for the event signal reconstruction at the sink with the minimum number of sensor nodes. DNRS enables sensor nodes to distributively decide whether it is a designated node or not according to its correlation with its neighbors and the event source. In addition, DNRS distributively regulates the sampling frequency rate of each sensor node to provide a certain reconstruction distortion with minimum energy consumption. Based on the immune network models, DNRS selects the appropriate sampling frequencies of sensor nodes according to the congestion in the forward paths and the event signal reconstruction distortion periodically calculated at the sink node by Adaptive LMS Filter.

The remainder of this chapter is organized as follows. In Section 2, after we briefly introduce the Biological Immune System, we introduce the relationship between Wireless Sensor Networks (WSN) and Immune System. In Section 3, we present the DNRS method. We firstly introduce the Distributed Node Selection scheme and then, introduce our Distributed Rate Selection mechanism. In Section 4, we evaluate the performance of DNRS. Finally, we conclude this chapter with Section 5.

2 Immune System and Wireless Sensor Networks

In this section, we first briefly introduce the immune system and its basic operation principles. Then, we discuss the similarities and the relation between the Immune System and Wireless Sensor Networks.

2.1 Biological Immune System

The human immune system is a complex natural defense mechanism. It has the ability to learn about foreign substances (pathogens) that enter the body and to respond to them by producing antibodies that attack the antigens associated with the pathogen [14]. The adaptive immune system is made up of lymphocytes which are white blood cells, B and T cells. Each of B-cells has distinct molecular structure and produces 'Y' shaped antibodies from its surfaces. The antibody recognizes antigen that is foreign material and eliminates it. This antigen-antibody relation is innate immune response [15]. Most antigens have various antigen determinants that are called epitope. In order to grab and latch onto antigen, antibody possesses a structure called paratope. Furthermore, each antibody has a unique epitope called idiotope. Note that different antibody types may have many epitopes in common. An epitope that is unique to a given antibody type is called an idiotope, hence the name idiotypic network for any scheme of regulation that works through the recognition of idiotopes [18]. An antibody type is thought to be stimulated when its paratopes recognize other types, and suppressed if its epitopes are recognized by other antibody paratopes.

The surface of a B-cell contains antibodies for that B-cell. When an antibody for a B-cell binds to an antigen, the B-cell becomes stimulated. The level of B-cell stimulation depends not only on the success of the match to the antigen, but also on how well it matches other B-cells in the immune networks. As the response of the antigen, the stimulated B- cells secrete the antibody to eliminate the antigen.

Immune system has the ability to process information, to learn and memorize, to discriminate between self and non-self, and to keep up harmony of the whole system. Because of these abilities, it gives the great inspiration to many kind of researches in computer sciences, robotics and signal processing. Therefore, Artificial Immune System (AIS) models have been introduced. In [18], an AIS model is proposed, which consists of a set of B-cells and the links between those B-cells. Each B-cell object can represent a data item which is being used for learning. According to this model, each B-cell is capable of responding to an antigen specified by a data item when it is stimulated by that antigen specific data item. In the natural immune system, the level of the B-cell stimulation relates that how well its antibody binds to the antigen and its affinity (or enmity) to its neighbors in the network. In [14], B-cell stimulation is modeled in terms of these three factors. The primary stimulus is defined as the affinity between the B-cell and the pathogen, i.e., how well they match. The second stimulus for a B-cell is the affinity to its neighbors. Third is the suppression (enmity) factor from the loosely connected neighbor of this B-cell.

The stimulated B-cells start to produce more lymphocytes (i.e., to clone) and to secrete free antibodies. To model the antibody secretion of the stimulated B-cell, in [19], idiotypic network hypothesis is proposed, which is based on mutually stimulus and suppression between antibodies. In [18], an analytical immune network model is proposed using the idiotypic network hypothesis introduced in [19] as follows

$$\frac{dS_i(t+1)}{dt} = \left(\frac{\alpha \sum_{j=1}^{N} m_{ij}s_j(t)}{N} - \frac{\alpha \sum_{k=1}^{N} m_{ki}s_k(t)}{N} + \beta g_i - k_i\right)s_i(t) \quad (1)$$

where m_{ij} is mutual stimulus coefficient of antibody i and j, g_i is affinity between antibody i and antigen, k_i is the natural extinction of the antibody i, α, β, η are constants. $s_i(t)$ is the concentration of antibody i at time t and given as follows

$$s_i(t+1) = \frac{1}{1+e^{(0.5-S_i(t+1))}} \quad (2)$$

In Section 3, we adopt the immune network model given in (1) and (2) to derive an effective communication model between sensor nodes and the sink. In the next section, we introduce some analogies between Immune System and Wireless Sensor Networks (WSNs).

2.2 Immune System Based Sensor Networks

Although Immune System seems different from the Wireless Sensor Networks (WSN), they have the great deal of analogies considering the tasks they accomplish. For example, when the Immune System encounters a pathogen, some B-cells are stimulated and secrete antibodies in different densities to eliminate the antigens produced by the pathogen. Similarly, when an event occurs in WSN environment, some sensor nodes referred to as designated node (DN) sense the event information and send this information to sink with a sampling frequency to achieve a certain event signal reconstruction distortion at the sink.

Table 1. Relationship between Immune System and Wireless Sensor Networks (WSN)

Immune System	Wireless Sensor Networks (WSN)
B-cells	Sensor nodes
Antibody	Sensor data
Antibody density	Sampling frequency rate f
Pathogen	Event source
Antigen	Reconstruction distortion
Natural extinction	Packet loss

In WSN, since multiple sensors record information about a single event in the sensor field, sensor observations are spatially correlated with respect to spatial location of sensor nodes. This results in transmission of highly redundant sensor data which is not necessary to reconstruct the event signal at the sink node. Therefore, it may not be necessary for every sensor node to transmit its data to the sink; instead, a smaller number of DN measurements might be adequate to communicate the event features to the sink within a certain reliability/fidelity level [8]. DNRS use the B-cell stimulation model given in [14] to determine the minimum number of DNs while meeting the application specific distortion bound at the sink.

For a fixed number of DNs, the minimum distortion can be achieved by choosing these nodes such that (i) they are located as close to the event source as possible and (ii) are located as farther apart from each other as possible [8]. Similarly, as given in [14], the stimulation of a B-cell depends on the distance between the pathogen and the B-cell and distance between the B-cell and its neighbors which stimulates or suppress it. That is, the B-cells which is the nearest to the pathogen and suppressed as little as possible by their neighbors become the stimulated B-cells. The stimulated B-cells start to secrete the antibodies in different densities to keep the antigen densities in a desired level. Similarly, after the selection of DNs they send the event information to sink with a sampling frequency to achieve the distortion constraint at the sink node. In Table 1, we summarize this relationship between Immune System and Wireless Sensor Networks (WSN). According to this relationship, DNRS distributively selects the DNs by using the B-cell stimulation model and distributively regulates the sampling frequency of the DNs based on immune network principles given in (1) and (2).

3 Immune System Based Distributed Node and Rate Selection

In this section, we first introduce an event signal reconstruction scheme used in DNRS. Then, we introduce DN selection scheme and the sampling frequency regulation mechanism in DNRS.

3.1 Event Signal Reconstruction

According to the application, the physical phenomenon can be modeled by a point source or by a field source. For example, while the case of object tracking can be modeled by a point source, the case of monitoring magnetic field and seismic activities can be modeled by field source. Here, in order to exploit the spatio-temporal correlation with Immune System based DNRS mechanism, we adopt the point source model given in [10] for event signal, and then we introduce a model for the event signal reconstruction. Note that, in this chapter, we use the point source based event signal reconstruction scheme, however, this scheme can be directly extended to the field source case.

In WSN applications, the event signal modeled by a point source is assumed to generate a continuous random process $f_S(t)$ having a variance σ_S^2 and a mean μ_S. Assuming that the point source is at the center of the coordinate axis [10], the sensor node situated at location (x, y) receives the signal given by

$$f(x, y, t) = f_S(t - \frac{\sqrt{x^2 + y^2}}{v}) e^{-\frac{\sqrt{x^2+y^2}}{\theta_s}} \qquad (3)$$

where v denotes the diffusion velocity of $f_S(t)$ and θ_s denotes the attenuation constant of event signal $f_S(t)$. Since $f(x, y, t)$ is the delayed and attenuated version of $f_S(t)$, the variance and mean of $f(x, y, t)$ can be expressed by

$$\mu_E(x, y) = (\mu_s e^{-\frac{\sqrt{x^2+y^2}}{\theta_s}}) \qquad (4)$$

$$\sigma_E^2(x, y) = (\sigma_s e^{-\frac{\sqrt{x^2+y^2}}{\theta_s}})^2 \qquad (5)$$

Using the observations which are attenuated and delayed version of the event signal, each sensor node generates its sample sets. The kth sample set generated by sampling the received signal $f(x_i, y_i, t)$ by sensor node n_i at location (x_i, y_i), i.e., $S_{i,k}$, is defined as

$$S_{i,k} = \begin{bmatrix} f(x_i, y_i, t_{kp}) & f(x_i, y_i, t_{kp+1}) & \cdots & f(x_i, y_i, t_{(k+1)p}) \end{bmatrix} \qquad (6)$$

where $S_{i,k}$ is $p \times 1$ vector, p and k denote the length of sample set and set number, respectively, $f(x_i, y_i, t_{kp})$ is a sample of event signal $f(x_i, y_i, t)$, taken at time t_{kp}.

In WSNs, after the detection of an event, sensor nodes sample the event signal and send the generated data packets to the sink. During this process, sensor circuitries add noise to its sample sets ($S_{i,k}$) as

$$S'_{i,k} = S_{i,k} + N_{i,k} \qquad (7)$$

where $S'_{i,k}$ is the sensor packet and $N_{i,k}$ is the observation noise, where $N_{i,k} \sim N(0, \sigma_N^2)$.

In addition to the additive noise, due to the constraint in cost, power, and communication, in WSNs, packet losses often arise during the communication between the sensor nodes and the sink. In this chapter, to model the packet losses, we assume an average packet loss probability λ_i which indicates the packet losses of sensor node i due to channel errors, collision, and congestion. The packet loss probability λ_i is computed by the sink by dividing the number of loss packets of sensor node i to the total number of packets transmitted by sensor node i. Hence, we define the lossy and noisy version of the sensor packet $(X_{i,k})$ as

$$X_{i,k} = \begin{cases} 0 & \text{with probability } \lambda_i \\ S'_{i,k} & \text{with probability } (1-\lambda_i) \end{cases}$$

Since uncoded transmission considerably outperforms any approach based on the separation paradigm when the measured physical phenomenon is reconstructed within some prescribed distortion level [11], we assume that the sensors perform uncoded transmission. When the sensor nodes transmit the uncoded observations, minimum mean square error (MMSE) estimation is the optimal decoding technique. Therefore, the decoded version of the sensor packet $S_{i,k}$, i.e., $Z_{i,k}$, can be expressed as follows

$$Z_{i,k} = \frac{\left(\sigma_E^2(x_i,y_i) + \mu_E^2(x_i,y_i)\right)(X_{i,k})}{(1-\lambda_i)\left(\sigma_E^2(x_i,y_i) + \mu_E^2(x_i,y_i) + \sigma_N^2\right)} \quad (8)$$

According to the decoded sensor packets $(Z_{i,k}, \forall i,k)$, sink generates the reconstructed event signal, i.e., $\hat{S}(\tau, f, M, p)$, as follows

$$\hat{S}(\tau, f, M, p) = \frac{1}{M}\left[\sum_{i=1}^{M} Z_{i,1} \sum_{i=1}^{M} Z_{i,2} \cdots \sum_{i=1}^{M} Z_{i,(\frac{\tau f}{p})}\right] \quad (9)$$

where τ, M, f denote the reconstruction interval, the number of DNs and the sampling frequency of sensor nodes, respectively. The sampling frequency of a sensor node is the number of samples taken from the sensed event signal. However, since in WSNs each data packet includes a number of samples (p), reporting frequency which is defined as the number of transmitted data packet per unit time is directly proportional with sampling frequency (f). Therefore, throughout this chapter, we mainly use the sampling frequency.

3.2 Distributed Designated Node Selection

As discussed in Section 2, the stimulation of a B-cell depends on three factors. First factor is the affinity between the B-cell and the pathogen. Second factor is the affinity between the B-cell and its neighbors which make it stimulated. Third factor is the affinity between the B-cell and its neighbors which suppress it [14]. According to the B-cell stimulation model given in [14], for a B-cell if the summation of these three factors specified by some equations is greater than a certain threshold value, this B-cell becomes stimulated. In WSN, the minimum distortion can be achieved by choosing the DNs such that (i) they are located as close to the event source as possible and (ii) are located as farther apart from each other as possible [8]. In fact, as B-cell stimulation, the selection of DNs depends on the three factors given as follows:

(i.) First factor is the affinity between the sensor node (B-cell) and event source (pathogen).
(ii.) Second factor is the affinity between the sensor node (B-cell) and its uncorrelated neighbor nodes (stimulating B-cells).
(iii.) Third factor is the affinity between the sensor node (B-cell) and its correlated neighbor nodes (suppressing B-cells).

To determine the correlated and uncorrelated neighbors of a sensor node, we define the application specific correlation radius r. For a sensor node, while the neighbor nodes in its correlation radius r are the correlated neighbors for this sensor node, the neighbor nodes which are not in correlation radius r are uncorrelated neighbors for this sensor node. The spatial correlation between the sensor nodes depends on the variance and the mean of the observed event signal, environment size, node density. Therefore, we assume that correlation radius r is an application specific value, which can be determined based on the environment size, node density and statistical properties of the physical phenomena observed in a given sensor network application [8]. We also assume that each sensor node is aware of its position and event source location by means of existing localization techniques [22]. Each sensor knows the relative position of its neighbors, as all nodes periodically broadcasts its position to neighbors.

For the determination of the affinity between the sensor node and its correlated and uncorrelated neighbor nodes, the correlations between the nodes and event source are needed. To express the correlations between event source and the nodes, we use the correlation coefficients, $\rho_{s,i}$ and $\rho_{i,j}$. $\rho_{s,i}$ states the correlation between the sensor node i sending information and the event source S. $\rho_{i,j}$ states the correlation between the sensor node i and j. We use the power exponential form [20] to model the correlation coefficients $\rho_{s,i}$ and $\rho_{i,j}$ and give these coefficients as follows.

$$\rho_{s,i} = K_\vartheta(d_{s,i}) = e^{(-d_{s,i}/\theta_1)^{\theta_2}}; \theta_1 > 0, \theta_2 \in (0, 2] \qquad (10)$$

$$\rho_{i,j} = K_\vartheta(d_{i,j}) = e^{(-d_{i,j}/\theta_1)^{\theta_2}}; \theta_1 > 0, \theta_2 \in (0, 2] \qquad (11)$$

where $d_{s,i}$ and $d_{i,j}$ denote the distances between event source S and sensor node i and between sensor node i and j, respectively. For $\theta_2 = 1$, the model becomes exponential, while for $\theta_2 = 2$ squared exponential. The covariance function is assumed to be nonnegative and decrease monotonically with the distance, with limiting values of 1 at $d = 0$ and of 0 at $d = \infty$.

Now, as the B-cell stimulation model given in [14], we mathematically model the DN selection. B-cell stimulation model given in [14] depends on three factors (i) the matching between the pathogen a B-cell (ii) the affinities between the B-cell and its neighbor B-cells (iii) the enmity between the B-cell and its neighbor B-cells. According to this model, when the summation of these three factors exceeds certain threshold for a B-cell, this B-cell becomes stimulated. Using this B-cell stimulation scheme, we model the DN selection depending on the similar three factors as follows:

(i.) We model the first factor which is the affinity between the sensor node i (B-cell) and event source s (pathogen) as $\rho_{s,i}$. Here, $\rho_{s,i}$ indicates the correlation between event source s and sensor node i. As the distance between sensor node i and event source s decreases, $\rho_{s,i}$ increases. Hence, it is more possible to become a DN for a sensor node nearest to the event source.

(ii.) We model the second factor which is the affinity between the sensor node i (B-cell) and its uncorrelated neighbor sensor nodes (stimulating B-cells) $j, \forall j$ as follows. Sensor node j is selected as the neighbor node which is not in the correlation radius of sensor node i.

$$\sum_j (1 - \rho_{i,j}) \qquad (12)$$

Here, $\rho_{i,j}$ indicates the correlation between sensor node i and j. As the distance between sensor nodes i and j increases, $\rho_{i,j}$ decreases. Therefore, $\sum_j (1-\rho_{i,j})$ is higher for a sensor node with the higher number of uncorrelated neighbors (out of r). Hence, it is more possible to become a DN for such sensor nodes.

(iii.) We model the third factor which is the affinity between the sensor node and its correlated neighbor sensor nodes (in r) $k, \forall k$ such that k is in the correlation radius of sensor node i and give as

$$\sum_k (-\rho_{i,k}) \qquad (13)$$

where $\rho_{i,k}$ indicates the correlation between sensor node i and sensor node k. As the distance between sensor node i and k decreases, $\rho_{i,k}$ increases. Therefore, $\sum_k (-\rho_{i,k})$ is small for a sensor node having the more correlated neighbors. Hence, it is the least possible to become a DN for such sensor nodes.

Here, we define the DN selection weight of sensor node i (T_i) as the combination of these three factors as follows:

$$T_i = \rho_{s,i} + \sum_j (1 - \rho_{i,j}) + \sum_k (-\rho_{i,k}) \qquad (14)$$

When T_i exceeds the certain threshold t_{dns} called the DN selection threshold, sensor node i becomes a DN. It is clear that while t_{dns} increases, the number of DN decreases because the number of nodes whose weight (T_i) exceed t_{dns} decreases. t_{dns} is determined at the sink node by using reconstruction distortion periodically computed by the Adaptive LMS Filter.

Adaptive LMS Filter is typically applied in environments where signals with unknown or non-stationary statistics are involved and appear for this reason particularly suited to be used in highly dynamic systems such as sensor networks [21]. Using the Adaptive LMS Filter, the sink node predicts the reconstructed version of the event signal sample $\hat{S}(k, \tau, f, M, p)$ given in (1) from the previous samples

$\hat{S}(k-r,\tau,f,M,p)$ $r = 1...m$, where m is the order of the filter and $\hat{S}(k,\tau,f,M,p)$ denotes the kth sample of the reconstructed event signal. Hence, the event signal reconstruction error in the Adaptive LMS Filter is expressed by

$$e[k] = \hat{S}(k,\tau,f,M,p) - \bar{\hat{S}}(k,\tau,f,M,p) \qquad (15)$$

where $\bar{\hat{S}}(k,\tau,f,M,p)$ denotes the prediction of $\hat{S}(k,\tau,f,M,p)$ with Adaptive LMS Filter. Depending on this error, $e[k]$, we give the reconstruction distortion D as follows

$$D = \frac{1}{\tau f} \sum_{k=1}^{\tau f} e[k]^2 \qquad (16)$$

To determine the minimum number of DN which can be sufficient to represent the all sensor nodes in the WSN environment, sink periodically increases t_{dns} because the number of DNs decreases while t_{dns} increases. To increase t_{dns}, the sink node adds a number (u) to t_{dns} in every decision interval t and broadcasts the updated designated node selection threshold ($t_{dns} + u$) to sensor nodes. As the t_{dns} increases in each decision interval t, the number of DN decreases. Unless the distortion level of Adaptive LMS Filter at the sink is increased, the sink continues to increase t_{dns} because the increase in the distortion means that henceforth, the selected DNs can not represent the event data gathering with all sensor nodes in the event area. Therefore, when the distortion is increased, the sink node stops increasing the t_{dns} and it sets t_{dns} to the last updated value of t_{dns} before the increase of the estimation distortion. After the determination of maximum allowable t_{dns}, the sink node broadcasts this t_{dns} to all sensor nodes. Therefore, each sensor node can distributively decides whether it is a DN or not. If $T_i > t_{dns}$, sensor node i is a DN. If $T_i < t_{dns}$, sensor node i is not a DN.

Now, we outline the entire DN node selection scheme of DNRS as follows

- When an event occurs, each sensor node computes its DN selection weight (T_i, $\forall i$) according to its correlated and uncorrelated neighbor nodes and the distance to the event location.
- Sink initially sets the DN selection threshold t_{dns} as low as possible value such that at the beginning all nodes in the event area become DN ($T_i, > t_{dns}, \forall i$).
- Sink node increases t_{dns} at each decision interval t without disturbing the reconstruction distortion D at the sink. It broadcasts the updated threshold value to all sensor nodes in each t. Thus, at each t, the number of DN decreases because the number of nodes which exceed the updated threshold t_{dns} decreases.
- Until D is increased, sink node continues to increase t_{dns} and to broadcast the updated t_{dns}.
- When the distortion is increased at a decision interval, the sink node stops increasing t_{dns} at this decision interval. Thus, sink node sets t_{dns} to the last updated value of t_{dns} before the increase of the reconstruction distortion and it broadcasts this t_{dns} to all sensor nodes.

Algorithm 3.1 DN Selection Algorithm in DNRS

foreach *sensor i* **do**
    computes T_i, $\forall i$
end
while *D is increased* **do**
    increases t_{dns} **foreach** *sensor i* **do**
        if $t_{dns} < T_i, \forall i$ **then**
            sensor i is a DN
        end
    end
end

- According to the final t_{dns}, the all sensor nodes distributively decide whether it is a DN not or not as follows.
- If $T_i > t_{dns}$, sensor node i is a DN.
- If $T_i < t_{dns}$, sensor node i is not a DN.

Thus, when an event occurs in WSN environment, DNRS distributively selects the DNs by using above algorithm. The selected DNs represent all nodes in the event area for this event. However, when the location of the event changes, DNRS again employs the DN selection algorithm given above to determine the new DNs. In Algorithm 3.2, we give a pseudo-code for DN selection algorithm of DNRS.

After the determination of the DNs, it is important to regulate their sampling frequency f for exploiting the temporal correlation in the event data. In the next section, we give our distributed frequency rate selection scheme in DNRS.

3.3 Distributed Frequency Rate Selection of Designated Nodes

The sampling frequency of a sensor node f is defined as the number of samples taken and hence packets sent out per unit time by that node for a sensed phenomenon. Hence, the sampling frequency f controls the amount of traffic injected to the sensor field while regulating the number of temporally-correlated samples taken from the phenomenon. This, in turn, affects the event signal reconstruction distortion [8]. For the temporally correlated data regions, less number of samples is sufficient to reconstruct the event signal while more samples are needed to reconstruct the event signal region which is not temporally correlated. Therefore, while smaller f is needed to reconstruct the temporally correlated regions in the event signal, higher f is needed to reconstruct the event signal regions which are not temporally correlated. However, while f increases, the contention at the forward path increases and packet losses arise at higher f. Since packet losses prevent to deliver the sufficient information for the reconstruction, it also affects the reconstruction performance such that while the packet loss rate increases, the reconstruction distortion increases. Therefore, to exploit the temporal correlation in the event signal the appropriate frequency rate selection scheme have to consider the packet losses at the forward path to provide the most appropriate sampling frequency to the sensor nodes such that the DNs having higher λ_i use small f while the DNs having less λ_i use higher f. This kind of frequency rate regulation scheme provides a new congestion control mechanism to sensor networks such that while at the data paths in which the packet losses arise, f is decreased, on the non-congested paths f is increased.

As discussed in Section 2.1, in human immune system when a pathogen enters a body, some B-cells are stimulated and they start to secrete the antibody with the appropriate density to eliminate the antigens produced by the pathogen. This natural mechanism is modeled with the immune network models given in (1), (2). Similarly, in WSN when an event occurs, some sensor node should be selected as the DNs and they should send the sensed information with the appropriate sampling frequency to sink node to reconstruct the event signal without the congestion in the forward path. In order to address the sampling frequency regulation and the congestion in the forward path, we exploit the relationship between the immune system and Wireless Sensor Networks (WSN) and we adopt the immune network equations given in (1), (2) to provide the sensor nodes a distributed rate control mechanism. Now, we give the relation between Immune System and WSNs, which is summarized in Table 1, as follows:

- We consider each sensor node as a B-cell which secretes only one kind of antibody.
- We consider the sensor data as the antibodies which are secreted by B-cells and we model the sampling frequency of the sensor nodes (f_i) as the antibody concentration given by s_i. f_i denotes the sampling frequency identifier of sensor node i and it is determined by mapping the actual sampling frequency of sensor node i to a number between 0 and 1.
- In (1), (2), the stimulus value of antibodies (S_i) manages the concentration of the antibody i according to the antigen concentration and the natural extinction of the antibodies. Based on the similarity summarized in Table 1, to manage the sampling frequency rate of DNs according to the reconstruction distortion (D) and the packet loss rates (λ_i), we consider S_i as the rate control parameter $F_i(t_k)$ and give as follows.

$$F_i(t_{k+1}) = F_i(t_k) + (\frac{1}{K} \sum_{j=1}^{K} f_j(t_k) + aD - b\lambda_i) f_i(t_k) \quad (17)$$

where

$$a = (1 - signum(\lambda_i - \lambda_c))(1 - signum(D - D_c))/2$$
$$b = (1 + signum(\lambda_i - \lambda_c))/2$$

λ_c and D_c are the application-specific packet loss rate and reconstruction distortion constraints, respectively, K is the number of DNs, t_k is the kth decision interval, $f_i(t_k)$ is the sampling frequency identifier of sensor node i at the decision interval t_k, and $F_i(t_{k+1})$ is the rate control parameter at the decision interval t_{k+1}. Here, a and b are the coefficients which controls the respective influence between D and λ_i. If $\lambda_i > \lambda_c$, $a = 0$ and $b = 1$. In this case, to reduce λ_i, DNRS neglects the reconstruction distortion ($a = 0$) such that after DNRS decreases λ_i, it deals with decreasing D if $D > D_c$. If $\lambda_i < \lambda_c$ and $D > D_c$, $b = 0$ and $a = 1$. In this case, since $\lambda_i < \lambda_c$ ($b = 0$), DNRS deals with decreasing D.

- We consider the reporting frequency of DN i at t_k ($f_i(t_k)$) as the concentration of antibody i (s_i), which is secreted by B-cell i, given in (2). Based on the rate control parameter $F_i(t_k)$ given in (17), the sampling frequency identifier $f_i(t_{k+1})$ can be given as follows

$$f_i(t_{k+1}) = \frac{1}{1 + e^{(0.5 - F_i(t_k))}} \qquad (18)$$

In each decision interval (t_k), each sensor node distributively evaluates these mechanism given in (17), (18) by means of the coordination between DNs and sink as follows:

- In each t_k, each DN broadcasts its sampling frequency identifier ($f_i(tk)$) to all DNs.
- In each t_k, each DN computes its packet loss rate (λ_i) by the feedback from the sink node.
- D is computed at the sink by Adaptive LMS Filter and it is broadcasted to all sensor nodes in each (t_k).

The operation of immune System based distributed rate control mechanism given in (17), (18), which is based on the coordination among the DNs, can be given as

- When an event occurs, some sensor nodes are selected as the DN through the DN selection scheme given in Section 3.2.
- The selected DNs start to sample the event signal with a sampling frequency.
- When for DN i, $\lambda_i > \lambda_c$, $a = 0$ and $b = 1$. In this case, F_i starts to decrease and this result in decrease in the sampling frequency of DN i (f_i). Since the decrease in the sampling frequency of DN i reduce the contention at the forward path of DN i, λ_i decreases.
- When for DN i $\lambda_i < \lambda_c$, $b = 0$. In this case, if $D > D_c$ ($a = 1$), DN i increases its sampling frequency ($f_i(t_k)$) to decrease D.
- In addition to rate regulation scheme, immune system based DNRS algorithm provides a kind of combinatorial behavior to DNs such that (i) when the over all packet loss rates increases, all sampling frequencies of DNs (f_i, $\forall i$) decrease more faster to reduce the overall packet losses (ii) when D becomes immensely large, all sampling frequencies of DNs (f_i, $\forall i$) increase more faster to reduce D.

DNRS also enables DNs to collaboratively converge a stable state in which the smaller packet loss rate and event signal reconstruction distortion and significant energy conservation can be provided to WSN by collaboratively regulating sampling frequency of DNs. In Algorithm 3.3, we give a pseudo-code for the rate selection algorithm in DNRS.

Algorithm 3.2 Rate Selection Algorithm in DNRS

foreach t_k **do**
    sink broadcasts D and $\lambda_i, \forall i$ to all DNs **foreach** *DN i* **do**
        broadcasts its sampling frequency $f_i(t_k)$ to all DNs computes $F_i(t_{k+1})$ using (17)
        computes $f_i(t_{k+1})$ using (18)
    end
end

4 Performance Evaluation

In this section, we present the simulation results of DNRS. In order to evaluate the performance of DNRS, we develop a simulation environment using MATLAB. In this environment, 100 sensor nodes in $100 \times 100 m$ sensor field were randomly positioned. We use CSMA/CA based MAC protocol and a routing protocol which finds the shortest path from sensor nodes to the sink. An event was generated with a point source randomly positioned in this environment. The parameters used in our simulation are given in Table 2.

Table 2. Simulation Parameters

Parameter	Value
Area of sensor field	$100 \times 100 m$
Number of sensor nodes (N)	100
Packet length	100 samples/sec.
Queue size of sensor nodes	20 packets
Radio range of a sensor node	20 m
Correlation radius r	3-15 m
Correlation coefficients θ_1	0.3
Correlation coefficients θ_2	0.3

DNRS determines the DN selection threshold t_{dns} at the sink node and the sink node broadcasts t_{dns} to all sensor nodes whereby each sensor node distributively decides whether it is a DN or not. To show the effect of the t_{dns} on the number of DN, we increase the t_{dns} by starting from a certain minimum value. In Fig. 1, the number of DNs is shown for varying t_{dns} with different r. While the t_{dns} increases, the number of DNs whose weights (T_i) can exceed t_{dns} decreases. Therefore, the number of DNs decreases. As r increases, the number of correlated neighbors of DNs increases and the weights of DNs (T_i) decreases. Therefore, as observed in Fig. 1, the number of DNs decreases while r increases.

In WSNs, energy consumption mainly depends on the number of source nodes (DNs) and the total number of samples injected to sensor field by DNs. Since sink can regulate the number of DNs by changing t_{dns}, it can also regulate the average total energy consumption by changing t_{dns}. To investigate the energy consumption of the

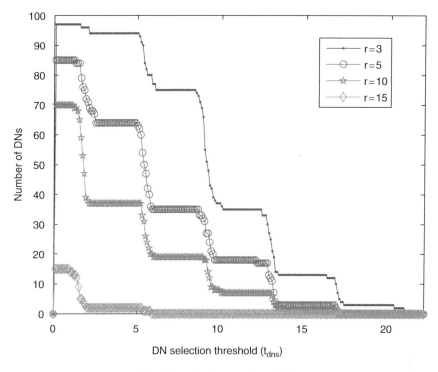

Fig. 1. t_{dns} for the number of DNs

DNRS, we assume that on average each DN consumes $20mW$ cumulatively for sensing the event and communicating the observed information to the sink node during the overall event monitoring period. In Fig. 2, the average total energy consumption is shown for varying t_{dns} with different r. As observed, while t_{dns} increases, the number of DNs decreases and therefore, the average total energy consumption decreases. Similarly, since increase in r results in decreasing the number of DNs, the average total energy consumption decreases as r increases. Hence, maximum r and maximum t_{dns} are observed to be imperative to provide the minimum energy consumption to sensor networks. However, selected r and t_{dns} must also provide the application specific event signal reconstruction reliability to sensor networks. Therefore, t_{dns} and r should be selected, which can provide the application specific reconstruction reliability and minimum energy consumption. In Fig. 3, D is shown for varying t_{dns} with different r. As observed, while t_{dns} increases, D decreases. However, after some certain values of t_{dns}, D starts to increase. This is mainly because the number of DNs starts to be insufficient to deliver the sufficient information needed for the reconstruction. The DN selection algorithm of DNRS uses these points in which D starts to increase to determine the most appropriate value of t_{dns}. Furthermore, since increase in r results in smaller number of DNs, as r increases, D starts to increase at the smaller t_{dns}. This prevents to decrease D even at the smaller t_{dns}.

Apart from the number of DNs, the sampling frequency of DNs affects the reconstruction performance. To show this effect over the reconstruction performance, in Fig. 4 D is shown for varying sampling frequency with different t_{dns}. Common method for increasing the reconstruction accuracy is the over sampling such that increasing sampling frequency, reconstruction accuracy can be improved. As observed in Fig. 4, while sampling frequency of DNs increases, D decreases. However, after the certain values of f, D starts to increase as f increase. Higher values of f increase the contention at the forward path and packet loss rate of DNs (λ_i). Therefore, for higher values of f, the sufficient information for the reconstruction cannot be delivered to the sink and this increases D as f increases. As t_{dns} decreases, the number of DNs and the contention at the forward path increases. This results in increase in λ_i. Therefore, as observed in Fig. 4, while t_{dns} decreases, D increases. Hence, the appropriate rate control mechanism must deal with decreasing the packet loss rate of DNs to provide minimum reconstruction distortion as well as minimum energy consumption.

In addition to the designated node selection, DNRS regulates the sampling frequency of the designated nodes to exploit the temporal correlation in the event data. As discussed in Section 3.3, temporally correlated event data can be inferred from its previous sample, less number of samples delivered to the sink by DNs is sufficient to reliably reconstruct the event signal at the sink. Therefore, DNRS enables the DNs to regulate their sampling frequency to exploit the temporal correlation in the

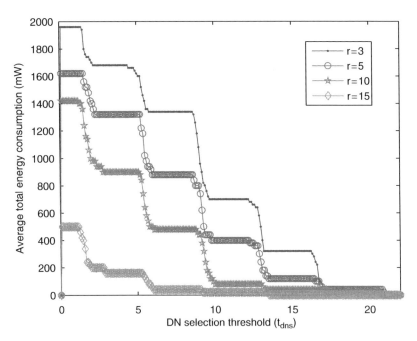

Fig. 2. t_{dns} for the average total energy consumption

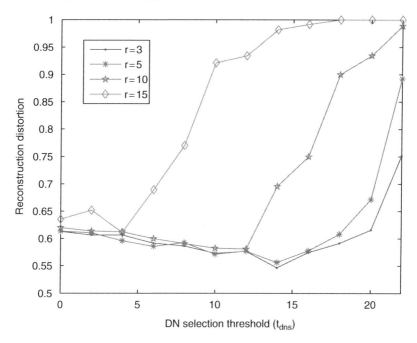

Fig. 3. D for varying t_{dns} with different r

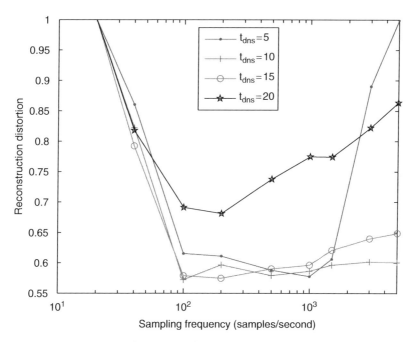

Fig. 4. D for varying f with different t_{dns}

event data such that when the temporally correlated event data is observed, DNRS reduce the sampling frequency of DNs and when the temporal correlation can not be observed, it increases the sampling frequency of DNs. Furthermore, DNRS reduces the sampling frequency of a DN when it has the higher packet loss rate (λ_i).

To evaluate the performance of the distributed rate control mechanism in DNRS we consider a scenario in which DNs sample the event signal and transmit the sampled event signal to sink during 10 $seconds$ time period and each DN updates its sampling frequency in every decision interval ($t = 0.5\ seconds$). We assume that $D_c = 0.85$ and $\lambda_c = 0.2$ and each DN initially sets its sampling frequency as 200 ($f_i(t_0) = 200\ samples/sec.\ \forall i$) and each DN can change its sampling frequency between 100 $samples/sec.$ and 1000 $samples/sec.$ such that 100, 200, 300, ...,1000. According to the scenario, in Fig. 5, D versus time is shown with different t_{dns}. As observed, for all t_{dns}, DNRS provides the reconstruction distortion which is below $D_c = 0.85$ and slightly constant after 2.5 $seconds$. Furthermore, while t_{dns} increases, D decreases. As t_{dns} increases, the contention at the forward path decreases together with the number of DNs. Therefore, for higher t_{dns}, DNRS manages and regulates the sampling frequency of DNs to provide minimum D more easily than the case for smaller t_{dns}, which provides more DNs.

To show the energy efficiency of the distributed frequency regulation mechanism in DNRS, in Fig. 6, using the same simulation parameters in Fig. 5, average total energy consumption versus time is shown with and without DNRS. For the case

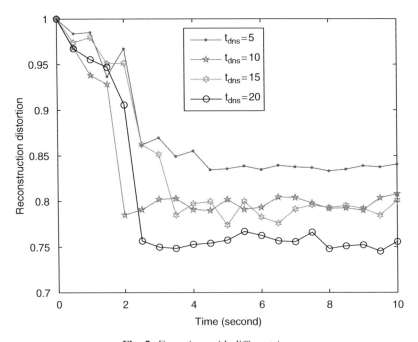

Fig. 5. D vs. time with different t_{dns}

without DNRS, all DNs use the same sampling frequency ($f = 500\ samples/sec.$). We assume that for all case, on average, each DN consumes $0.01mW$ cumulatively to deliver a sample of the event signal to sink node. As observed in Fig. 6, in terms of energy consumption, the case with DNRS completely outperforms the case without DNRS. This is mainly because DNRS can regulate the sampling frequency of DN according the temporal correlation in the event signal and the packet loss rate of DNs.

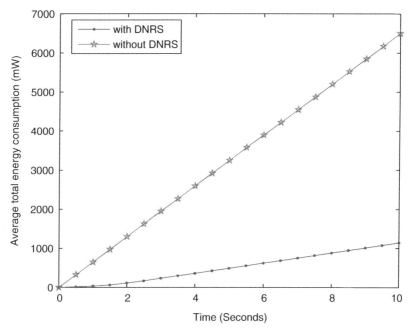

Fig. 6. Average total energy consumption vs. time with and without DNRS

5 Conclusion

In this chapter, Distributed Node and Rate Selection (DNRS) algorithm, which based on the principles of natural Immune System, is introduced. Based on the B-cell stimulation in immune system, DNRS selects the most appropriate sensor nodes that send samples of observed event, are referred to as designated node (DN). Using the method, each sensor node can distributively decide whether it is a designated node or not according to its correlation with its neighbors and event source. In addition to the determination of the DNs, to exploit the temporal correlation in the event data based on the immune models, DNRS selects the appropriate sampling frequencies of sensor nodes according to the congestion at the forward paths and the event signal reconstruction distortion periodically calculated at the sink node by Adaptive LMS

Filter. With the selection of the minimum number of designated nodes and the regulation of the sampling frequency of designated nodes, DNRS provides the significant energy saving to Wireless Sensor Networks (WSN).

References

1. I. F. Akyildiz, W. Su, Y. Sankarasubramaniam, and E. Cayirci, (2002) Wireless Sensor Networks: A Survey, Computer Networks (Elsevier) Journal 38:393–422,
2. J. Kusuma, L. Doherty, K. Ramchandran, (2001) Distributed compression for sensor networks, in: Proceedings of the IEEE Image Processing 1:82–85.
3. H. Gupta, V. Navda, S. R. Das, V. Chowdhary, (2005) Efficient Gathering of Correlated Data in Sensor Networks, in: Proc. ACM MOBIHOC, pp. 402-413, Urbana-Champaign, Illinois, May 2005.
4. P. V. Rickenbach, R. Wattenhofer (2004) Gathering Correlated Data in Sensor Networks, in. Proc. ACM DIALMPOMC'04, Philadelphia, Pennsylvania, USA, October 2004.
5. R. Cristescu, B. Beferull-Lozano, M. Vetterli (2004) On network correlated data gathering, in Proc. IEEE INFOCOM 2004, pp. 2571-2582, March 20.
6. R. Cristescu and M. Vetterli (2005) On the optimal density for real-time data gathering of spatio-temporal processes in sensor networks, in Proc. IPSN 2005, April 2005.
7. D. Ganesan, R. Cristescu, B. Beferull-Lozano (2004) Power-efficient sensor placement and transmission structure for data gathering under distortion constraints, in Proc. IPSN 2004, pp. 142-150, April 2004.
8. M. C. Vuran, O. B. Akan, and I. F. Akyildiz (2004) Spatio-Temporal Correlation: Theory and Applications for Wireless Sensor Networks, Computer Networks Journal (Elsevier) 45, no. 3: 245-261, June 2004.
9. M. C. Vuran, I. F. Akyildiz (2006) Spatial Correlation-based Collaborative Medium Access Control in Wireless Sensor Networks, IEEE/ACM Transactions on Networking, June 2006.
10. M. C. Vuran, O. B. Akan (2006) Spatio-temporal Characteristics of Point and Field Sources in Wireless Sensor Networks, in. Proc. IEEE ICC 2006, Ýstanbul, June 12-15.
11. M. Gastpar, M. Vetterli (2003) Source-Channel Communication in Sensor Networks, in Proc. IPSN'03, Palo Alto, USA, April 2003.
12. S. S. Pradhan, K. Ramchandran (2003) Distributed source coding using syndromes (DISCUS): design and construction, IEEE Transactions on Information Theory 49: 626-643, March 2003.
13. C. Guestrin, P. Bodi, R. Thibau, M. Paski, S. Madde (2004) Distributed regression: an efficient framework for modeling sensor network data, in. Proc. IPSN'04, pp. 26-27, April 2004, Berkeley, California USA.
14. J. Timmis, M. Neal, J. Hunt (2000) An artificial immune system for data analysis, Elsevier, BIOSYSTEMS 2000, pp. 143-150.
15. J. H. Jun, D. W. Lee, K. B. Sim (1999) Realization of Cooperative Strategies and Swarm Behavior in Distributed Autonomous Robotic Systems using Artificial Immune System, in. Proc. IEEE SMC'99, pp. 614-619.
16. J. Timmis, M. Neal, J. Hunt (1999) Data analysis using artificial immune systems, cluster analysis and Kohonen networks: some comparisons in. Proc. IEEE SMC'99 Systems Man and Cybernetics.
17. S. Forrest, A. S. Perelson, L. Allen, R. Cherukuri (1994) Self-Nonself Discrimination in a Computer, in Proc. IEEE Symposium on Security and Privacy.

18. J. D. Farmer, N. H. Packard, A. S. Perelson (1986) The immune system, Adaptation, and Machine Learning, Physica 22D: 187-204, North-Holland, Amsterdam.
19. N.K. Jerne (1984) Idiotypic Network and Other Preconceived Ideas, Immunological Rev 79: 5-24.
20. J.O. Berger, V. de Oliviera, B. Sanso, (2001) Objective bayesian analysis of spatially correlated data, J. Am. Statist. Assoc 96:1361-1374.
21. S. Santini, K. Römer, An Adaptive Strategy for Quality-Based Data Reduction in Wireless Sensor Networks, Institute for Pervasive Computing ETH Zurich.
22. J. Hightower, G. Borriello (2001) Location systems for ubiquitous computing, IEEE Computer, Aug. 2001.
23. B. Atakan, O. B. Akan, (2006) Immune System Based Distributed Node and Rate Selection in Wireless Sensor Networks, in: Proc. IEEE/ACM BIONETICS 2006, Cavalese, Italy, December, 2006.

A Bio-Inspired Architecture for Division of Labour in SANETs

Thomas Halva Labella and Falko Dressler

Autonomic Networking Group
Dept. of Computer Science 7, University of Erlangen-Nuremberg
Martensstr. 3, 91058 Erlangen, Germany
`hlabella@ulb.ac.be, dressler@informatik.uni-erlangen.de`

Summary. Division of labour is one of the possible strategies to efficiently exploit the resources of autonomous systems. It is also a phenomenon often observed in animal systems. We show an architecture that implements division of labour in Sensor/Actuator Networks. The way the nodes take their decisions is inspired by ants' foraging behaviour. The preliminary results show that the architecture and the bio-inspired mechanism successfully induce self-organised division of labour in the network. The experiments were run in simulation. We developed a new type of simulator for this purpose. Key features of our work are cross-layer design and exploitation of inter-node interactions. No explicit negotiation between the agents takes place.

1 Introduction

Sensor/Actuator Networks (SANETs) are interesting research objects. They are hybrid systems consisting of networked sensor nodes (also called *motes*) and mobile robots (wheeled, legged, flying, and so on). Robots and motes (henceforth referred to as *agents*) can communicate together via wireless communication for further cooperation, e.g. to achieve a common goal.

SANETs might be used for fire fighting in forests, or to assist a rescue team after an earthquake or similar natural disasters. Motes are deployed to sense the environment. Their output can then be directed to a base host which elaborates the data and sends out commands to the robots. A more challenging issue is to have robots responding autonomously to the motes' output.

Akyildiz et al. [1] cites as unique features of a SANET: node heterogeneity, realtime requirements, different deployment strategies for motes and robots, mobility and co-ordination paradigm—mote/robots and not only mote/sink as typically in Wireless Sensor Networks (WSNs). Research issues include power management, routing, co-ordination algorithms, design, and many other topics.

This work addresses one common problem that arises in autonomous systems: *division of labour*. Suppose that the agents in a SANET have to perform different tasks. Considering the mentioned rescue example, the tasks would probably be: report the state of the environment (temperature, pictures, sounds, etc.) and help victims. There are obviously a number of interesting applications that might fit this scenario. We focus only on coverage [2]. An autonomous SANET has to decide which and how many agents should perform each task, in such a way that tasks do not overlap in the area. This is what we mean by division of labour. The aim of division of labour is to efficiently exploit the available resources of the network.

Not all authors agree on our definition of division of labour. On the contrary most of them call the problem of deciding which and how many agents should perform a task as *task allocation* (we discussed on this in [3]). We think this is mostly due to different interpretations of the word "task". In most cases, it is intended as an operation associated with a starting and a duration (or terminating time). The task "lives" less than the overall system. In our case, a task is something that "lives" as long as the SANET itself. The examples provided in Sec. 3 make this definition clearer.

Our previous work [3, 4] demonstrated how a simple learning mechanism could be useful to improve the efficiency of a group of autonomous robots. The learning, inspired by ants' foraging, was also able to introduce self-organised division of labour in the group. The discussions and results presented in this paper extend our work in the context of SANETs.

We describe here an architecture for division of labour in SANETs. The architecture is based on probabilistic decisions. During the lifetime of the SANET, the agents adapt the probability to execute one task among a given set. The architecture exploits the interactions between agents, but only within a limited range. The local interactions are however enough to induce division of labour at the global level, i.e. to provide a self-organising behaviour [5]. No particular knowledge of the environment or of the other nodes' activity is required. Additionally, the architecture is based on a cross-layer design. Application and network layers are both responsible for the division of labour.

We could not find any related work about division of labour in SANETs. There are however a few about task allocation, where "task" is meant as a short-living set of actions to be performed. Gerkey et al. [6] proposed a market-based task allocation schema. The agents bid to acquire a task based on the estimation of their capabilities. The authors' auction involves the whole SANET. They are aware that this might be a problem and envisage to solve it by running only localised actions. Clustering methods provide the inherent capability of performing operations in a local context. Younis et al. exploited this behaviour for task allocation in WSNs [7]. Batalin et al. [8] used a greedy policy to allocate tasks. Every task is allocated to the best available agent. Low and co-workers [9] used a bio-inspired task allocation mechanism. It is based on the threshold model [10] and is used to allocate the task of tracking objects in the network. As in the architecture we describe below, their agents adapt during the network lifetime. Jesi et al. [11] also use an adaptive threshold model for the purpose of selecting superpeers in an overlay network. They employ a different formula to adapt the system behaviour. It will be part of future work to compare the different formulae in detail.

We tested our architecture in a proof-of-concept using a simulated environment. There are still some minor things that have to be improved before porting to real hardware. We took particular care in designing a reliable simulator. Our previous experience with similar simulations make us confident in a successful port on real hardware and to more complex environments in the future.

Section 2 briefly describes the simulator that we developed for our experiments. Sections 3 and 4 describe the application layer (for task selection) and the network layer (for message routing) of our architecture. Section 5 shows the results of the experiments, and Section 6 draws our conclusions.

2 Simulation

Good simulation tools can be found for WSNs as well as for robots. The former focus on the simulation of the communication layers and environments, the latter on the physical environment and how the robots perceive it. Something that does both is, to the best of our knowledge, missing.

On the side of network simulations, OMNeT++[1] is becoming more and more popular. OMNeT++ is a modular discrete event simulator. It can be extended by means of additional components. The Mobility Framework (MF) simulates mobile nodes that communicate through wireless communication. The environment in which the nodes move is quite simple. Nodes are simulated as points in a rectangular two-dimensional area. Most of the simulations of wireless networks can be reliable and effective also in such a simple environment.

On the side of robot simulation, a number of libraries have appeared for the simulation of rigid bodies. Open Dynamics Engine (ODE)[2] is one of them. A body is defined by its position, mass, velocity, orientation, and momentum of inertia. More bodies can be connected through different types of joints. Joints constrain the movement of one body w.r.t. another. Each body might also be given a shape. Shapes are used by the library to detect when and where two bodies collide. ODE solves the equations of dynamics to obtain the trajectory of the bodies.

We extended OMNeT++ to include in it the rigid-body simulation provided by ODE. We called the resulting simulator BARAKA. Unfortunately, there is not enough place here to describe BARAKA in details. The interested reader can refer to [12]. Fig. 1 shows the outcome of our hybrid simulator. BARAKA is based on he principles set forth by Jakobi et al. [13]. These principles make more likely a successful port of the algorithms on real hardware.

3 Task Selection

We first describe the application layer, which is in charge of selecting the tasks to perform. Agents have the capability to perceive whether their actions are successful

---
[1] http://www.omnetpp.org/
[2] http://www.ode.org

or not, either by directly sensing the environment, or by receiving a feedback from other nodes (we discuss more on this in Sec. 4.3). Before choosing a new task, the application layer adapts its parameter on the base of the outcome of the previous tasks.

Robots and motes know *a priori* the list of possible tasks that they can perform. They have generally different sets of tasks. In our experiments we used four tasks both for robots and motes, therefore we can indicate the set of tasks as $T_{\text{agent}} = \{T_1, T_2, T_3, T_4\}$. The robots' and motes' four tasks are described below.

Each agent k associates a task to a real number τ_i^k, with $i \in T_{\text{agent}}$. At the moment of selecting a task to perform, the agent chooses randomly between the tasks. The probability to choose task i is

$$P(i) = \frac{(\tau_i^k)^{\beta_{\text{task}}}}{\sum_{k \in T_{\text{agent}}} (\tau_k^k)^{\beta_{\text{task}}}}, \quad (1)$$

with $\beta_{\text{task}} \geq 1$ (in the experiments of Sec. 5, $\beta_{\text{task}} = 3$). This equation is like the one used to model how ants choose one path among the several that bring to a food source [14]. In the original formulation, τ_i^k was the concentration of pheromone on path i. The parameter β was introduced to increase the exploitation of good paths.

The agent initialises $\tau_i^k = \tau_{\text{init}}$, $\forall i \in T_{\text{agent}}$. If the agent is successful in performing task i, then

$$\tau_i^k = min\{\tau_{\text{max}}, \tau_i^k + \Delta\tau\} \quad (2)$$

and

$$\tau_i^k = max\{\tau_{\text{min}}, \tau_i^k - \Delta\tau\} \quad (3)$$

if it is unsuccessful.

The agents in our experiments have four tasks. For three of them, the behaviours of robots and sensors are the same. We suppose the agents are used to sense the environment and to report to a base host. The tasks are:

(i.) measure the temperature locally and send it to the base;
(ii.) record the sound in the surroundings and send it to the base;
(iii.) record a video of the place and send it to the base.

Task T_1 ends immediately because it creates and sends only one small packet. Tasks T_2 and T_3 occupy the node for more time, because they generate a stream of packets, which is usually a big load for the network. T_2 differs from T_3 because it generates less packets than the latter. T_1, T_2 and T_3 are thus the sensing task that both motes and robots have to perform. They are successful if the packets are not rejected by a node on the route to the host (Sec. 4.3).

Motes' and robots' behaviours are different in the case of the fourth task:

(iv.) motes broadcast help requests, robots answer to them and travel where they are needed.

In a general application, motes might decide that they need help by analysing their data, or after instructions from the base. In our experiments, motes' help needs are modelled with a stochastic process. Robots listen to incoming help requests and answer to them by travelling where they are required. Robots and motes co-ordinate their actions through a well defined protocol. In the context of this paper, it is sufficient to say that motes consider a failure if they do not need to send any help request, a success otherwise (a mote's success does not depend on robots' action). A robot considers the task successful if it can reach the requesting mote, a failure if there are no pending requests or if it can not reach its destination.

Assuming that the robots can know the direction of a nearby mote[3], the robots can use the route discovery capability of the network layer to find a path to their destination. In WSNs, the topology of the network usually corresponds to the topology of the environment. Robots do not need a map of the environment because the network topology can be seen as a simple map.

4 Networking

The literature proposes several routing protocols for ad hoc networks with mobile nodes [15]. We focused our attention on *AntHocNet* [16] which is a self-organising routing algorithm, inspired by the food searching behaviour of ants. There are two main reasons why we chose this algorithm: firstly, it is inspired by ants' behaviour and perfectly fits in the context of our work; secondly, and more important, Di Caro et al.[16] showed that it performs better than AODV [17], which is the common reference point for *ad hoc* networks. We modified the original algorithm to fit into our work. We describe below the algorithm used for our experiments. We are going to explicit the differences w.r.t. the original algorithm during our exposition.

Each message coming from the application layer belongs to different classes, one for each type of task. The network layer of node i keeps routing information in a number of tables $_c\mathbf{R}^i$ for each class c. There is an additional class that is used for the messages originated by the network layer, like delivery error notifications or route discovery responses.

Each entry $_c\mathbf{R}^i_{nd}$ is a tuple $\langle t, h, e, s, m, v \rangle$[4] that records some statistics about the path from node i to node d using node n as next hop for a message belonging to class c. The entry t is the estimation of the transmission time, h is the number of hops in the route, e is the energy required for transmission, s is the minimal signal-to-noise ratio of all the links in the path, $m \in \{\text{true}, \text{false}\}$ denotes if the next hop n is mobile, and $v \in \{\text{true}, \text{false}\}$ whether it is still valid for routing. $_c\mathbf{R}^i_{nd} \in \mathbf{R} = \mathbb{R}^3 \times \mathbb{N} \times \{\text{true}, \text{false}\}^2$.

---

[3] It is not the purpose of our work to address this problems, but it might be done, e.g., by triangulating the signal emitted by a node, by using directional antennas, or by means of a vision system.

[4] In *AntHocNet*, there is only one routing table, because it does not differentiate messages into classes. Each entry is just one real number: the estimation of delivery time for a message.

Storing several statistics allows the node to use different routing strategies. Every node might have different objectives to maximise. It might choose, for instance, the route with higher minimal signal-to-noise ratio, increasing the reliability of the message delivery. The node might also choose to use different criterion in different moments. If it detects important information to be sent, it might decide to use a route that minimise the end-to-end delay. If the node has low power level, it might decide to send the message to a near node to reduce transmission power.[5]

When the application layer wants to send a message to a host for which there is no known path, the network layer starts a route discovery process. When one or more routes have been found, the network layer routes the data on the newly discovered paths. The processes of route discovery and routing are done independently for each message class. If a route is known for a class c_1 but the network layer receives a message of class c_2, a new route discovery takes place. This was introduced to address problems, like anycast communication and spatial load balancing of the routes, that we do not address in this paper.

To describe the routing algorithm, we need to describe the route discovery process (Sec. 4.1) and the route selection (Sec. 4.2) separately. We then describe an additional feature of our system in Sec. 4.3: a packet filtering mechanism. Sec. 4.4 describes the additional packet types used by the network layer.

4.1 Route Discovery

If a node wants to send packets to a destination for which it does not know any route, it broadcasts a ROUTEDISCOVERY packet, containing the address of the desired destination d. The request is treated by the other nodes as a normal data packet.

The packet is transmitted to the node's neighbours. They might know a route that reaches d, or not. If a neighbour knows how to reach the destination, it randomly chooses a next hop n to relay the request. The node uses a function $r : \mathbf{R} \rightarrow \mathbb{R}^+$ to transform the statistics about the route to d in a positive real value.[6] The probability $\mathcal{P}_{nd}^i$ at node i to choose n as next hop to reach d is:

$$\mathcal{P}_{nd}^i = \frac{r(_c\mathbf{R}_{nd}^i)^{\beta_{\mathrm{disc}}}}{\sum_{j \in N_d^i} r(_c\mathbf{R}_{jd}^i)^{\beta_{\mathrm{disc}}}} . \qquad (4)$$

where N_d^i is the set of neighbours for which a path to d is known. β_{disc} is a parameter that can control the exploratory behaviour of the algorithm, in the same fashion as for task selection. For route discovering, it is however set to 1.

---

[5] The distance of a node can be estimated by the receiving power of the message, as measured by the antenna, that is, by the physical layer. It should be noted however that it is not really important in this case to know the real distance, but the power required for the transmission.
[6] *AntHocNet* does not need the function $r(\,\cdot\,)$ since the routing table contains already positive real numbers.

The actual $r(\cdot)$ used in our experiments is:

$$r_{\text{time}}(_c\mathbf{R}_{nd}^i) = \begin{cases} \frac{1}{t} & \text{if } t \neq 0, \\ H & \text{if } t = 0. \end{cases}$$

$$r(_c\mathbf{R}_{nd}^i) = \begin{cases} 0 & \text{if } n \text{ is not a valid hop}, \\ H & \text{if } n \text{ is a valid hop and } n = d, \\ \frac{r_{\text{time}}(_c\mathbf{R}_{nd}^i)}{2} & \text{if } n \text{ is a valid hop}, n \neq d \\ & \text{and } n \text{ is a mobile hop}, \\ r_{\text{time}}(_c\mathbf{R}_{nd}^i) & \text{otherwise.} \end{cases}$$

where H is a high-value constant, and t is the estimated transmission time. As it can be seen, this function increases the probability to route packets through nodes which are not mobile, i.e., the sensors. This is because a link to a mobile node is likely to break soon, while the sensors can form a sort of stable backbone.

If a neighbour does not know anything about d, it broadcasts the incoming request again. Due to broadcasting, the discovery messages can proliferate quickly and follow different paths in the network.

The ROUTEDISCOVERY stores the path travelled so far. If a node receives several requests originating from the same node, it compares the path of each packet with the shortest known path, the distance being measured in number of hops. We apply the same filters as [16]: only packets that have not travelled over very bad paths are let through.

We assume that the paths are symmetric: if node A can directly communicate with node B, than node B can directly communicate with node A. On arrival to destination, the receiving node generates a ROUTEDISCOVERYRESPONSE packet. The ROUTEDISCOVERYRESPONSE is sent back with high priority along the same path of the incoming packet. During its travel back, it collects the statistics about the path that will be used to update $\mathcal{R}_{nd}^i$. The statistics are not collected and stored only in the ROUTEDISCOVERYRESPONSE packets, but in all the packets that the network layer receives. In this way, other nodes can passively set up routes to other hosts when they receive a message from it.[7]

At arrival of a new packet with data about the route to d through n, the network layer updates the routing table using a custom function $\oplus : \mathbf{R}^2 \to \mathbf{R}$. If $\mathbf{r}_{nd}^i$ is the information obtained from the incoming packet,

$$_c\mathbf{R}_{nd}^i =_c \mathbf{R}_{nd}^i \oplus \mathbf{r}_{nd}^i$$

performs a weighted sum of the real values t, h, e, s and sets m to the new value. All occurs only if both the previous information and the new information are valid.[8]

---

[7] In *AntHocNet* only the returning packets set up the routes. *AntHocNet* keeps the routing table up-to-date by means of a proactive strategy. While communicating with the destination, *AntHocNet* generates new ROUTEDISCOVERY packets to find new routes. Our solution avoids this, at the cost of slightly bigger network packets.

[8] In our experiments, it does happen that sensors become mobile, and thus changing m might seem useless. It could happen in other applications that some robot decides to "become" a

Once a node starts the route discovering process, it waits for a response and buffers the data to send. If it did not receive any response after some time (5 s), it starts the discovering process again. The node repeats the process for a maximum number of times (5) before giving up. Once a route has been found, it is kept in the routing table for 120 s and then removed.

If a node does not find a route to the destination, it takes one of two actions according to who originated the message: if the message came from the application layer, the network layer notifies it about the failure; if the message came from someone else, it sends a ROUTEFAILURE packet to the origin.

4.2 Routing

Once one or more routes have been found, the network layer starts sending the data. The node chooses randomly the next hop for each packet. The probability for each hop is calculated with (4), only using another exponent, β_{route}, higher than β_{disc} (in our experiments, $\beta_{\text{route}} = 2$). The higher exponent results in a greedier behaviour w.r.t. good paths. The probabilistic data routing leads to data load spreading, relieving congested paths.

4.3 Packet Filtering

Most of the packets travelling though the SANET are likely to contain data coming from sensor readings. In order to reduce the congestion of the network, a node might decide, instead of routing a packet, to drop it, for instance, if it contains redundant information. The rejection of others' messages plays an important role in the division of labour. It is the source of the competition that is required by the agents to specialise. If an agent can successfully send a message associated to a task, it increases the probability to perform the task again. If it is not successful, someone has taken over it, and thus it decreases its probability to perform the task again. We already showed [3] that similar interactions play a key role in division of labour.

Agents do not need to signal to their neighbours the task they are currently executing, because their output is likely to be read anyway by nearby agents. This is why it is reasonable to directly use this 'free' source of information as base mechanism for agents' differentiation.

Each node remembers the last packet that it received of a given class and for a given destination.[9] Upon arrival of a new packet to route, the network layer compares

---

sensor, that is, not to move. The robot could decide it because the network in that point is particularly under load, or because there was a failure of one sensor and the network lost connectivity.

[9] This implementation might require much memory and might not well scale. Other solutions can be used on devices with limited memory. We could use a limited array for recording the last messages. If the array is full and a new message should be stored, then the oldest element can be deleted. This system would have the effect only to weaken the interaction between nodes. This effect could be compensated by some other means, like increasing ΔQ—see later.

it with the one previously stored. It then randomly decides whether the packet should be routed or not. In case the node decides to route the packet, the node increases the probability to route again packets of the same class and decreases its τ_i^k related to the packet class using (3). If the node rejects the packet, it decreases the probability to route packets of the same class, sends a ROUTEFAILURE message to the source and its application layers increases τ_i^k using (2).

Packet filtering does not take place in the following cases:

(i.) the packet is broadcast (it might be some important message to spread in the network);
(ii.) the packet is not the first packet of a stream generated by T_2 or T_3 (streams are interrupted at the beginning, but not when the connection with the destination has already been established);
(iii.) it is a packet belonging to T_4 (this task requires a strict co-ordination between robots and motes, thus it should not be interrupted);
(iv.) the packet comes from a source further than a given hop-count threshold D (packets from near sources have correlated content, and thus they can be dropped without losing much information).

Each node i keeps a table $_c\mathcal{Q}^i$ of values $_c\mathcal{Q}_d^i \in [\mathcal{Q}_{\min}, \mathcal{Q}_{\max}]$ for known destinations d and packet class c. The probability $_c\mathcal{P}_d^i$ to route a packet is

$$_c\mathcal{P}_d^i = \begin{cases} _c\mathcal{Q}_d^i & \text{if this is the first packet of class } c \text{ to } d \text{ seen, or} \\ \alpha_1\alpha_2(_c\mathcal{Q}_d^i - \mathcal{Q}_{\min}) + \mathcal{Q}_{\min} & \text{otherwise.} \end{cases},$$

$$\alpha_1 = \left(1 - e^{-\gamma_1 h}\right), \qquad \alpha_2 = \left(1 - e^{-\gamma_2 \Delta t}\right),$$

where h is the number of hops travelled by the incoming packet, Δt is the elapsed time from the previous known message and the incoming packet, $0 < \mathcal{Q}_{\min} < \mathcal{Q}_{\max} \leq 1$, $\gamma_2 > 0$ (in our experiments, $\gamma_2 = 0.01 \text{ s}^{-1}$). The coefficients α_1 and α_2 decrease the probability to route a packet for near sources and for information recently transmitted. We want that the effect of α_1 smoothly decreases for $h \to D$. Because $\alpha_1 \approx 1$ when the exponent $\gamma_1 h \approx 5$, we set $\gamma_1 = 5/D$.

Every $_c\mathcal{Q}_d^i$ is first initialised to $\mathcal{Q}_{\text{init}}$. If a node decides to reject a packet, it updates $_c\mathcal{Q}_d^i$ using

$$_c\mathcal{Q}_d^i = max\{_c\mathcal{Q}_d^i - \Delta\mathcal{Q}, \mathcal{Q}_{\min}\},$$

and uses

$$_c\mathcal{Q}_d^i = min\{_c\mathcal{Q}_d^i + \Delta\mathcal{Q}, \mathcal{Q}_{\max}\},$$

if it decides to route the packet.

The threshold D grants that the interactions between agents are localised. Each agent looks only at neighbours no more than D hops away and adapts accordingly. Therefore, agents decide only based on local information, but their decisions have an effect on the whole system, as we show in Sec. 5.

4.4 Other Packet Types

For the correct functioning of the network layer, the nodes need some other information to be exchanged. In addition to ROUTEDISCOVERY, ROUTEDISCOVERYRESPONSE and ROUTEFAILURE (used for route discovery), the network layer generates DATA packets. They contain the information coming from the application layer. The information is routed as explained in Sec. 4.2 and using the routing table corresponding to the message class.

The network layer regularly broadcasts to all its neighbours a HELLOMESSAGE. After a node receives a HELLOMESSAGE, it expects to receive it regularly. If it does not occur within a given amount of time, the neighbour and all its associated routes are deleted from the routing tables.

The HELLOMESSAGE rates are different for robots and motes.[10] The robots use a higher sending rate (30 s) than the motes (120 s). The nodes use also two different timeouts in case the node is mobile (180 s) or not (600 s). The information about the mobility of the nodes is contained in the HELLOMESSAGE.

5 Experiments

We simulated SANETs in a squared area, whose side is 500 m. Twenty-five motes were placed in a grid that covers the environment. This is a likely placement if the motes are dropped from above and if the purpose is to uniformly cover the environment. To simulate however a real deployment process, the motes were randomly placed in an area 5 m around the actual grid point. Robots were placed at the corners of the environment. We tested 1, 2 and 3 robots at each corner, for a total group sizes of 4, 8 and 12 robots. Figure 1 shows the set-up, from the point of view both of the network and of the simulated three-dimensional world.

The remaining parameters of our architecture were set as follows: $Q_{min} = 0.01$, $Q_{max} = 1.0$, $Q_{init} = 1.0$, $\Delta Q = 0.02$, $D = 2$, $\tau_{min} = 0.1$, $\tau_{max} = 10$, $\Delta \tau = 0.5$ and $\tau_{init} = 3.0$. The sound stream size was 40 kB and the video stream size 400 kB. The choice of the values was based on empirical experience.

The base host is at the top left corner of Fig. 1(a). The resulting set-up is symmetrical along one of the diagonals.

Robots and motes did not know the address of the base. The application layer of the base broadcast regularly a packet with its address. The agents started working only after the arrival of this message. Given the importance of this information, it was replicated and broadcast also in each node's HELLOMESSAGE. Apart from broadcasting its address, the application layer of the base only received and recorded the messages it had received. The network layer of the base was the same as the other nodes.

---

[10] *AntHocNet* uses one rate for all the nodes. It should be noted however than in the networks studied in [16] the nodes are all homogeneous. The different rates are then a natural extension to face heterogeneous nodes.

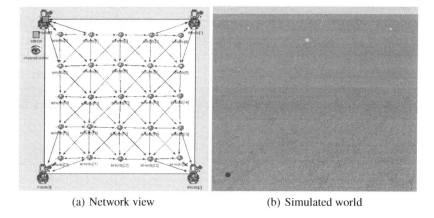

(a) Network view (b) Simulated world

Fig. 1. Set-up of the experiments: views of the network and of the simulated three-dimensional world. In the bottom view, the real dimension of the motes, usually few centimetres, were increased to make them visible. For comparison, the robot in the bottom left corner in the simulated world view is 42 cm tall

The MAC and physical layers simulated the IEEE 802.11 protocol. We used the modules included in the MF.

The help requests were generated by a stochastic process. Each mote decided every second whether it required help or not. We call *help density* the probability per second that a mote requires help. In this set of experiments, help densities were constant during each run. Each newly generated help request was put in a queue in the mote. When motes executed T_4, they checked if the queue was empty or not. If it was empty, they considered it a failure and adapted $\tau_{T_4}^k$ using (3). If there were pending requests, they broadcast a HELPREQUEST packet and adapted $\tau_{T_4}^k$ using (2). We used help densities $12.5 \ 10^{-6}\text{s}^{-1}$, $25 \ 10^{-6}\text{s}^{-1}$ and $50 \ 10^{-6}\text{s}^{-1}$. Each combination of robot group size/help density was tested in forty different runs. Each run was described by a seed for the random number generator and the misplacement of the motes. Another random number generator, initialised with a different seed, was used to generate the help-request events during the simulation.

5.1 Evaluation

The intuitive way of measuring the division of labour is to count how many nodes are involved in a task. This is the method we followed for instance in [3]. This method has one problem: it does not consider the spatial positions of the nodes. Two neighbouring nodes performing the same task would be treated in the same way as two nodes far away. If the task were measuring the temperature, the neighbouring nodes would be doing redundant work.

To account also for the spatial distribution of the tasks, we used the *hierarchic social entropy* [18]. Each agent was represented as a point in a 5-dimensional space.

The co-ordinates of each point were given by the (x, y) co-ordinates of the node in the arena (normalised by the arena side), and by the probability of performing T_1, T_2 and T_3. There is no point to consider also the probability to perform T_4 because it is dependent on the other three tasks.

Then, we fixed the value of a parameter d, and we clustered all the points that were no further away than d in the 5-dimensional space. The number of clusters depends on d, thus we denote it with $C(d)$. The total number of agents is A. A cluster i contains $I(i,d)$ agents. Picking up an agent randomly, it has probability $p_i = I(i,d)/A$ of belonging to cluster i. The entropy of the system [18] can be measured by:

$$H(d) = -K \sum_{i=1}^{C(d)} p_i \log_2 p_i ,$$

where K is an arbitrary constant, which we set to 1. The hierarchic social entropy is defined as

$$S = \int_0^\infty H(d)\, dd . \qquad (5)$$

Note that, thanks to the integral, S is not dependent on d, but only on the positions of the points in the 5-dimensional space. Note also that $\exists M : \forall d > M, C(d) = 1$ and $H(d) = 0$. Balch[18] shows that S increases when the system becomes more and more heterogeneous and the agents differentiate themselves. In our case, S increases either if two neighbours have different probabilities of performing the same task, or if the nodes have the same probabilities but are far away.

5.2 Results

There might several interesting data to show about our simulations, like delivery rate, end-to-end delays, and so on. Unfortunately we have no space for all of them. We focus then only on those measurements that are strictly related with the division of labour.

At the beginning of the experiments, the agents of the SANET had uniform probability of performing each task. The hierarchic social entropy was thus at its minimum since the only source of differentiation was given by the positions in the arena of the nodes.

Figure 2 shows how the hierarchic social entropy developed through time during the experiments. As a reference, we report the median value of the distribution of hierarchic social entropy at 0 s (bottom segments in Fig. 2) and the median of estimated upper bounds (upper segments). The hierarchic social entropy at 0 s was not constant because of the stochastic placement of the motes, thus the median value represents a sort of approximated lower bound. The estimated upper bounds were calculated by keeping fixed the motes, and finding the value of robots' positions and agents' task probabilities that maximised S. S is however complex to maximise, so we searched for the local optimum reached from random initialisation of agents'

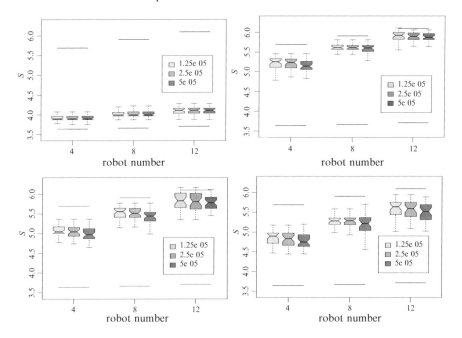

Fig. 2. Evolution of hierarchic social entropy in the SANET during the experiments. The first plot refers to 120 s after the beginning of the experiments (top left plot), the others to 1200 s (top right), 2400 s (bottom left) and 3600 s (bottom right). The three groups of bars in each plot correspond to a group size (on the x axis). Each bar in a triplet refers to a help density (as shown in the legend) and summarises the results of 40 runs. The notches of the bars correspond to the median results. The bars extend to the first to the third quartile of the distribution of results. The whiskers extend till the points that are no more than 1.5 times the inter-quartile distance. The horizontal segments show the approximated upper and lower bounds of the hierarchic social entropy. See the text for discussion

probabilities and robots' positions. We repeated the process for each motes' initial position. The segments in Fig. 2 are the medians of the local optima we found.

The results in Fig. 2 show that the agents in the SANET specialised themselves. The level of S increased after the beginning of the experiments, meaning that they agents tended either to be near and perform different tasks, or to perform the same tasks but far one from the other. The level of differentiation reaches the maximum value around 1200 s, keeps it till 2400 s and then slowly decreases.

The hierarchic social entropy can increase because the agents are more apart in the environment, because they have different task probabilities, or a combination of both. Given that most of the agents are motes, and that the robots start from outside and move inward the network, the physical distance component can only be reduced. Therefore the hierarchic social entropy increases because of a more heterogeneous distribution of probabilities. We can see that the agents indeed differentiate their activities by looking at the distribution among the agents of the probabilities

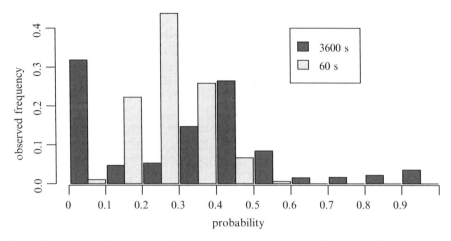

Fig. 3. Distribution of the probabilities to perform task T_2 among the agents (12 robots, help density $50\ 10^{-6}s^{-1}$). On the x-axis there are ten intervals of probability. On the y-axis, we report the fraction of agents that have been observed having probability to perform T_2 in the corresponding x range at two snapshots of the system (see the legend)

to perform each task. They look for every robot group size, help density and for tasks T_1, T_2 and T_3 like Fig. 3.[11]

Shortly after the beginning (60 s), there is only one peak. Agents' probabilities were all initialised to the same value, and are now spreading. At 3600 s, three peaks appear. About 30% of the agents had low probability of performing a given task. Less than 30% had probability between 0.4 and 0.5. Finally, there is a small peak also for probability between 0.9 and 1. We are granted that two agents with same probability of performing a task were not likely to be neighbours by the high values reached by S (Fig. 2).

The situation was different for task T_4. The distributions of the probability to perform T_4 looked like Fig. 4. The left peak grew for decreasing help density. The high number of agents that did not perform task T_4 is explained by the relative low help density. The distribution expanded to the right for fixed help density and increasing robot number. This is simply due to the fact that the higher the number of robots, the higher is the probability that a robot answers to a help request. Thus, the higher is the number of successful ends of this task, and the higher the probability that the same robot repeats the task.

The development of the distribution of performing a task is summarised in Fig. 5. We reproduce only the result for twelve robots, but those for four and eight are similar. What struck our attention is that, while the hierarchic social entropy does not change between 1200 s and 2400 s, the distributions still do. There are new peaks arising (corresponding to the stripes for $P \in (0.4, 0.5]$ that become darker and darker in in Fig. 5) even after 1500 s. If more and more agents get similar probability of

---

[11] Due to the limited space, we do not show all the plots.

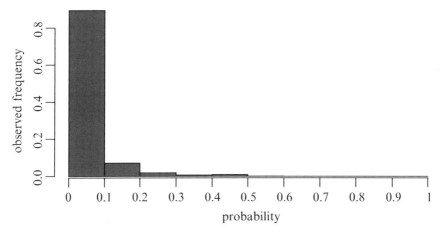

Fig. 4. Distribution of the probabilities to perform task T_4 among the agents (12 robots, help density $50\ 10^{-6} s^{-1}$)

performing a task, they become more "alike" and thus the hierarchic social entropy should decrease. This does not occur only in the case the nodes become more "alike" in the task probability space but they are further away in the physical space. That is, if two nodes have the same high probability of performing a task, they are far one from the other in order to keep the same distance in the clustering space and maintain the same hierarchic social entropy.

It is also possible to note that more and more agents set their probability of performing a task next to the central peak value. This is shown by the stripes that become darker and darker beside the main one ($P \in (0.3, 0.4]$ and $P \in (0.5, 0.6]$), starting at about 2300 s. We think that this phenomenon can explain the slow decrease of the hierarchic social entropy at the end of the experiments. At 2300 s in fact, a consistent number of agents has probability in $(0.4 \in 0.5]$, but they are relatively far from each other. When other agents get probability in $(0.3, 0.4]$ and $(0.5, 0.6]$ (and become similar in the task space), they fill the positions between the previous agents. They are therefore also physically nearer, and this is why the hierarchic social entropy decreases.

In Fig. 6, we show a typical snapshot of the distribution of tasks in the SANET. The plot refers to T_3. It can be seen, that when a node had high probability of performing T_3, its neighbours were likely to have a low one. The routes that were used to send the data to the base host are depicted in Fig. 7. The network was split in two halves: there were few links between the top right triangle and the bottom left triangle. Figures 6 and 7 do not represent the steady state of the SANET. The network reached a dynamic equilibrium, where things continually changed. This is especially true for the routes in Fig. 7, since the routing tables entries were removed after a while, and new discoveries took place.

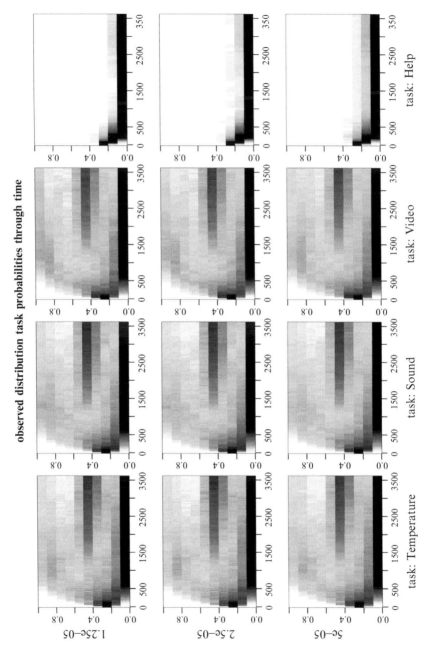

Fig. 5. Dynamics of the distribution of task probabilities (twelve robots). Each plot in the array refers to a combination of tasks (one per column) and help densities (one per row). The task probabilities are on the y axes, the time from the beginning of the experiments on the x axes. The darkness of a cell in position (t, p) is proportional to the number of agents with probability of executing the task equal to p after t seconds from the beginning of the experiments

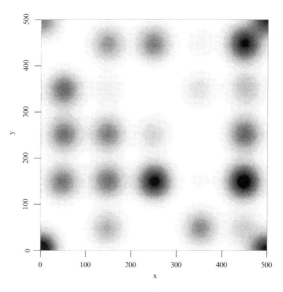

Fig. 6. Distribution of task T_3 among the agents. Each agent is represented by a fuzzy circle. The picture refers to a run with 8 robots, help density $25\ 10^{-6}\mathrm{s}^{-1}$ and depicts the state at 2490 s after the beginning. The darker the circle, the higher is the probability that an agent performs T_3

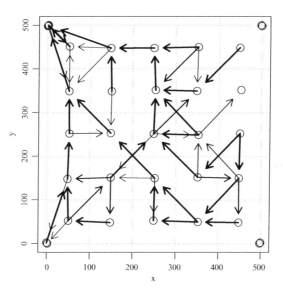

Fig. 7. Routes to deliver the output of T_3 to the base host (in upper left corner). The situation is the same as in Fig. 6. The arrows show for every node the known next hops. The thickness of the arrows is proportional to the probability of choosing a node as next hop

6 Conclusions

This paper illustrated an architecture for division of labour in SANETs. The agents make use of solutions inspired by ants' behaviour. The control architecture is based on strong inter-layer and inter-agent interactions. The latter are local, meaning that they occur only between agents within a given range, smaller than the experimental area. The local interactions are however enough to induce self-organised division of labour in the SANET.

The degree of division of labour can be still improved. It could be argued that it is possible to achieve the maximum degree of division by pre-programming the nodes to do only one task and to spread them alternatively. This solution however introduces other problems that do not touch our solution. Firstly, the deployment process becomes more complex, since the right nodes must be placed in the right place to cover the environment. Secondly, the SANET can not adapt to changes in the environment. Nodes require to be reprogrammed for a new situation. Reprogramming might be driven by the base host, which sends the new programs to the nodes. The drawback is that the network has to sustain a bigger load. Another solution would be to use the robots to go and reprogramme individual nodes. The drawback here is that a number of resources, the mobile robots, are taken away from other tasks only for the maintenance of the network. Algorithms similar to those used by our SANET have shown to adapt well to changes in the environment [16, 3, 14]. Thirdly, one needs to know in advance the characteristics of the environment and the number of motes and robots to find the optimal division of labour. This *a priori* knowledge might not be correct or difficult to retrieve.[12] This knowledge is not required by our architecture.

Future work will test other adaptation rules both at the application layer and at the network layer in order to improve the division of labour. We will also test how our system behaves under dynamic environments. To improve the response time to changes in the environment, we think it will be important to base the packet filtering mechanism also on the content of the messages. Suppose that two neighbouring nodes are measuring the temperature of the environment, but they measure two very different values. It is important that both values are reported, since it might mean that a fire broke out.

Acknowledgements

This work was funded by the Deutscher Akademischer Austausch Dienst (DAAD), grant number 331 4 03 003.

---

[12] Additionally, if we already knew the environment, it would make little sense to use a WSN to sense it.

References

1. Akyildiz, I., Kasimoglu, I.: Wireless sensor and actor networks: research challenges. Ad Hoc Networks **2** (2004) 351–367
2. Cardei, M., Wu, J.: Coverage in Wireless Sensor Networks. In Ilyas, M., ed.: Handbook of Sensor Networks. CRC Press, West Palm Beach, FL (2004)
3. Labella, T., Dorigo, M., Deneubourg, J.L.: Division of labor in a group of robots inspired by ants' foraging behavior. ACM Transactions on Autonomous and Adaptive Systems **1** (2006) 4–25
4. Labella, T., Dorigo, M., Deneubourg, J.L.: Efficiency and task allocation in prey retrieval. In Ijspeert, A., Murata, M., Wakamiya, N., eds.: Biologically Inspired Approaches to Advanced Information Technology: First International Workshop, BioADIT 2004. Volume 3141 of Lecture Notes in Computer Science., Springer Verlag, Heidelberg, Germany (2004)
5. Dressler, F.: Self-Organization in Ad Hoc Networks: Overview and Classification. Technical Report 02/06, University of Erlangen, Dept. of Computer Science 7 (2006)
6. Gerkey, B., Matarić, M.: A market-based formulation of sensor-actuator network coordination. In Sukhatme, G., Balch, T., eds.: Proceedings fo the AAAI Sping Symposium on Intelligent Embedded and Distributed Systems, AAAI Press, San Jose, CA (2002) 21–26
7. Younis, M., Akkaya, K., Kunjithapatham, A.: Optimization of Task Allocation in a Cluster-Based Sensor Network. In: Proceedings of the Eighth IEEE Symposium on Computers and Communications (ISCC 2003), IEEE Computer Society, Los Alamitos, CA (2003) 329–334
8. Batalin, M., Sukhatme, G.: Using a sensor network for distributed multi-robot task allocation. In: Proceedings of the IEEE International Conference on Robotics and Automation (ICRA2004). Volume 1., IEEE Press, New York, NY (2004) 158–164
9. Low, K., Leow, W., Ang, M.: Autonomic mobile sensor network with self-coordinated task allocation and execution. IEEE Transactions on Systems, Man and Cybernetics, Part C **36** (2006) 315–327
10. Bonabeau, E., Theraulaz, G., Deneubourg, J.L.: Quantitative study of the fixed threshold model for the regulation of division of labor in insect societies. Proceedings of the Royal Society of London, Series B-Biological Sciences **263** (1996) 1565–1569
11. Jesi, G., Montresor, A., Babaoglu, O.: Proximity-aware superpeer overlay technologies. In Keller, A., Martin-Flatin, J.P., eds.: Proceedings of SelfMan'06. Volume 3996 of Lecture Notes in Computer Science., Springer Verlag, Heidelberg, Germany (2006) 43–57
12. Labella, T., Dietrich, I., Dressler, F.: BARAKA: A hybrid simulator of sensor/actuator networks. In: Proceedings of the Second IEEE/Create-Net/ICST International Conference on COMmunication System softWAre and MiddlewaRE (COMSWARE 2007), Bangalore, India (2007)
13. Jakobi, N., Husbands, P., Harvey, I.: Noise and the reality gap: the use of simulation in evolutionary robotics. In Moran, F., Moreno, A., Merelo, J., Chacon, P., eds.: Advances in Artificial Life: Proceedings of the Third European Conference on Artificial Life. Volume 929 of Lecture Notes in Computer Science., Springer Verlag, Heidelberg, Germany (1995) 704–720
14. Bonabeau, E., Dorigo, M., Theraulaz, G.: Swarm Intelligence: From Natural to Artificial Systems. Oxford University Press, New York (1999)
15. Abolhasan, M., Wysocki, T., Dutkiewicz, E.: A review of routing protocols for mobile ad hoc networks. Ad Hoc Networks **2** (2004) 1–22

16. Di Caro, G., Ducatelle, F., Gambardella, L.: AntHocNet: An adaptive nature-inspired algorithm for routing in mobile ad hoc networks. European Transactions on Telecommunications, Special Issue on Self-organization in Mobile Networking **16** (2005) 443–455
17. Perkins, C., Royer, E.: Ad hoc On-Demand Distance Vector Routing. In: 2nd IEEE Workshop on Mobile Computing Systems and Applications, IEEE Computer Society, Los Alamitos, CA (1999) 90–100
18. Balch, T.: Hierarchic social entropy: An information theoretic measure of robot group diversity. Autonomous Robots **8** (2000) 209–238

A Pragmatic Model of Attention and Anticipation for Active Sensor Systems

Sorin M. Iacob[1], Johan de Heer[2] and Alfons H. Salden[3]

[1] Telematica Instituut, PO Box 589, 7500 AN, The Netherlands,
sorin.iacob@telin.nl;
[2] T-Xchange, Twente University,
PO Box 217, 7500 AE, Enschede, The Netherlands,
johan.deHeer@txchange.nl;
[3] Almende, Westerstraat 50, 3016 DJ Rotterdam, The Netherlands,
alfons@almende.com

Abstract. Mammals have acquired highly efficient mechanisms for information processing that facilitate quick and adequate responses to environmental changes, and high robustness against distracting signals. Selection of attention is such a fundamental mechanism that leads to a prioritization of sensor information analysis and actuator information synthesis tasks. Different models for selection of attention try to explain this mechanism through the integration of, or interaction between top-down and bottom-up signals and processes. How exactly this integration or interaction occurs in mammals' brain is still an open problem. We propose in this work a pragmatic model of attention that combines bottom-up (data-driven) analysis with top-down controls, at different levels of sensor and actuator signal processing. The aim of this model is not to explain how attention works in biological systems, but rather to provide a practical design for artificial systems that mimic the behavior of their biological archetype. The main functional elements of this pragmatic model are the selection and fixation of attention, which result from the optimization of an objective function that includes both data-related and knowledge or task-related quantities. The working of the model is illustrated through a generic architecture for complex computer vision tasks, and demonstrated on two applications for detecting and recognizing multiple patterns in complex visual scenes.

1 Introduction

Attention is a fundamental mechanism by which a limited-resource system optimizes its goal-directed behavior in an arbitrarily complex environment. Attentive systems are characterized by a set of sensors and actuators, and a data analysis system which maps sensed data on actuator controls, based on some behavior optimization criteria embedded in the analysis system. Depending on the complexity of the analysis system, the biological agent displays different levels of behavioral complexity.

Various attention processes have been identified, such as the human ability to divide attention over multiple tasks and stimuli, and to selectively focus on task-relevant information, in addition to how top-down and bottom-up factors influence the way that attention is directed within and across sensory and actuator modalities [1]. This contextualization of attention processes ensures a graceful degradation of a human's performance with a raise in stimulus complexity. Explanatory models of attention have been devised that could give accounts for the functional or computational behavior of biological attentive systems.

However, we do not believe that artificial attentive systems will come about just by a simple translation of these explanatory models devised in cognitive neuroscience research into manually handcrafted knowledge-based information, computing and communication technologies. Rather, we plea for a new research paradigm for developing artificial intelligent systems, in which biological cognitive systems and artificial intelligent systems co-evolve, interact and are aligned for the purposes of various problem domain areas. We believe that this research paradigm shift will boost cross disciplinary cognitive and computer science research resulting in new scientific community networks, will encourage new research areas, and will strengthen the scientific and methodological foundations of cognitive sciences.

We address the design and implementation of artificial agents whose apparent behaviors mimic that of biological systems. The main purpose of attention and anticipation in an artificial agent is to optimize the performance of complex data analysis tasks, while ensuring graceful degradation of performance and no catastrophic failure. The methodological journey that we take is analyzing explanatory models (empirical or theoretical, functional or computational) of attention and isolating those properties, functions, or mechanisms that seem to have some high generality with respect to data acquisition, processing, and analysis. This leads to a pragmatic model of attention and anticipation from which we devise an architecture for an artificial system that is capable of displaying the apparent (i.e. observable) behavior of a biological system described by the explanatory model.

Our paper is organized as follows. In Section 2 we focus on explanatory theories of attention and anticipation and sketch its challenging problem space. In section 3 we outline a pragmatic model of visual attention and anticipation for which we devised the Proactive Attentive Support System (PASS) architecture described in section 4. In essence, PASS takes a specialized task, namely focusing on relevant information in huge amounts of data (selective attention) based on sensible predictions in the course of action (anticipation). Finally, in Section 5 we showcase the applicability of PASS.

2 Explanatory Models of Attention and Anticipation

In the previous century there have basically been two main periods, which illustrate that attention theory was central to research. The first period was around 1900, in particular the work by William James [2] in which he stated that "Every one knows what attention is. It is the taking possession by the mind, in clear and vivid form, of one of what seem simultaneously possible objects or trains of thought... It implies withdrawal from some things in order to deal effectively with others." The fundamental property of attention is "to make the contents of consciousness appear clearer."

The second period was in the 1950s. In particular the work initiated by Donald Broadbent [3, 4] was driven by real insight into practical problems, such as man-machine system design. Broadbent's Filter theory of attention tried to explain how the human brain coped with the information overload caused by having many senses receiving information simultaneously. Strongly influenced by communication theory and computer technology he put forward the concept of a single-channel, limited capacity information processing system, in which an observer can block or weaken the strength of the incoming messages to the brain by a 'filter'. In addition, another way in which attention can act was acknowledged, namely as biasing the interpretation of information (to a certain extent acknowledging the ideas by James). Thus someone expecting to see a dog in the clouds will see one, while a person expecting to see a car will see a car. This kind of bias was understood to be set by the probability of events, contextual information derived from latest inputs and memories, and their subjectively perceived value.

More than a dozen theories of attention have been proposed since Broadbent's Filter Theory, such as Response selection theory of attention [5]; Capacity Theory [6]; Resource Theory [7, 8, 9, 10]; Treisman's theories of attention [11, 12] and the Feature Integration Theory (FIT) [13]; Van der Heijden's [14], Allport's [15] and Neumann's [16] unlimited capacity to process theories; and computational theory of attention [17]. Even though a unifying theory has not yet emerged, it is widely accepted that two processes of attention co-exist. First, a bottom-up process, which assumes an exploratory behavior, which is more or less synonym with Broadbent's Filter theory and variants thereof. This bottom-up process is successful in explaining the organization of the sensory apparatus up to the level of the primary sensory cortex, but is not very convincing in explaining the higher-level cognitive functions. Nevertheless it is useful for implementing unsupervised functions for low-level pattern detection. Second, a top-down process that requires intentionality, which

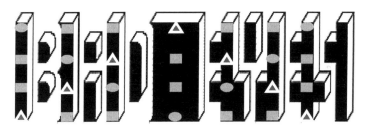

Fig. 1. [1]Empirical indication of variable top-down contributions in pattern detection. Without specifying any task (i.e. no specific top-down information), the triangular elements pop out (pre-attentive detection), then a small top-down signal suffices to fit these triangles to an arrowhead shape. Then, as one continues to observe, more top-down influences lead to noticing that some of the light-grey dots have different shapes. Finally, one will apply more complex strategies for making sense out of the remaining lines and patches. Top-down signals (such as being told that the pattern reads "bad eyes") will slowly lead to choosing the appropriate scale at which the fragments can be integrated to fit something that 'makes sense'

---

[1] Adaptation of the example found on http://www.michaelbach.de/

resembles the second function of attention (biasing) described above. This top-down process successfully explains selection of attention; useful for artificial support systems, were human operators specify the high-level goals for the artificial agent.

Note that these attention processes underlying perception, processing and actuation do not only occur independently for different perceptual modalities (vision, hearing etc), but may also transcend modalities to ensure a globally optimal performance of the perception-actuation cycles. At the same time, multi-sensory and multi-actuator integration phenomena appear to be sensitive to attentional states [18].

An example showing that both attentional processes can become apparent is illustrated in Figure 1. The question is *how* these two processes, independent of modality, sum up the hybrid model of attention in which the top-down (or knowledge-driven) and bottom-up (or data-driven) processes are somehow integrated. Unfortunately, there is no agreement regarding the relative contribution of the two types of processes to cognition, the level where integration is produced, whether integration results as a competitive or cooperative process, or whether some intertwining and coupling occurs. Current models of integration propose either that top-down processes constraint bottom-up processes, or the other way around. For example, [19] argues for the need to integrate top-down and bottom-up attentional influences by finding the optimal top-down influence on bottom-up processes. However, their integration architecture rely on a distinct (explicit) integration function, which computes weight factors for the relative contribution of top-down and bottom-up information, according to some arbitrary algorithm.

Another model, grounded in neurobiology, dealing with top-down predictions with bottom-up inputs, proposed by [20] argues that large scale dynamics, expressing contextual influences and knowledge stored in the system, can influence local processing. These authors indicate that top-down predictions might be embodied in the temporal structure of both stimulus-evoked and ongoing activity, and that synchronous oscillations are particularly important in the process: "Incoming afferent signals induce some patterning of activity in cortical areas that arises from local computations operating on the input. These local patterns, however, are constantly subject to modulation by specific synchronizing and desynchronizing influences [...] through long-range interactions both from other cell populations in the same area and from assemblies that are activated in other areas" (p. 714). The different top-down modulation signals generated in cortical areas that were activated by the current inputs lead to "resonant states" which will amplify certain patterns and attenuate others. The amplified patterns will further influence a larger number of cortical areas, which in turn will add their own modulation signals to the bottom-up signals. This way, different assemblies of cortical areas resonate for a given set of stimuli and stored patterns: "The synchronizing influences from a large number of cortical areas that exploit different binding criteria could compete, and the binding solution that is compatible with maximal resonance could win" (ibid.).

In other words, a self-organizing process leads to the temporal coordination of input-triggered responses and to their binding into functionally coherent assemblies.

In our pragmatic approach we do not intend to find the *right* answer, but to derive a *useful* model, where this integration leads simultaneously to an appropriate bottom-up selection of relevant input data, and a top-down tuning of the selection algorithms. In order to distinguish between these two functions of attention we refer to the concept of attention in the case of bottom-up processes and anticipation in the case of top-down processes. Dennett [21] suggests that "the trade-off between top-down and bottom-up processes is a design parameter." But is that parameter arbitrary? Or static? We argue that the integration is itself an optimization process, rather than a mere weighting of the relative contributions of top-down and bottom-up information streams, and as such, bottom-up and top-down processes are both intertwined and coupled [22] rather than combined through some explicit convolution operator.

3 A Pragmatic Model of Attention for Artificial Vision

Visual information analysis is one of the most challenging tasks in terms of data volume and complexity. Computer vision applications are typically confronted with resource limitations in real-time pattern analysis of high-resolution video sequences. We propose a pragmatic model of selective attention that continuously optimizes the completion of some pattern detection and analysis task for a given set of processing resources.

In this section we first look at some of the previous works concerned with the practical implementation of visual attention mechanisms that rely on the integration of bottom-up and top-down processes. Further, we state and explain the functional requirements for the pragmatic model. Next, the key functions of the model (selection, fixation and anticipation) are explained.

3.1 Related Work

The necessity of integrating top-down and bottom-up processes in complex computer vision tasks has been identified by many authors. Itti, Koch and Niebur proposed in [17] an attention model based on the computation of bottom-up features salience through top-down multiplicative gain. In essence, this model builds upon the guided search model proposed earlier by Wolfe [23]. A salience map is produced starting form a linear combination of three conspicuity maps, which in turn result from a range of multi-scale feature maps that combine features like color, intensity, and edge orientations. The combined saliency map indicates the most salient image location, which should be attended first. The model, however, also accounts for the temporal variations in saliency, and therefore the focus of attention (FOA) is selected by a 2D winner-take-all (WTA) neural network. The first element that fires (the "winner") indicates the image location of the FOA. After firing, an inhibitory signal resets all neurons in the WTA network and provides a local inhibition signal to the currently selected FOA, which would prevent the immediate return to this location. Although this model does not include an explicit top-down signal derived from some pre-defined properties of the system (such as goals or knowledge), its extension in this direction is straightforward.

A subsequent refinement of this model is proposed in [19], where top-down saliency maps are also produced from learned statistical properties of the features. Moreover, this model distinguishes between target and distractor features, which are both accounted for in the production of bottom-up and top-down saliency maps. The integration between the bottom-up and top-down processes results from the multiplication of the two categories of salience maps. An adaptive version of this strategy has been proposed by Choi et al. [24], where the top-down multiplicative gains for modulating the bottom-up saliency map are learned through a supervised technique, which leads to an improved relevance of the selected FOA.

Rasolzadeh et al. [25] propose a further improvement of this model by optimizing the top-down weighting factors based on task and context information. For each region of interest (ROI) in the bottom-up saliency map an error measure is computed that shows the distance between the actual top-down value and the optimal one. This measure is calculated dynamically, such that the selection of attention results from a dynamic linear combination between the two saliency maps.

All these models rely on some explicit (linear) combination of the individual top-down-and bottom-up information. A consequence of this approach is that the combination operator must be defined arbitrarily, i.e. based on some a priori design decision. This issue is somewhat less critical in [25], due to the dynamic computation of the operator's coefficients.

3.2 Functional Requirements

The major functional requirements for the pragmatic model of attention are:

(i) Detection of triggers of attention through bottom-up (data-driven) analysis;
(ii) Selection and prioritization of subsets from the input data, based on the detected triggers of attention and the given analysis task;
(iii) Switching between different foci of attention, based on the results of the analysis task.

The large dimensionality of input video information has to be efficiently dealt with in the first function. In order to ensure a fast processing, this function must implement simple algorithms that operate autonomously, without any feedback from the subsequent analysis functions.

The second function analyzes, in a sequential manner, small subsets of the input data, to detect those subsets that qualify for further analysis. The decision of selecting or rejecting a certain subset is based on the presence of certain features, which are characteristic for the searched patterns. The top-down information used in this function comprises the definition of the relevant features.

The data set to be analyzed by the third function already benefits from a normalized, or even canonic representation in terms of features strengths, given the specific features of the searched pattern. Such an input set of features is then classified, with a certain confidence level, as belonging to a particular class. As the bottom-up information flow changes through continuous observation, the confidence level of the classification increases until it eventually reaches a certain threshold. This leads to generating a top-down control signal that causes switching the attention focus.

A more complex behavior results when the confidence threshold is dynamically computed based on some constraints derived from the analysis task (top-down information). Switching between different foci of attention may occur in this case before a reliable classification of the current pattern is reached. Such a decision for switching may be taken when, for instance, the priority of the data subset currently analyzed becomes significantly lower. This mechanism of premature attention shift will be referred to as anticipation, and described in more detail in section 3.5.

The functional architecture sketched in Fig. 2 shows how top-down information influences bottom-up data processing at different levels, with varying contributions from the two information pathways. This is consistent with the observation of Engel et al. that "the extent of top-down [contribution] clearly increases as one moves up in the cortical hierarchy" [20].

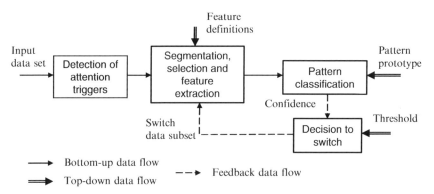

Fig. 2. Functional diagram of top-down and bottom-up integration in attention selection and fixation

3.3 Selection of Attention through Local Optimization

The first two blocks in Fig. 2 implement the selection of attention. The detection of attention triggers is performed by algorithms that run in a 'pre-attentive' manner, i.e. without any top-down feedback. Such algorithms would typically detect simple generic features that are likely to appear in all possible input data sets (i.e. lines, gradients, etc.), being thus relevant for all possible tasks of the system.

Assuming that the task of the system is fixed (i.e. feature definitions do not change), the selection of the relevant subsets and the feature extraction algorithms operate without requiring top-down feedback signals. Indeed, at a given moment (i.e. for a given input data set), the selection of a particular subset would only depend on data's properties.

After feature extraction each subset can be described by a real-valued vector of feature strengths. The next task is to prioritize the selected subsets, i.e. to decide which of the selected subsets shall be fed to the next level of analysis. Generally, this ordering cannot be based on the feature vector values themselves, since these only represent a necessary condition for detecting a relevant pattern. For instance, if the task is *'avoid impact with incoming cars'*, then some relevant visual features could be formulated as *'car shape'*, *'apparent position'*, and *'apparent motion vector'*. At this stage, however, it is impossible to distinguish between a nearby incoming toy car, and a remote incoming automobile. For this reason, the prioritization strategy should rely on some criteria that are independent of the particular task. A very good example of such a prioritization strategy is given in [19], where a signal to noise ratio is attached to each of the detected features to indicate how well the feature can be discriminated from a distracting background. This meta-information is then used to produce a 'salience map'. Another strategy could try to maximize the relative size of each analyzed subset, such that fewer iterations would be required for analyzing the complete incoming data set. The choice for particular optimization criteria and procedures is a design parameter.

3.4 Fixation of Attention

An active sensing system produces a reliable detection of targeted patterns through continuously analyzing input data streams. The fact that pattern analysis does not rely on a mere snapshot of the observed environment calls for a dynamic mechanism for pattern detection.

Fixation of attention assumes that the 'pattern classification' function in Fig. 2 is fed a feature vector extracted from a subset of the input data. As input data is continuously refreshed, the vector's values will vary as well. The pattern classification algorithm will perform several iterations, until a stable classification is reached. After each iteration a confidence measure is attached to the classification result. The classifier stops when a sufficiently reliable decision regarding the class membership of the current pattern is reached, i.e. the overall confidence is larger than a predefined threshold. The way the consecutive confidence values are combined to produce the final decision is an additional design parameter.

Fixation of attention based on a static confidence threshold may lead the system into undesirable stable states where poor quality data prevent the classification function from reaching the required level of reliability. The system will then fail to further analyze the other relevant data subsets. To avoid this, an objective function should be defined that combines the confidence level reached at a given moment, $c(t)$, and the time when the analysis of the current pattern started, t_s. We call this the *utility measure* that indicates how useful a further analysis of the current pattern is under an absolute time constraint. A possible (empirical) definition of this function is

$$u(t) = \frac{1}{[1 + c(t)] \cdot \alpha(t - t_s)}, \qquad (1)$$

where α controls the temporal decay of the utility function. The decision to end the analysis for the current pattern can be taken when u goes for the first time below a certain (new) utility threshold.

3.5 Anticipation through Dynamic Contextualization

Generally, a utility function should incorporate task-specific measures, not just generic constraints. Indeed, when the task at hand requires accurate classification of a single pattern (e.g. "is that airplane friend or foe?"), a rigid temporal constraint is not appropriate. On the contrary, when the task requires spotting a specific pattern in a fast-changing sequence, timeliness becomes more important than accuracy.

Besides the task-specific (top-down) constraints, the utility function may incorporate data-specific (bottom-up) constraints as well, such that it adapts the behavior of the attentive system to the environmental conditions as well. For one and the same task, for instance, the utility of analyzing a pattern may be influenced by the presence and properties of other possible patterns in the input data set. This influence of the environment (i.e. other possibly relevant patterns) upon the analysis of the current pattern results in an anticipative behavior, as explained further in Section 4.

4 The PASS Architecture

The Proactive Attentive Support System (PASS, see Fig. 3) illustrates the principles exposed earlier in Section 3. The system is designed to recognize specific targets in a complex visual scene, and relies on the assumption that target patterns display some scale-invariant features. This enables the "pre-attentive" processing module to work with a low-resolution version of the input video data stream, which ensures a real-time operation.

The low resolution video can be obtained through uniform sub-sampling. The lower resolution limit is to be established separately for each recognition task, depending on the scale invariance properties of the features to be detected.

The 'feature extraction' algorithm detects rectangular areas in the input video that contain typical features indicating the presence of the searched pattern. It does not produce a feature vector, but rather a 'region of interest' (ROI) which will be further analyzed by the 'post-attentive processing' algorithms.

The short-term memory has several roles. On the one hand, it ensures temporal independence (asynchrony) between the pre-attentive processing functions and the rest of the system, such that ROIs are detected continuously, independent of the particular state of the attention mechanism (selection, fixation), or the progress of the post-attentive pattern analysis. On the other hand, the short term memory allows the feedback from the ROI analysis function (utility measure u_1) to be correctly interpreted as reinforcement signal for previously detected ROIs. This feedback can trigger some adaptation of the feature detection algorithm.

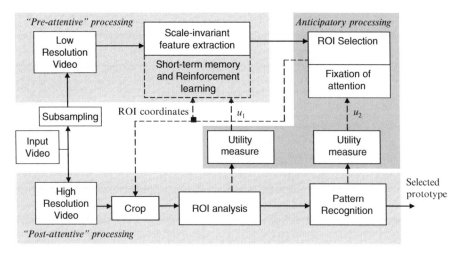

Fig. 3. PASS architecture for visual analysis of complex scenes

All the detected ROIs in each video frame are then analyzed by the selection function, which assigns to each of them a priority, based on the strength of the detected features. The relative coordinates and sizes of the ROI with the highest priority are communicated to the 'crop' function, which extracts from the full-resolution video a ROI with the same relative position and sizes. This segment is then analyzed to detect the presence of more complex patterns, and to produce a utility measure indicating the strength of the detected patterns. This value, and the coordinates of the selected ROI are used by the reinforcement learning function, as explained earlier. The pattern recognition function calculates a distance between known target patterns and those detected by the subset analysis function. A second utility measure is derived from the recognition confidence, and used for locking the attention until the expected pattern is reliably detected.

To avoid possible stable states, the anticipation algorithm monitors the realization of the global optimum, and overrides, as needed, the focus or fixation of attention. The anticipation process is driven in this example by the overall performance requirement of analyzing as many patterns as possible in the given scene. This global optimum is achieved by including in the second utility measure, u_2, a dependency on the number of total ROIs detected at a given moment, and the average analysis time, under the assumption that each ROI will only remain in the observed scene for a limited time:

$$u_2(r_k, t_i) = \frac{\min\left\{(t_e - t_i), \frac{t_{avg}}{N_r}\right\}}{[1 + c(r_k)] \cdot (t_i - t_s)}, \qquad (2)$$

where r_k is the ROI being analyzed at time t_i, t_e is the earliest time to exit from the scene of one of the detected ROIs, t_{avg} is the average processing time calculated for the last few regions, N_r is the number of regions pending analysis, and t_s is the moment when the analysis of the current region has started. Since this measure is only calculated after performing the first recognition iteration, t_i is always strictly larger than t_s. The confidence of pattern recognition, $c(r_k)$, has to be defined in accordance with the classification, or matching algorithm. The analysis of the current region stops when the value calculated in (2) decreases below a certain limit. Since this limit has no clear physical meaning, it has to be established it experimentally.

5 Experimental Implementations

Based on the PASS architecture, several applications to video surveillance have been developed (see also [26, 27]). In a first implementation the objective was to detect multiple moving objects in a relatively large observation area. For each moving object a high resolution snapshot has to be captured and stored, but no pattern recognition has to be performed.

The implemented system consists of two video cameras. A fixed camera produces an overview video of the whole observation area, which is fed to the (pre-attentive) motion detection algorithm. For each region ROI_i a motion vector $m_i(x_{i0}, y_{i0}, t) = [\dot{x}_{i0}(t), \dot{y}_{i0}(t)]$ is calculated for the center (x_{i0}, y_{i0}) of the region at each moment. The selection is straightforward, based on the earliest time to exit from the scene of each region, $t_{ei} = \min\{t_{eix}, t_{eiy}\}$, where t_{eix}, t_{eiy} are the times to exit for the horizontal and vertical components of the motion vector, calculated under the assumption of constant speed as:

$$t_{eix} = \begin{cases} \frac{D_x - x_0}{\dot{x}_{i0}}, & \dot{x}_{i0} > 0 \\ \frac{-x_0}{\dot{x}_{i0}}, & \dot{x}_{i0} < 0 \\ t_{e\max}, & \dot{x}_{i0} = 0 \end{cases}, \quad t_{eiy} = \begin{cases} \frac{D_y - y_0}{\dot{y}_{i0}}, & \dot{y}_{i0} > 0 \\ \frac{-y_0}{\dot{y}_{i0}}, & \dot{y}_{i0} < 0 \\ t_{e\max}, & \dot{y}_{i0} = 0 \end{cases} \quad (3)$$

where D_x, D_x, are the width and height of the low-resolution video, and $t_{e\max}$ is a predefined maximum waiting time. In order to ensure that each region will be handled within the given maximum amount of time, the ordering of the ROIs is performed on the values

$$t_i = \min\{t_{ei}, t_{\max} + t_{i0} - t\}, \quad (4)$$

where t_{i0} is the time at which ROI_i was detected, and t is the current time. After each selection (that is, for each video frame) the t_i values are recalculated.

A second camera relies on a pan-tilt-zoom mechanism to select from the natural scene an area that corresponds to the currently selected ROI from the overall video. The coordinates and sizes of the ROIs are mapped onto pan, tilt, and zoom values.

In a next implementation, a face detection algorithm was used in place of the motion detection (see Fig. 4). The face detection algorithm (based on Viola and Jones [28]) is applied both to the overview and to the detail video streams. The pattern detection strategy is based on analyzing the pixels within a sliding window, at all possible scales, in each video frame. In order to speed up the pre-attentive processing the search space for the face detection algorithm is reduced to only fine scales, under

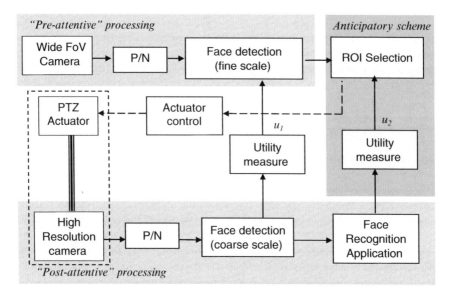

Fig. 4. Attentive system for face recognition

the assumption that faces will have small apparent sizes in the overview video. For the post-attentive processing, a similar reduction of the search space is applied, but now only coarse-scale sliding windows are used, under the assumption that, at the current position and viewing angle, the second video will only contain one face. The result of this second detection is used to produce the utility measure u_1 that indicates how well the detection algorithm performed at low resolutions. This utility value is set to 1 if a face was detected in the second video stream, otherwise it will be 0. The corresponding region, previously detected in the overview video and stored in the short-term memory, will be stored as a positive or negative example, respectively, on the permanent storage. At a later stage, the face detection kernel will be retrained with this additional set of samples.

The face area detected in the second video is then passed on to a face recognition module based on a back-propagation neural network, which can perform both verification and identification. The utility measure u_2 is implemented as in equation (2), with $c(r_k)$ calculated as the cosine of the output ($\mathbf{v}$) and target ($\mathbf{t}$) vectors of the neural network:

$$c(r_k) = \frac{\mathbf{v} \cdot \mathbf{t}}{|\mathbf{v}| \cdot |\mathbf{t}|}. \tag{5}$$

6 Conclusions and Future Work

Starting from several explanatory theories of attention based on the integration of top-down and bottom-up signals and processes, we argued in this work that this integration occurs at different levels of processing. Next, we provided several key functional requirements that allowed the formulation of a pragmatic model of attention selection and fixation that mimics the behavior of biological systems, but allows for a relatively simple computational model. The essence of integration relies on the optimization of an objective function that includes both data-related (bottom-level) and knowledge or task-related (top-level) quantities.

A generic architecture that illustrates the working of the proposed model has been proposed that supports the implementation of complex computer vision tasks. The different functions of the attention and anticipation model have been implemented and tested in two applications for complex computer vision tasks. The performance of the attention mechanism in these applications is very promising, in that the complex task of detecting and attending to multiple (moving) objects is performed with a relatively low computational load.

In short, we have shown how explanatory models of natural intelligence proposed by cognitive scientists can inspire new computational models of artificial intelligent systems.

Several immediate extensions of the current system can be envisaged, such as distributing the workload of the post-attentive analysis algorithms across multiple computers taking video inputs from different pan-tilt-zoom cameras. A more interesting problem, however, is the extension of the top-down and bottom-up integration mechanism to the higher level of task definition. Not only would a dynamic task definition increase the flexibility of the system, but would also allow multiple instances of such a system to collaborate or compete in task acquisition and completion, and eventually lead to emergent behavior. Bio-inspired emergent computing mechanisms for selection of attention and anticipation valid at evolutionary time scales manifest themselves as dynamical features of co-evolving and interacting biological cognitive systems and artificial intelligent systems. For this reason, a possible approach to extending the proposed attention and anticipation mechanism could rely on the dynamic scale-space paradigm proposed in [22].

7 Acknowledgement

This research has been supported in part by the Dutch Freeband Communication Research Programme (AWARENESS project) under contract BSIK 03025.

References

1. Kramer AF, Wiegmann DA and Kirlik A (2006) Attention: from theory to practice. Oxford University Press
2. James W (1890) Principles of Psychology. New York
3. Broadbent DE (1958) Perception and Communication. Pergamon Press, London
4. Broadbent DE (1971) Decision and Stress. Academic Press, London
5. Deutsch JA and Deutsch D (1963) Attention: Some theoretical considerations. Psychological Review, 70, pp 80–90

6. Kahneman D (1973) Attention and Effort. Englewood Cliffs, Prentice-Hall, NJ
7. Navon D and Gopher D (1979) On the economy of the human processing system. Psychological Review, 86, pp 214–255
8. Norman DA and Bobrow DG (1975) On data-limited and resource-limited processes. Cognitive Psychology, 7, pp 44–64
9. Wickens CD (1980) The structure of attentional resources. In: R.S. Nickerson (ed) Attention and Performance 8, Academic Press, New York
10. Wickens CD (1984) Processing resources in attention. In: Parasuraman R and Davies DR (eds) Varieties of Attention. Academic Press, New York
11. Treisman AM (1960) Contextual cues in selective listening. Quarterly Journal of Experimental Psychology, 12, pp 242–248
12. Treisman AM (1964) Verbal cues, language, and meaning in selective attention. American Journal of Psychology, 77, pp 206–219
13. Treisman AM (1988) Features and objects: The XIVth Sir Frederic Bartlett memorial lecture. Quarterly Journal of Experimental Psychology, 40A, pp 201–237
14. Van der Heijden AHC (1992) Selective Attention in Vision. Routledge and Kegan Paul, London
15. Allport DA (1987) Selection for action: Some behavioral and neurophysiological considerations of attention and action. In: Heuer H and Sanders AF (eds) Perspectives on Perception and Action. Erlbaum, Hillsdale, NJ
16. Neumann O (1987) Beyond capacity: A functional view of attention. In: Heuer H and Sanders AF (eds) Perspectives on Perception and Action. Erlbaum, Hillsdale, NJ
17. Itti L, Koch C, and Niebur E (1998) A model of saliency-based visual attention for rapid scene analysis. IEEE Transactions on Pattern Analysis and machine Intelligence, 20(11), pp 1254-1259
18. Calvert GA, Spence C and Stein BE (eds) (2004) Handbook of Multisensory Processing, MIT Press
19. Navalpakkam V, Itti L (2006) An Integrated Model of Top-down and Bottom-up Attention for Optimal Object Detection. In: Proc. IEEE Conference on Computer Vision and Pattern Recognition, pp 2049-2056
20. Engel AK, Fries P and Singer W (2001) Dynamic Predictions: Oscillations and Synchrony in Top-down Processing. Nature, October 2001, Vol 2, pp 704–716
21. Dennett DC (1994) Cognitive Science as Reverse Engineering: Several Meanings of "Top Down" and "Bottom Up". In: Prawitz D & Westerstahl D (eds) International Congress of Logic, Methodology and Philosophy of Science. Kluwer, Dordrecht http://cogsci.soton.ac.uk/~harnad/Papers/Py104/dennett.eng.html
22. Salden AH, and Kempen MH (2004) Sustainable Cybernetics Systems - Backbones of Ambient Intelligent Environments. In: Remagnino P, Foresti GL, and Ellis T (eds) Ambient Intelligence. Springer
23. Wolfe JM (1994) Guided Search 2.0: A Revised Model of Visual Search. Psychonomic Bulletin & Review, 1 (2), pp 202–238
24. Choi S, Ban S, Lee M, (2004) Biologically Motivated Visual Attention System Using Bottom-Up Saliency Map and Top-Down Inhibition. Neural Information Processsing Letters & Review 2 (1)
25. Rasolzadeh B, Björkman M, Eklundh J-O (2006) An Attentional System Combining Top-Down and Bottom-Up Influences. International Cognitive Vision Workshop (ICVW06), Graz, Austria
26. Iacob SM and Salden AH (2004) Attention and Anticipation in Complex Scene Analysis – An Application to Video Surveillance Systems. Proc IEEE International Conference on Systems, Man and Cybernetics, The Hague, The Netherlands.

27. Iacob SM, Salden AH, and de Heer J (2006) The PASS demonstrator: Detailed design and experimental results. Telematica Instituut, Enschede, The Netherlands https://doc.telin.nl/dscgi/ds.py/Get/File-64559/INES_PASS_D3.pdf
28. Viola P, Jones M (2001) Rapid Object Detection using a Boosted Cascade of Simple Features. Proc IEEE Computer Society Conference on Computer Vision and Pattern Recognition (1)

Part IV

Search and Optimization

Self-Organization for Search in Peer-to-Peer Networks

Elke Michlmayr

Women's Postgraduate College for Internet Technologies (WIT),
Institute of Software Technology and Interactive Systems,
Vienna University of Technology,
Favoritenstrasse 9-11/E188,
1040 Vienna, Austria
http://wit.tuwien.ac.at/people/michlmayr
michlmayr@wit.tuwien.ac.at

Summary. This chapter presents the design and evaluation of an ant-based approach to query routing in peer-to-peer networks. After pointing out how to employ the ant metaphor for query routing, we evaluate the impact of different settings for the configurable parameters present in ant algorithms on the performance values. In particular, the focus is on the effects of setting the ratio between ants exploiting the option currently known as the best one, and ants exploring the search space with the aim of finding improved options. We show that the exploitation-exploration dilemma can be avoided by an adequate design of the exploring option.

1 Introduction

The principles of self-organization and emergence have received a lot of interest in the research community recently. In particular, the trail-laying and trail-following behavior observed from foraging ants has been employed for solving diverse problems in computer science. Although the ant metaphor has been successfully applied to routing of data packets both in wireless networks [1] and fixed networks [2], little is yet known about its adequacy for the task of query routing in peer-to-peer networks. The challenges for the latter are the following: Each peer is connected via outgoing links to some other peers which are called its neighbor peers. If a peer issues a query or receives a forwarded query from one of its neighbor peers, it has to decide based on its *local knowledge* which neighbor peer to send the query to. Since ant-based methods rely on local knowledge and indirect communication only, they are suitable for this task.

In *reputation learning* approaches [3] to query routing, the local information of a peer is gained by (1) continuously observing the queries and answers that pass the local node and by (2) recording which kind of queries its neighbor peers are able to answer. The recorded data must be accumulated and stored in an appropriate way to support the neighbor selection process. Based on the recorded data, the peer chooses the neighbor peer which is most likely to store results itself, or has neighbor peers that are likely to store such resources.

One of the advantages of using the ant metaphor for reputation learning is that it is readily applicable. Ant algorithms are inspired by the collective foraging behavior of specific ant species, which is referred to as the trail-laying and trail-following behavior. Trail-laying means that each ant drops a chemical substance called pheromone when moving. Trail-following means that each ant senses its environment for existing pheromone trails and their strength in order to decide which path to follow. In the case of reputation learning, the trail-laying and trail-following is applied as follows: The pheromone trails store the accumulated data about the successful queries in the past, and the queries are represented as ants.

This work is part of our ongoing efforts [4, 5] to design and evaluate SEMANT, a reputation learning-based algorithm for query routing in peer-to-peer networks compliant with the *Ant Colony Optimization* meta-heuristic [6]. One of the challenges of our work is that ant algorithms include various configurable parameters for which appropriate value settings must be set. In particular, there is a parameter that influences a basic decision each ant has to make before selecting an outgoing link. It can either

- *exploit* the best results known so far for path selection, or it can
- *explore* a path that is not currently known as the best one in order to possibly find an improved solution to the problem. If it succeeds, this will enhance the performance of the system.

The question of which one of the strategies to select according to its desirability in the current context of the ant is referred to as the *exploitation-exploration dilemma* [7, 8]. The dilemma occurs not only in ant algorithms, but in reinforcement learning [9] in general. It has been discussed in several contexts, such as in the case of foraging bees [10], in economic systems [11], or in software product development [12].

In ant algorithms, and also in SEMANT, the decision about which strategy to use is performed based on a parameter that specifies a pre-defined ratio between exploring and exploiting. What we will show in the following is that the SEMANT algorithm is self-configuring in the sense that the overall efficiency of the search process is the same no matter which ratio is employed, and also the performance of the system as perceived by individual users is not affected by the ratio. This chapter is organized as follows. Sec. 1.1 defines the problem and states the assumptions which were made. Sec. 2 contains a detailed description of all aspects related to the SEMANT algorithm. Sec. 3, next to showing the performance of the algorithm in comparison to other approaches, presents experimental results that prove the claims we made above. Sec. 4 discusses related work on ant algorithms in peer-to-peer networks.

1.1 Problem Description

A peer-to-peer network is a network consisting of interconnected nodes in which each node manages an information repository containing a certain number of resources. Every peer (1) offers its resources to the other nodes in the network and (2) issues queries to the network. All nodes collaborate to answer the queries and therefore – as a whole – implement a distributed search engine. For each query, the shortest path through the network must be found that leads from the querying peer to one or more answering peers offering one or more resources that are appropriate for satisfying the query.

In order to make it possible to concentrate on the problem of query routing, the following assumptions are made about the application scenario:

(i.) Each peer has an unique address that can be used as an unique identifier.
(ii.) The resources at each peer can be uniquely identified using an existing resource identifier (such as a filename) together with the peer identifier.
(iii.) All links between peers are bi-directional, that is, can be used in both directions.
(iv.) The network topology already exists. Each peer already knows which neighbor peers it is connected to. Peer discovery is not within the scope of this work.
(v.) The network topology and the content distribution in the network are considered static.

Resources can be any kind of files that are annotated with metadata. The metadata are composed of name-value pairs, also called *elements*. These elements are used for specifying additional information about a certain resource. A meta-data schema defines the name and the meaning for each of the elements the schema is comprised of. Since the query routing procedure is the same no matter which element is employed, only one element is considered in the following. The values that can be used for annotation originate from a controlled vocabulary, e.g, the concepts of a taxonomy or an ontology. The query vocabulary is the same as the metadata vocabulary used for annotating resources. The queries do not consider the actual content of a resource, but rather the metadata that describes its content. A query Q consists of a concept c from the controlled vocabulary. A resource is an appropriate result for query Q if it is annotated with concept c.

2 Specification of the Algorithm

This section specifies the components of the SEMANT algorithm. Sec. 2.1 documents the data structures that need to be stored at each peer. Sec. 2.2 describes the query routing procedure and all aspects related to facilitating ant algorithms in a peer-to-peer network. In Sec. 2.3, the mechanisms to select outgoing links in the query routing procedure are covered. Sec. 2.4 explicates how routing tables are updated after successful queries. Sec. 2.5 describes the activities executed locally at the peers. Sec. 2.6 explains the bootstrapping mechanism.

2.1 Data Structures

Pheromone trails are maintained in a table τ at each peer P_i. Table τ is of size $C \times n$, where C is the size of the controlled vocabulary that defines the allowed keywords

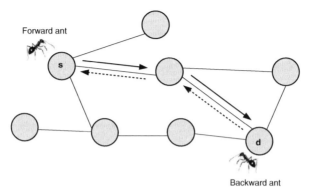

Fig. 1. The concept of forward and backward ants. Node s is the querying peer and node d is the answering peer. The solid lines show the path of the forward ant, the dotted lines that of the backward ant

in a query and n is the number of peer P_i's outgoing links to neighbor peers. Each τ_{cu} stores the amount of pheromone corresponding to concept c dropped at the link from peer P_i to peer P_u, for each concept c and each neighbor peer P_u. All entries in table τ are initialized with the same value $\tau_{init} = 0.009$. This is necessary to prevent divisions by zero in the evaporation feature (see Sec. 2.5) and in link selection (see Sec. 2.3).

2.2 Query Routing

The concept of forward ants and backward ants (see Fig. 1) from the *AntNet* algorithm [2] is employed as the foundation of the query routing procedure. In addition, *AntNet*'s mechanism for preventing cycles is adopted. Whereas the latter can be adopted without any changes, the concept of forward ants and backward ants needs to be adapted to the application purpose of query routing. In the following, the necessary adaptations are discussed.

Queries

In the SEMANT algorithm, queries are represented as ants. This approach has two advantages. Firstly, no additional traffic is created in the network. Secondly, representing queries as ants guarantees that the degree of optimization for certain query keywords directly depends on the popularity of a given keyword. The more often a query keyword is requested, the better its paths will be optimized in terms of indicating the way through the network to the most appropriate peers. Instead of sending out forward ants at regular intervals from random peers like in *AntNet*, a forward ant is created for each query that occurs in the network. This ant is created at the peer which issued the query, and it is responsible for answering it.

Link Costs

Ant algorithms usually consider two different types of information in the link selection process. Firstly, there is the pheromone distribution of the outgoing links. This information is built incrementally by the ants. Secondly, there is problem-specific information that defines the cost of every link. In the case of query routing, this would be the time it takes to travel from one peer to another. Since this feature is already integrated in ant algorithms, including support for different link costs in the SEMANT algorithm would be easily possible. It would also be useful because in a real-world setting, each link between two peers has a certain latency, a certain bandwidth, and a certain throughput depending on the hardware of the connection.

However, there are two reasons that make the evaluation of this feature cumbersome. Firstly, there is no appropriate test data composed of a real-world network topology together with defined latency/bandwidth/throughput properties available. Secondly, even if such data would be available, there is no other approach to search in peer-to-peer network that considers different link costs. This makes it impossible to conduct performance evaluation by comparison to reference values. For these reasons, the algorithm only considers the pheromone distribution of the outgoing links, but no link costs.

Time-to-live Parameter

Next, it is necessary to define at which point in time the forward ant should stop its travel. In *AntNet*, forward ants terminate their travel through the network when they arrive at their well-defined destination peer. This behavior can not be transferred, since the forward ant's task is to find an unknown destination peer that in the worst case does not even exist. Instead, a stop point for forward ants must be defined to prevent forward ants from running infinitely if no results can be found. The simplest solution is to use a time-to-live (TTL) parameter ttl_{max} like introduced in Gnutella (see [13]). Each time an ant travels one hop to reach another peer, it decrements ttl_{max} by one. The stop point is reached if $ttl_{max} = 0$.

MinResources Variation and MaxResults Variation

After a forward ant has found a result and creates a backward ant, there are two possibilities for proceeding further. Either the forward ant (1) terminates its travel, or (2) it continues until the maximum time-to-live parameter is reached. The idea behind the latter approach is that if forward ants are allowed to go on after they found the first peer that stores results, they can increase the absolute number of results found by detecting other appropriate peers. Keeping in mind that query routing in peer-to-peer networks is a special kind of optimization problem, the choice between these two options determines the optimization goal of the algorithm:

- *MinResource* **variation.** If the ants are terminated after they found the first result, the optimization goal is to use the minimum amount of network resources. Therefore, the forward ant strategy of stopping after the first result is found is referenced to as the *minResource* variation of the SEMANT algorithm.

- **MaxResults variation.** In the other case, where the ants use the maximum time-to-live parameter, the ants use approximately the same amount of network resources for each query and the optimization goal is to maximize the number of results that are found for a query. The strategy of using the maximum time-to-live parameter is referenced to as the *maxResults* variation of the SEMANT algorithm.

In practice, both of these variations are valid, because both of the optimization goals are desired at the same time. Using the *minResource* variation is of benefit for the performance of the entire network, since the algorithm saves as many network resources as possible. Using the *maxResults* variation is of benefit for the individual users of the network, since the algorithm tries to find as many results for a single query as possible. It is possible to combine these variations by using a weight that defines the ratio between employing the *minResource* variation and employing the *maxResults* variation.

Step-by-step Description of the Query Routing Procedure

Now the complete procedure for answering a query is laid out. Consider a query q issued at a peer P_q. For simplicity, the assumption is that query q is a simple query containing exactly one keyword c. The extensions to the algorithm for supporting complex queries are described in [14]. The following seven steps are necessary for answering query q.

(i.) Check the resource repository of peer P_q. If any results are found, present them to the user. If the number of results found is less than r_{max}, go to step 2. If the number of results found is greater than r_{max}, terminate the algorithm.

(ii.) Create a forward ant F_q with timeout ttl_{max} at peer P_q. Add the identifier of peer P_q to F_q's stack of already visited peers $s(F_q)$.

(iii.) Use the link selection procedure described in Sec. 2.3 to select the neighbor peer(s) P_{j_x} ($x \in [1..n], n \in \mathbb{N}$) the forward ant F_q should choose. Send ant F_q to peer P_{j_1}. For every peer P_{j_x} where $x \in [2..n]$, create a clone F_q^c of forward ant F_q and send ant F_q^c to peer P_{j_x}.

(iv.) For every forward ant F_q that arrives at a peer P_j, check if peer P_j was already visited by a clone F_q^c. If so, terminate ant F_q. Otherwise, check the resource repository of peer P_j for resources r that are results for query Q. If there are no results, continue at step 6. Otherwise, add the identifiers of all resources r to the set R and continue at step 5.

(v.) Generate a backward ant B_q. Pass it R, the identifier of the peer P_j that stores R, and the stack of already visited peers $s(F_q)$. Send B_q back to the querying peer P_q using the procedure described in Sec. 2.4. In case the *minResources* variation is used, terminate the forward ant F_q. Otherwise, continue at step 6.

(vi.) Add the identifier of peer P_j to the stack of already visited peers $s(F_q)$.

(vii.) If $ttl_{max} > 0$, decrement ttl_{max} by 1, and let F_q continue at step 3. Otherwise, terminate F_q.

As soon as a backward ant B_q arrives at the querying peer P_q, the results $r \in R$ are presented to the user. In case the user decides to download a resource r, a direct connection between peer P_q and the peer P_j that stores r is established and resource r is retrieved from peer P_j.

2.3 Link Selection

Now the selection of outgoing links by the forward ants is described. In ant algorithms, this selection is made by applying a so-called transition rule. The transition rule designed for the SEMANT algorithm is based on the transition rule from the *Ant Colony System* algorithm [15], which consists of two strategies that supplement each other.

- In the exploiting strategy, the ant determines the quality of the links depending on the amounts of pheromone and always selects the link with the highest quality.
- The exploring strategy encourages the ants to discover new paths. This is achieved by deriving goodness values for the neighbor peers according to the amounts of pheromone on the links that lead to them, and by probabilistically selecting a subset of the peers in proportion to their goodness values.

As discussed in Sec. 1, the decision for one of the strategies is based on a parameter $w_e \in [0, 1]$. For example, if parameter w_e is set to 0.85, the forward ants will employ the exploiting strategy in 85% of the cases. Each time a forward ant has to select an outgoing link, it individually decides for a strategy by applying the *roulette wheel selection technique* [16] together with parameter w_e as an input value.

In case the exploiting strategy is used, a forward ant F_q located at a certain peer P_i selects the neighbor peer P_j currently known as the most appropriate one by applying the transition rule shown in Eq. 1.

$$j = arg\ max_{u \in U \wedge u \notin s(F_q)} \tau_{cu} \qquad (1)$$

In Eq. 1, U is the set of neighbor peers of peer P_i, and $s(F_q)$ is the set of peers already visited by F_q.

In case a forward ant F_q utilizes the exploring strategy, the transition rule shown in Eq. 2a and Eq. 2b is applied *for each* neighbor peer P_j in order to decide whether peer P_j should be selected. Note that this is an adaptation of the roulette wheel selection technique: Each p_j is *separately* placed on the continuum between 0 and 1, and for each p_j a random value q is calculated for deciding if peer P_j should be selected. This mechanism allows more than one peer to be selected in order to account for the fact that there are multiple possible destination peers which contain answers for a query. To ensure that at least one peer will be selected, the algorithm falls back to the exploiting strategy in case applying the exploring strategy does result in not selecting any peer.

$$p_j = \frac{\tau_{cj}}{\sum_{u \in U \wedge u \notin s(F_q)} \tau_{cu}} \qquad (2a)$$

In Eq. 2a, the assumption is that a forward ant F_q is located at a certain peer P_i, U is the set of neighbor peers of peer P_i, and $s(F_q)$ is the set of peers already visited by F_q.

$$GO_j = \begin{cases} 1 \text{ if } q \leq p_j \wedge j \in U \wedge j \notin s(F_q) \\ 0 \text{ else} \end{cases} \quad (2b)$$

In Eq. 2b, q is a random value, $q \in [0,1]$, and the sum of all goodness values

$$\sum_{j \in U \wedge j \notin s(F_q)} p_j = 1.$$

If $GO_j = 1$, the forward ant F_q creates a clone F_q^c of itself and sends the clone F_q^c to peer P_j.

2.4 Routing Table Updates

A description of the mechanism used by the backward ants for updating the routing tables follows. A backward ant is created at a certain peer P_r storing a set of results R. Each backward ant B_q is in possession of a copy of the stack data containing a list of all visited peers $s(F_q)$ recorded by its corresponding forward ant F_q. Based on this information, the backward ant B_q calculates the number of hops h_{qr} between the querying peer P_q and the answering peer P_r. After that, it travels back hop-by-hop to the querying peer P_q according to the information stored in $s(F_q)$. At each intermediate peer, ant B_q is responsible for dropping pheromone by applying the pheromone trail update rule shown in Eq. 3a and Eq. 3b. The amount of newly added pheromone depends on the goodness of the found path, which is determined by comparing the number of resources found and the length of the path to pre-defined reference values. For the reference solution, the value for a path's total length is set to $\frac{1}{2} \cdot ttl_{max}$, and the number of resources is set to r_{max}.

$$\tau_{cj} \leftarrow \tau_{cj} + Z, \quad (3a)$$

where

$$Z = w_d \cdot \frac{|R|}{r_{max}} + (1 - w_d) \cdot \frac{ttl_{max}}{2 \cdot h_{qr}} \quad (3b)$$

In Eq. 3b, parameter w_d weights the influence of resource quantities and path length.

2.5 Peer Activity

Each peer performs management procedures on its local routing table. It applies the evaporation rule shown in Eq. 4 in predefined intervals t_e for each link to neighbor peer P_u and each concept c. The amount of pheromone that evaporates in every interval is controlled by parameter $\rho \in [0,1]$. Evaporation is one of the constituents of pheromone management in ant algorithms. It is useful for preventing that paths become dominant.

$$\tau_{cu} \leftarrow (1-\rho) \cdot \tau_{cu} \qquad (4)$$

This rule is adopted from the evaporation part of *Ant Colony System*'s global pheromone update rule [15].

2.6 Bootstrapping

In the start-up phase, directly after the initialization of the routing tables, all pheromone trails are of equal strength. Therefore, the ants make their decisions completely randomly. This means that the performance values of the algorithm will be rather bad in the beginning. Informed decisions are made as soon as an appropriate number of ants found results, and updated the pheromone trails accordingly. After a certain time, the pheromone trails will be near to the optimum. When reaching this point, the performance values do not improve significantly any longer. One possibility to speed up the time to reach this point is to send additional ants in the start-up phase. Since the exploring strategy already provides for selecting more than one outgoing link, sending additional ants is most useful in the exploiting strategy.

A simple criteria to detect if the algorithm is in the start-up phase is to check the amounts of pheromone on the outgoing links before link selection is performed in order to find out if all amounts for the keyword of the query are of equal size. If this is the case, one possibility would be to send an ant to every outgoing link, but this is equal to flooding. Another option would be to randomly choose two outgoing links, but this still creates a high amount of network traffic. Instead, a parameter $p_{sendAdditional} \in [0,1]$ is used. It defines the probability that two outgoing links are chosen in case (1) the pheromone trails on all outgoing links are equal and (2) the exploiting strategy is employed.

3 Results

This section describes the experimental evaluation of the the SEMANT algorithm. After presenting the setup used in Sec. 3.1, the metrics and the rationale behind choosing them are discussed in Sec. 3.2. Sec. 3.3 shows the performance in comparison to the standard k-random walker approach, and Sec. 3.4 discusses the exploitation-exploration dilemma in context.

3.1 Setup

For evaluating the design of the algorithm, a peer-to-peer system needs to be simulated. The SEMANT algorithm is independent from a certain application scenario. It can be used in every situation in which groups of individuals want to share annotated resources. The application scenario chosen for the evaluation is to support the cooperation between researchers in the field of computer science by allowing them to share scientific articles.

The three most important characteristics of a peer-to-peer system that have an impact on the performance of the search algorithm used are the (1) network topology used, the (2) distribution of the content within the network, and (3) the distribution of queries within the network. The specifications for all three of them are described below.

Network topology. Since the use case concerns a community of researchers, and therefore a social network, a *small world* [17] network topology is selected for the experiments in order to choose a realistic model for social networks. The size of the network is set to 1024 peers, and the clustering coefficient of the network is set to 1.

Content distribution. The content is modeled by utilizing the ACM Computing Classification System [18] as the underlying meta-data vocabulary. The ACM Computing Classification System is a taxonomy consisting of 1473 topics for the domain of scientific literature in the field of computer science. Each resource is annotated with one keyword, i.e., each resource is an instance of one leaf topic from the taxonomy. In total, the ACM Computing Classification System taxonomy contains 910 leaf topics. In order to represent each research topic equally, the same number of resources is created for each leaf topic. Each peer stores on average 30 resources, and the total number of resources in the network is 30940 resources.

In real-world settings, the available resources in a peer-to-peer network are not randomly distributed among the peers. Instead, patterns in the data can be observed since the interests of the peers are not uniform, but each peers concentrates on one or a few areas it has special interests in. In the application scenario of researchers, the special interests are certain research areas. Modeling research areas with the leaf topics from the taxonomy would lead to very narrow interests. Therefore, the subtopics of a third-level topic of the taxonomy are facilitated for modeling a research area. There are 177 third-level topics in the taxonomy. The assumption is that each peer is an expert on a certain research area and for this reason, on average 60% of the resources in his or her repository are instances of one particular research area. On average, another 20% of the resources are related to another research area. The remaining 20% of the resources are instances of random leaf topics. Note that the hierarchical relationships between the topics in the taxonomy are used only for modeling the research areas, but not for query routing. This topic has been addressed in [19], together with an evaluation of the impact of different content distributions on the performance of the algorithm.

Query distribution. An uniform query distribution is employed. For uniformly distributing the queries within the network, a ticker clock at each peer is used. The probability that a peer issues a query within one time unit is set to 0.1. Each query consists of a randomly selected leaf topic from the ACM Computing Classification System taxonomy.

3.2 Metrics

Several scientific articles about peer-to-peer systems use the metrics of precision and recall [20] known from information retrieval for their evaluation. The drawback of relying on these metrics is that they do not include the traffic created in

ρ	evaporation factor	0.07
ttl_{max}	timeout of forward ants	15
w_e	weight of exploiting vs. exploring strategy	0.85
r_{max}	maximum number of resources	10
w_d	weight of resource quantity vs. link costs	0.5
$p_{sendAdditional}$	bootstrapping parameter	0.0

Table 1. Parameter values chosen for the SEMANT algorithm

the network. Our aim is to create an algorithm for query routing which has an optimal ratio between network traffic and quantity of results. Therefore, we do not rely on precision and recall, but employ the following metrics instead:

- *Resource usage* is defined as the number of links traveled for each query within a given period of time.
- *Hit rate* is defined as the number of resources found for each query within a given period of time.
- *Efficiency* is the ratio of resource usage to hit rate. If we divide the number of links traveled by the number of resources found, we get the average number of links traveled to find one resource, which is the most practical metric.

Obviously, these metrics have the drawback that the recall – which measures the ratio between resources found and resources present in the network – is not known. The value for precision, which measures the ratio between correctly found resources and false positives, is always 100% since the SEMANT algorithm does not produce false positives.

3.3 Performance Evaluation

For performance evaluation, the SEMANT algorithm is compared against the well-known *k-random walker* approach [21]. A random walker is similar to a forward ant, except from the fact that it does not rely on routing tables, but instead makes a random decision about which outgoing link to choose in the link selection procedure. If resources are found, the walker sends a message back to the querying peer (similar to a backward ant), and walks on. There are two configurable parameters: k is the number of walkers per query, and a time-to-live parameter TTL defines the timeout for the walkers.

The time horizon for the experiment is set to 5000 time units. The parameter values shown in Table 1 are used for the configurable parameters of the SEMANT algorithm. Both the *minResource* variation and the *maxResults* variation are evaluated. In order to provide for comparison fairness,

- both algorithms use the same setup as described in Sec. 3.1,
- the time-to-live parameters are set to an equal value, and
- the parameter settings for the algorithms are set in such a way that – in total – the agents of both algorithms are allowed to travel a comparable number of links.

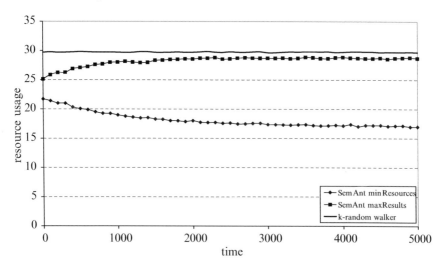

Fig. 2. Resource usage comparison between SEMANT using the *minResources* variation, SEM-ANT using the *maxResults* variation, and *k-random walker*

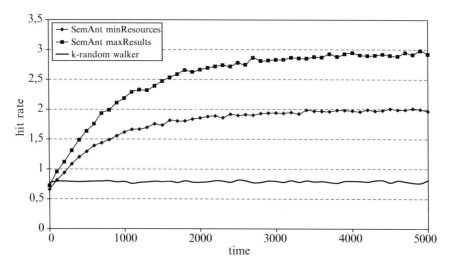

Fig. 3. Hit rate comparison between SEMANT using the *minResources* variation, SEMANT using the *maxResults* variation, and *k-random walker*

Consequently, the parameters for the *k-random walker* algorithm are set to $k = 2$ and $TTL = 15$. The results of the comparison are shown in Fig. 2 to Fig. 4. Note that Fig. 2 includes the resource usage of both forward and backward ants/agents for both algorithms.

These figures show that both variations of the SEMANT algorithm outperform the *k-random walker* approach. On average, the latter needs 29.7 hops for retrieving 0.79 results per query. This means that approximately 37.3 hops are necessary for

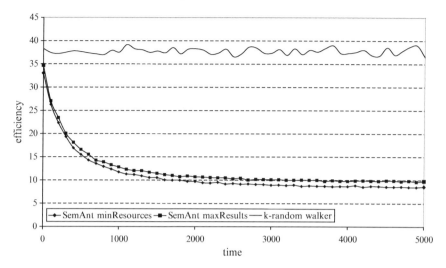

Fig. 4. Efficiency comparison between SEMANT using the *minResources* variation, SEMANT using the *maxResults* variation, and *k-random walker*

retrieving one appropriate resource. Since the *k-random walker* approach does not include any kind of optimization, the performance values do not improve but stay nearly constant over time. On the contrary, the SEMANT algorithm – which in the start-up phase, where all the pheromone trails store the same amount of pheromone, shows results which are only slightly better than those of the *k-random walker* reference – optimizes its performance. Particularly between time unit 0 and time unit 1000, the performance values improve significantly. About 1500 time units are necessary to reach a converged phase, after which the results are nearly stable and show only slight improvements.

The figures clearly indicate that, depending on which variation of the SEM-ANT algorithm used, two different optimization problems are solved. In case of the *maxResults* variation, the hit rate is maximized. After 5000 time units, on average 2.9 resources per query are found (Fig. 3). However, the resource usage values are only slightly better than those of the *k-random walker* reference. After 5000 time units, on average 28.7 hops are necessary for query routing (Fig. 2). On the other hand, employing the *minResource* variation minimizes the resource usage of the ants. In this case, after 5000 time units 16.9 hops are needed on average (Fig. 2) for finding on average 1.9 resources per query (Fig. 3). Note that in Fig. 2, the dotted curve for resource usage between time unit 0 and time unit 2500 in the *minResource* variation does not seem to decline as rapidly, but considering that (1) the hit rate increases at the same time and that (2) hit rate has an effect on resource usage, the decline of the curve is significant. The effects of minimizing resource usage and maximizing hit rate depending on the variation are even more present if a higher value for parameter ttl_{max} is used (see [14] for figures). However, employing a higher value for ttl_{max} has a negative effect on performance.

Since hit rate and resource usage depend on each other, the best way of comparing the variations against each other is to use the metric of efficiency (Fig. 4). It can be seen that in the *minResource* variation less hops are necessary to retrieve one resource in comparison to the *maxResults* variation. These results show that for the overall performance of the system, the *minResource* variation has a higher benefit in terms of a lower number of hops traveled for retrieving one resource. The average difference over the complete time span is approximately 1.13 hops per query (12.51 hops on average/11.38 hops on average). The cause for this difference can be explained as follows. In the *maxResults* variation, when the forward ants continue searching after they found the first appropriate peer, they tend to stay in the neighborhood of the peer they already found and try to go back to it, because the pheromone trails indicate that there is an appropriate peer nearby. However, this wastes resources, since forwards ants are not allowed to visit the same peer twice. For this reason, the *minResource* variation of the SEMANT algorithm is more favorable than the *maxResults* variation.

3.4 Exploration versus Exploitation

Now we address the exploitation-exploration dilemma by acquiring the simulation results for several different settings for parameter w_e. Since it turned out to perform better in the previous experiment, the *minResource* variation is employed. For the parameters other than w_e, the settings described in Sec. 3.3 are used. The results of the comparison are given in Fig. 5 to Fig. 7. Fig. 5 shows that in the start-up phase, the resource usage is proportional to the ratio between exploring and exploiting strategy. The more forward ants employ the exploring strategy, the higher the amount of messages in the network. Consequently, for the overall load of the system it is better to employ a low rate of exploring forward ants, but also a high rate, e.g., $w_e = 0.05$, is feasible. Although in this case the resource usage at time unit 0 is much higher (168.4 hops on average) than in case of a low rate (e.g., 16.7 hops on average for $w_e = 0.95$), still not even half as many resources are consumed as when using broadcast with a time-to-live parameter of 4. The hit rate is dependent on parameter w_e not only in the start-up phase, but also in the converged phase (Fig. 6). The more exploring forward ants in the network, the better the pheromone trails and, consequently, the higher the hit rate. The best result can be reached for setting $w_e = 0.05$. After 5000 time units, on average 2.24 resources are found in this case. The worst possible result (1.92 resources on average after 5000 time units) is obtained for setting $w_e = 0.95$. However, the difference in absolute numbers is marginal and the curves converge to the same limit over time.

Combining the two metrics shows that, independently from the ratio between exploring and exploiting strategy, the overall efficiency in terms of number of hops necessary for answering a query is similar for every setting of parameter w_e. Although a higher rate of exploring ants gives slightly less efficient results in the start-up phase, at the same time it moderately improves the performance in the converged phase. Fig. 7 highlights this by depicting the efficiency results for $w_e = 0.05$ and $w_e = 0.95$ plotted in log-log scale. Table 2, showing the average efficiency

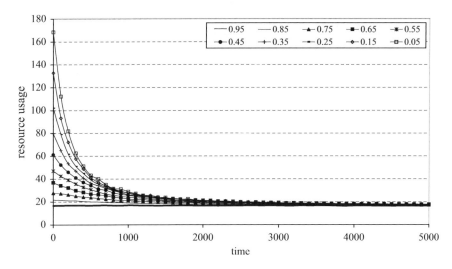

Fig. 5. Resource usage of SEMANT using the *minResources* variation when varying the value used for parameter w_e. Between time unit 0 and time unit 1500, resource usage is dependent on parameter w_e

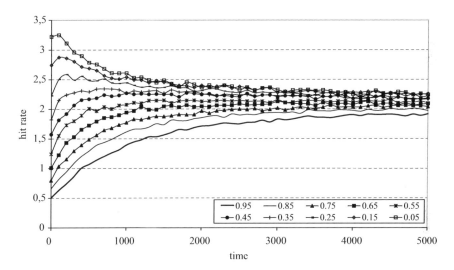

Fig. 6. Hit rate of SEMANT using the *minResources* variation when varying the value used for parameter w_e. The curves converge to the same limit

over the complete time span for every setting of parameter w_e, reveals that those two effects are in balance with each other. The mean of all values is 11.39 hops per query, and the low standard deviation of 0.087 indicates that there are no significant differences in average efficiency when varying the ratio between exploring and exploiting strategy. The reason for this effect lies in the design of the exploring strategy. It can

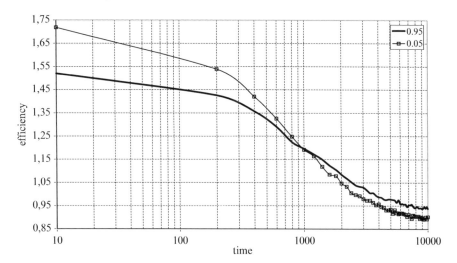

Fig. 7. Log-log plot of the efficiency of SEMANT using the *minResources* variation for parameter $w_e \in [0.05, 0.95]$. A higher rate of exploring ants results in slightly less efficient results in the start-up phase, but improves the performance in the converged phase. The curves for the other settings of parameter w_e lie in between those shown and are omitted for clarity. Note that the time axis is stretched by factor 2

0.05	0.15	0.25	0.35	0.45	mean
11.43	11.31	11.33	11.45	11.25	11, 38619209
0.55	0.65	0.75	0.85	0.95	standard deviation
11.35	11.37	11.42	11.38	11.57	0, 087454343

Table 2. Average efficiency of SEMANT using the *minResources* variation when varying the value used for parameter w_e

be explained by the fact that, although the forward ants choose outgoing links that are not currently known as the best ones, they still make this decision in proportion to their desirability.

4 Related Work

The most well-known practical approach towards employing ant algorithms in peer-to-peer networks is the Java-based open source framework *Anthill* [22] for the design, implementation, and evaluation of ant algorithms in peer-to-peer networks. An *Anthill* system is an overlay network of interconnected nests which correspond to peers. Nests provide services like document storage, routing table management, topology management, and generation of ants upon user requests. In addition, the *Anthill* API provides a basic set of actions for ants that enables them to travel from nest to nest, and to interact with the services provided by them. The main difference between *Anthill* and SEMANT is that the *Anthill* project focuses on the development of the framework. It does not specify ant algorithms. This task is left to the users of the framework according to the application scenario. However, there are two applications based on the *Anthill* framework which were built by the *Anthill* developers for demonstration and evaluation purposes.

Gnutant [22] is a file-sharing application. In this application, each file is identified by an unique file identifier and associated with meta-data comprised of textual keywords. The query routing algorithm is based on the ant metaphor, but not on the principles defined by the Ant Colony Optimization meta-heuristic. Three different types of ants are responsible (1) for constructing a distributed index that contains URLs pointing to shared files and (2) for managing routing tables. If a user adds a new file to a nest, one *InsertAnt* is generated for each keyword of the file. Insert-Ants propagate the presence of new files to the network by updating the distributed index. Gnutant utilizes hashed keyword routing based on the Secure Hash Algorithm (SHA). Each index entry contains the hash value of a keyword together with a set of nests that are likely to store files associated with the given keyword. *SearchAnts* are generated upon user queries and exploit the information stored in the routing tables in order to find files that match the queries' keywords. If no index entry that exactly matches the query's hash value exists, the SearchAnt selects the hash value that most closely matches the hash value of the query. If a SearchAnt localizes an appropriate file, it generates a *ReplyAnt* that immediately returns to the source nest and informs the user about the result of his or her query. The SearchAnt itself continues traveling until it reaches a defined time-to-live (TTL) parameter, that is, the maximum number of hops the ant is allowed to move. Once the TTL parameter is reached, it returns to its source nest via the same path it traveled before and updates both routing tables and distributed index. The second application that was built using *Anthill* is called Messor [23]. It is not intended for search, but for load-balancing in peer-to-peer networks based on the necrophoric behavior of ants.

Schelthout et al. [24] evaluate whether the principle of synthetic pheromone can be employed for coordination in distributed agent-oriented environments. Their framework is based on the idea of objectspaces known from concurrent computing. Similar to *Gnutant*, Schelthout et al. create pheromone trails for each query and allow agents to follow trails that represent only one out of the multiple keywords in the query. Evaporation is used and the evaporation factor is set to 0.1.

In addition to the directly related work, there are three areas of distantly related work. Firstly, there is work addressing the applicability of biological processes other than ant-based methods to search in peer-to-peer networks. For example, Babaoglu et al. [25] apply the principles of proliferation to search in unstructured overlay networks. Secondly, there are approaches to peer-to-peer which are inspired by stigmergy, but do not derive from the Ant Colony Optimization meta-heuristic. For instance, Handt [26] tackles the problem of search in peer-to-peer networks by inverting it. Instead of peers issuing queries, the annotated resources move through the network and lay/follow pheromone trails that guide them. In addition, the peers re-order according to their interests which are determined by creating node profiles out of the meta-data their locally stored resources are marked-up with.

Thirdly, there are ant-based approaches to peer-to-peer tasks other than search. The open-source project *MUTE* [27] implements a peer-to-peer system that relies on the ant metaphor for user discovery. Ants are used to track which neighbor connections are associated with particular sender addresses. Distributed search is based on controlled flooding to locate files by name based on free-form query strings. Wu and Aberer [28] use swarm intelligence to create a model for the dynamic interactions between Web servers and users, which provide relevance feedback by browsing Web pages. This model is used for ranking Web documents. A swarm intelligent module is added to the Web server architecture. The trail laying feature used is composed of an accumulation feature that increases the pheromone amount when a user visits a page, and of a spreading feature in which pheromone is diffused to the pages that link to a certain page. The spreading feature does not comply with the Ant Colony Optimization meta-heuristic. Labella et al. [29] present a foraging-inspired approach for division of labor between a group of robots performing object search and retrieval. The goal is to distributively determine which agent is suited best for carrying out a certain task using indirect communication only.

5 Conclusion

In this paper we evaluated the effectiveness of the self-organizing and emergent behavior observed from natural ants for query routing in peer-to-peer networks. We specified two possible variations of the SEMANT algorithm based on the *Ant Colony Optimization* meta-heuristic and experimentally evaluated the performance of both variations in order to identify the superior one. After that, we addressed the exploration-exploitation-dilemma by showing that the overall efficiency of the SEM-ANT algorithm's search process is not influenced by the parameter value chosen for the ratio between exploring and exploiting forwards ants. This finding is important, because it simplifies the configuration of the algorithm and makes its behavior more easily predictable.

References

1. Gianni Di Caro, Frederick Ducatelle, and Luca Maria Gambardella, "AntHocNet: An Ant-based Hybrid Routing Algorithm for Mobile and Ad Hoc Networks," in *Proceedings of Parallel Problem Solving from Nature*. September 2004, vol. 3242 of *Lecture Notes in Computer Science*, Springer.
2. Gianni Di Caro and Marco Dorigo, "AntNet: Distributed Stigmergy Control for Communications Networks," *Journal of Artificial Intelligence Research (JAIR)*, vol. 9, pp. 317–365, July 1998.
3. Sam Joseph and Takashige Hoshiai, "Decentralized Meta-Data Strategies: Effective Peer-to-Peer Search," *IEICE Transactions on Communications*, vol. E86-B, no. 6, pp. 1740–1753, June 2003.
4. Elke Michlmayr, Arno Pany, and Sabine Graf, "Applying Ant-based Multi-Agent Systems to Query Routing in Distributed Environments," in *Proceedings of the 3rd IEEE Conference On Intelligent Systems (IEEE ISŠ06)*, September 2006.

5. Elke Michlmayr, "Ant Algorithms for Search in Unstructured Peer-to-Peer Networks," in *Proceedings of the Ph.D. Workshop, 22nd International Conference on Data Engineering (ICDE 2006)*, April 2006.
6. Marco Dorigo and Gianni Di Caro, *New Ideas in Optimization*, chapter The Ant Colony Optimization Meta-Heuristic, pp. 11–32, McGraw-Hill, 1999.
7. John H. Holland, *Adaptation in Natural and Artificial Systems*, MIT Press, 1992.
8. Stewart W. Wilson, "Explore/Exploit Strategies in Autonomy," in *From Animals to Animats 4: Proceedings of the 4th International Conference on the Simulation of Adaptive Behavior*, 1996, pp. 325–332.
9. Richard S. Sutton and Andrew G. Barto, *Reinforcement Learning: An Introduction*, MIT Press, 1998.
10. Yael Niv, Daphna Joel, Isaac Meilijson, and Eytan Ruppin, "Evolution of Reinforcement Learning in Uncertain Environments: A Simple Explanation for Complex Foraging Behaviors," *Adaptive Behavior*, vol. 10, pp. 5–24, 2002.
11. Lilia Rejeb and Zahia Guessoum, "The Exploration-Exploitation Dilemma for Adaptive Agents," in *Proceedings of the 5th European Workshop on Adaptive Agents and Multi-Agent Systems (AAMASŠ05)*, March 2005.
12. Mikael Holmqvist, "Experiential Learning Processes of Exploitation and Exploration Within and Between Organizations: An Empirical Study of Product Development," *Organization Science*, vol. 15, no. 1, pp. 70–81, 2004.
13. Matei Ripeanu, "Peer-to-Peer Architecture Case Study: Gnutella Network," in *Proceedings of the 1st IEEE International Conference on Peer-to-Peer Computing (P2P2001)*, August 2001.
14. Elke Michlmayr, "Specification of the SemAnt Algorithm," Tech. Rep., Vienna University of Technology, March 2006.
15. Marco Dorigo and Luca Maria Gambardella, "Ant Colony System: A Cooperative Learning Approach to the Traveling Salesman Problem," *IEEE Transactions on Evolutionary Computation*, vol. 1, pp. 53–66, 1997.
16. David E. Goldberg and Kalyanmoy Deb, "A Comparative Analysis of Selection Schemes Used in Genetic Algorithms," in *Proceedings of the 1st Workshop on Foundations of Genetic Algorithms*, July 1990, pp. 69–93.
17. Jon M. Kleinberg, "Navigation in a small world," *Nature*, vol. 406, pp. 845, August 2000.
18. Association for Computing Machinery, "ACM Computing Classification System (ACM CCS)," 1998.
19. Elke Michlmayr, Arno Pany, and Gerti Kappel, "Using Taxonomies for Content-based Routing with Ants," in *Proceedings of the Workshop on Innovations in Web Infrastructure, 15th International World Wide Web Conference (WWW2006)*, May 2006.
20. Ricardo Baeza-Yates and Berthier Ribeiro-Neto, *Modern Information Retrieval*, Addison Wesley Longman Publishing Co. Inc., 1999.
21. Qin Lv, Pei Cao, Edith Cohen, and Scott Shenker, "Search and Replication in Unstructured Peer-to-Peer Networks," in *Proceedings of the 16th ACM Conference on Supercomputing*, June 2002, pp. 84–95.
22. Ozalp Babaoglu, Hein Meling, and Alberto Montresor, "Anthill: A Framework for the Development of Agent-Based Peer-to-Peer Systems," in *Proceedings of the 22nd International Conference on Distributed Computing Systems (ICDCS 02)*. July 2002, IEEE.
23. Alberto Montresor, Hein Meling, and Ozalp Babaoglu, "Messor: Load-Balancing through a Swarm of Autonomous Agents," in *Proceedings of the 1st International Workshop on Agents and Peer-to-Peer Computing*, July 2001.

24. Kurt Schelfthout and Tom Holvoet. A Pheromone-Based Coordination Mechanism Applied in Peer-to-Peer. In *2nd International Workshop on Agents and Peer-to-Peer Computing (AP2PC 2003)*, volume 2872 of *Lecture Notes in Computer Science*, pages 71–76. Springer, July 2003.
25. Ozalp Babaoglu, Geoffrey Canright, Andreas Deutsch, Gianni Di Caro, Frederick Ducatelle, Luca Maria Gambardella, Niloy Ganguly, Márk Jelasity, Roberto Montemanni, and Alberto Montresor, "Design Patterns from Biology for Distributed Computing," in *Proceedings of the European Conference on Complex Systems (ECCS05)*, November 2005.
26. Arne Handt, "Self-Organizing Information Distribution in Peer-to-Peer Networks," in *Proceedings of New Trends in Network Architectures and Services: International Workshop on Self-Organizing Systems (IWSOS 2006)*, September 2006.
27. Jason Rohrer, "MUTE: Simple, Anonymous File Sharing," `http://mute-net.sourceforge.net/`, 2005.
28. Jie Wu and Karl Aberer. Swarm Intelligent Surfing in the Web. In *Proceedings of the 3rd International Conference on Web Engineering*, July 2003.
29. T.H. Labella, M. Dorigo, and J.-L. Deneubourg. Division of labour in a group of robots inspired by ants' foraging behavior. *ACM Transactions on Autonomous and Adaptive Systems*, 1(1):4–25, 2006.

A Bio-Inspired Location Search Algorithm for Peer to Peer Networks

Sachin Kulkarni[1], Niloy Ganguly[1], Geoffrey Canright[2] and Andreas Deutsch[3]

[1] Department of Computer Science and Engineering, Indian Institute of Technology Kharagpur, India
`sachindkulkarni@gmail.com,niloy@cse.iitkgp.ernet.in`
[2] R&I Telenor, Oslo, Norway `geoffrey.canright@telenor.com`
[3] Center for Information Services and High Performance Computing (ZIH), TU-Dresden, Dresden 01062 `andreas.deutsch@tu-dresden.de`

Summary. In this paper, we propose a new $p2p$ network based location search algorithm. The algorithm is built upon the concept of a gradient search and is applicable to unstructured networks. It is inspired by a biological phenomenon called haptotaxis. The algorithm performs much better than the random walk algorithm and is also more efficient than flooding. We also present some mathematical reasoning to explain the superiority of the algorithm.

1 Introduction

A p2p based system is a network that relies on the computing resources available from all the participants in the network instead of concentrating on a small number of centralized servers. This kind of network allows to build a server-less infrastructure, where participating nodes cooperate to find the desired resources. Moreover many p2p architectures offer inherent scalability and robustness as the network reorganizes itself in case of dynamic entry and removal of nodes. However there is a cost associated with these desirable properties. The latency of locating the desired resource in the network increases as we use a $p2p$ network. Optimizing this lookup time is one of the primary challenges of p2p systems.

In this chapter, we propose a new algorithm based on a *gradient search*, which is a key-based algorithm. It performs a guided search for a desired key in the entire search space. *The algorithm is motivated by a biological phenomenon called haptotaxis, hence named "hapto-search"* [2]. We have previously developed algorithms for search, which are inspired by the natural immune system and chemotaxis [4], [5]. Both haptotaxis and chemotaxis are used in biological interacting cell systems to regulate cell migration [3]. The regulated cell migration phenomenon inspires us to build a gradient-based search algorithm.

2 Related Work

Based on the type of neighborhood relationship between the nodes, p2p systems can be classified as structured and unstructured/semistructured. For unstructured p2p networks, one of the simplest approach is flooding, which is adopted by Gnutella [9]. Although flooding quickly finds out the desired location, it is inefficient as it produces a huge number of message packets. [1] describes random walk strategies in power law networks. It shows that k-walker random walk is much more scalable search method than flooding but at the expense slight increase in the average number of hops. It also discusses replication-based search strategies and its analysis shows that the search cost expected in such strategies is $\frac{n}{r}$, where n is the network size and r is the number of replicas. Our algorithm, hapto-search is also a replication-based algorithm. However we observed that the hapto-search algorithm is far more time efficient.

Our algorithm to some extent resembles the algorithm used in loosely structured p2p systems such as freenet [10].

In structured p2p systems, the network topology is tightly controlled and the placement of the resources is done not at random nodes but at specific locations. This helps the queries to locate the desired resources efficiently. The highly structured p2p systems like CHORD and CAN use precise placement algorithms to achieve performance bound of $\log N$, where N is the number of nodes in the network. However the imposition of the structure on the topology restricts their ability to operate in presence of extremely unreliable nodes. In hapto-search, we assume totally unstructured network, hence is extremely robust.

3 Definition of the Hapto-search Algorithm

In this section, we discuss the hapto-search algorithm, its inspiration and details of how it works.

3.1 Design Goal and Basic Idea

Main design goal of this algorithm is to attain a speed of location search comparable to DHTs (i.e. $\log(N)$) but to make no assumption about the structure of the network. Structured networks are more prone to failure and incur overhead to achieve robustness while robustness is a natural property of unstructured networks.

The algorithm defines a gradient-based search, in which every node is assigned a k-bit key unique within our p2p network. The key represents the identity of the node (Node ID, IP Address). A node distributes its key to a fixed number of peers in the network at random. When a peer wants to search for a node, it tries to search for any node, which has the key of the destination. To do so, a query traverses in the network and the next destination of the query is chosen in such a way that it ensures that the query always moves "closer" to the destination than the current position.

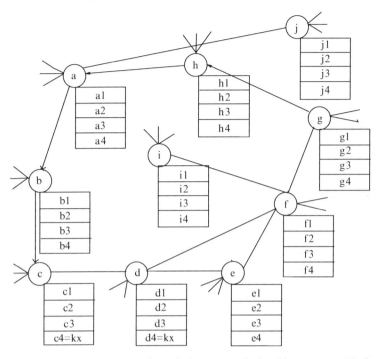

Fig. 1. Sketch of a network showing how the hapto-search algorithm operates. Each node is hosting 4 keys received from random peers. This diagram illustrates a case where node g wants to find node x (Node x is not shown in the figure). Nodes c and d have the key of x. Node g forwards the query to a neighbor, which has a key closer to k_x (node h in this case). This way a query traverses in the network and ultimately reaches node c. The path indicated by arrows shows a possible query traversal

3.2 Our Inspiration-Haptotaxis

Migration of cells within our body is a fundamental process for example in tissue development, tumor metastasis or wound healing. To achieve appropriate physiological outcomes, cells must adjust direction and speed of their movement in response to environmental stimuli. There are cell adhesion proteins at the outer surface of the cells, which bind to the adhesion ligands present in the extracellular matrix (ECM). This extracellular matrix is a substratum that surrounds the cells in a tissue. Cells tend to move in the direction where the adhesion between ECM ligands and cell receptors is larger. It is believed that the magnitude of this adhesion influences cell speed and random turning behavior, whereas a gradient of adhesion affects the resultant direction of cell movement. This phenomenon of a guided and gradient based movement of cells is called haptotaxis. Haptotaxis is influenced by the presence of the ECM consisting of immobilized molecules. Alternatively, cell migration towards a gradient of diffusible and soluble substances is known as chemotaxis.

In our example, the ECM resembles a p2p network. Then, a query path in the p2p network can be viewed as a haptotactic cell movement in the ECM. Accordingly, the concept of haptotaxis gives an idea of a guided search towards the destination so that one can always make sure that at every step, the moving entity is getting "closer" to its destination.

3.3 Detailed Algorithm

In this section, we introduce the detailed algorithm, which is key-based. It allows nodes to retrieve location information of each other. The key assigned to each node is unique across the network. This key can be generated by hashing the user id into a n-bit length code. Every node distributes its key, location information (IP Address) to a fixed number of nodes in the network. This distribution is done at random. A node (say) A knows the location information of another node (say) B, if it has the key of B. Thus if one node wants to search the location information of the destination, it tries to search for any node, which has the key for this destination. Once it finds such node, it can get the desired location information from it.

If a node doesn't have the key required, it routes the query to the neighbor, which is closest to the destination. A neighbor is said to be closer to the destination if it has a key which is closer to the destination key (in terms of Hamming distance) than the current node. The closest neighbor is selected to forward the query. This way the algorithm ensures that the query always moves "closer" to the destination as the Hamming distance between destination key and the closest key present at the current node always decreases, or in the worst case remains the same. The search ends when the query reaches a node hosting the destination key (i.e. Hamming distance becomes zero). Figure 1 illustrates the steps of the algorithm.

We have developed two versions of this algorithm, *simple* and *restricted*. In the simple version, the next hop to forward the query is chosen regardless of whether it has already been visited during the same search query or not. As it doesn't remember the path, it may happen that a query goes into the loop and keeps on moving in the circular path because of its deterministic nature. In the case of the restricted version, we maintain the path vector storing all the nodes visited during the search in that order. The query is forwarded to such a neighbor which is closer to the destination and which has not already been visited during the search. Thus loops are avoided and the query is guaranteed to reach the destination within N hops.

Dead end: It may happen that the current node itself is closer to the destination than all its neighbors. We can call this a *dead end* as there is no closer neighbor to forward the query. In such case, we continue the search by sending the query to the farthest neighbor, i.e. the node for which the Hamming distance of its key from the destination key is largest among all the neighbors of the current node.

The occurrence of dead ends implies the presence of local maxima in the search space. Dead ends will slow down the speed of the search algorithm. However, occurrence of dead ends depends upon the number of times a key is replicated as well as

Algorithm 3.1 Searching a key in an unstructured peer to peer network using the hapto-search algorithm. The input to the algorithm is the key for the destination and the output is the location information of the destination. The diagram shows both the simple and the restricted version of the algorithm. In the simple version, query is forwarded to a neighbor with shortest possible distance (and shorter than the current node's shortest distance) where as in the restricted version, besides choosing shortest of the shortest, a path vector is maintained to store all the nodes visited during the search and query is forwarded to a neighbor only if it is not present in the path vector.

Algorithm: Hapto-search

input : Key of the destination node
output: Location information of destination node

current ←source **while** *current node doesn't have the destination key* **do**
  for *all the neighbors of current node in the network* **do**
    shortest ← GetShortestHammingDistance
    (neighbor,destination); **if** *neighbor's shortest distance < my shortest distance* **then**
    | Add the neighbor to the sorted list of prospects
    end
    if *neighbor's shortest distance > farthest* **then**
    | Store neighbor and neighbor's distance as farthest
    end
  end
  while *List of prospects is not empty* **do**
    node ← take a node from the sorted list with minimum distance   /* In the restricted version, this if condition is necessary whereas in the simple version, no such check is performed.   */
    if *node not already visited in the search process* **then**
    | current ← node break
    end
  end
  if *no such node available* **then**
    /* This is the condition of dead end   */
    current ← choose the neighbor with the largest distance
  end
end

the average network degree. The details of both the simple and the restricted version of the hapto-search algorithm are illustrated through algorithm 3.1 whereas algorithm 3.2 depicts the simple random-search algorithm in which, the next hop to forward the query is chosen at random among the list of neighbors of the current node. The random-search algorithm is used as a benchmark to measure the efficiency of the hapto-search algorithm.

Algorithm 3.2 Searching a key in an unstructured peer to peer network using the random-search algorithm. In this algorithm, next hop is chosen at random from the list of neighbors.

Algorithm: Random search

input : Key of the destination node
output: Location information of destination node

current ← source **while** *current node doesn't have the destination key* **do**
  node ← Choose a node at random from the list of neighbors of current node
  current ← node
end

4 Analysis of Hapto-search

In this section, we present analysis of the hapto-search algorithm. We will be considering the simplified version of the algorithm for the analysis and the underlying network is assumed to be an E-R graph. We will be mainly interested in finding an upper bound for the average number of hops.

4.1 Average Number of Hops

Consider an E-R graph of N nodes with average node degree M. Now in our algorithm, every node distributes its key to say, R peers at random such that every peer has approximately R keys. Without loss of generality, we can assume the key of the destination to be 0 (all n bits are zero). In the simple hapto-search algorithm, a node not having the destination key, forwards the query to a neighbor which is "closer" to the destination or at least as close as the current node.

To compute the average number of hops, we need to derive formula for the rate of change in Hamming Distance (ΔHD) per hop traversal of the query. We need to know the probabilities of the current node being a local maximum and a global maximum. This can be represented by P_{lm} and P_{gm} respectively. We also need to introduce two more parameters, Average Forward Move (AFM) and Average Backward Move (ABM), where AFM is defined as the average change in HD of the current node from the destination in a single hop if the current node is not a maximum. Similarly ABM is defined as the average change in HD of the current node from the destination in a single hop if the current node is a local maximum. Using these parameters, we can calculate ΔHD as the following theorem states.

Theorem 1. *In the simple hapto-search algorithm, the expected change in HD (ΔHD) in a single hop traversal of query is given as:*

$$\Delta HD = P_{lm} * (-ABM) + (1 - P_{lm} - P_{gm}) * AFM$$

Proof: Consider a scenario where the current node is at HD i from the destination (destination key is assumed to be 0). Now there are three cases.

(i.) If the node is a global maximum (a node having the replication of destination key): In this case, search algorithm ends. So $\Delta HD = 0$
(ii.) If the node is a local maximum (dead end): In this case, it forwards the query to the farthest possible neighbor. Here HD is increased by the value ABM. So $\Delta HD = ABM$
(iii.) Otherwise: The current node finds a neighbor closer to the destination than itself to forward the query. In this case, HD is reduced by the value AFM. So $\Delta HD = AFM$

Therefore we can say that,

$$\Delta HD = P_{lm} * (-ABM) + (1 - P_{lm} - P_{gm}) * AFM$$

The average number of hops can be directly calculated from ΔHD. The following theorem states that.

Theorem 2. *Given that ΔHD is the expected change in HD in a single traversal of the query and IHD is the HD of the source (query generating) node from the destination, average number of hops taken by the query to reach the destination is given by the equation:*

$$Hops = \frac{IHD}{\Delta HD}$$

Now we will derive a formula for the probability of a local maximum, AFM and ABM.

4.2 Probability of a Local Maximum

A local maximum is a node, which doesn't have the destination key and it has no neighbor, which has a key closer to the destination key than the best key available at the node itself. Without loss of generality we can assume that the destination key is 0. In such case, there will be ^nC_i nodes in the network, which may have their key at HD i from the destination, where n is the number of bits used to define the key of a node. So the number of possible keys at HD i from the destination ($h(i)$) can be given as,

$$h(i) = {}^nC_i \tag{1}$$

Let $H(i)$ denote the number of keys at HD $> i$ from the destination. Then $H(i)$ can be computed by the following equation

$$H(i) = \sum_{k=i+1}^{n} (h(k)) \tag{2}$$

To compute the probability of a local maximum, let us first derive the formula for the probability that a node is at HD i from the destination.

Theorem 3. *Given the destination key as 0, the probability that a node is at HD i from the destination is given by:*

$$P_i = \sum_{j=1}^{R} \left(\frac{{}^{h(i)}C_j * {}^{H(i)}C_{R-j}}{{}^{2^n}C_R} \right)$$

Proof: A node will be at HD i from the destination if it has at least one key which is at HD i from the destination and no other key is closer than i. Each node has R keys in its storage. So there are R ways of it, where in each case the current node has j keys at HD i ($1 \leq j \leq R$). So the total number of ways a node can have at least one key at HD i can be given as,

$$n_i = \sum_{j=1}^{R} \left({}^{h(i)}C_j * {}^{H(i)}C_{R-j} \right)$$

Now the total sample set consists of the number of ways of choosing R keys from all possible keys. Total number of possible keys is 2^n. So there are ${}^{2^n}C_R$ ways of choosing R keys for a node. Hence the probability that a node at HD i can be given as,

$$P_i = \sum_{j=1}^{R} \left(\frac{{}^{h(i)}C_j * {}^{H(i)}C_{R-j}}{{}^{2^n}C_R} \right)$$

Theorem 4. *The probability that a node is a maximum is given as,*

$$P = \sum_{h=0}^{n-1} \left(P_h * \left(\sum_{j=h+1}^{n} P_j \right)^M \right)$$

Proof: The probability of a maximum is the probability that the best key available at the current node has HD less than the best key available at any of its neighbors. This can happen in many ways. A node can have best key at HD 0 (P_0) and keys available at all its M neighbors are at HD $1..n$ (n is the number of bits in a key). The probability of this case can be given as $(P_0 * (\sum_{j=1}^{n} P_j)^M)$. Alternatively, a node can have the best key at HD 1 and all the keys available at its neighbors are at HD $2..n$ and so on. Hence the probability of a maximum is summation of probabilities of all such cases. Hence P can be given as,

$$P = \sum_{h=0}^{n-1} \left(P_h * \left(\sum_{j=h+1}^{n} P_j \right)^M \right)$$

This equation gives us the probability of a maximum in the network, which includes both global and local maximum. When the query reaches a node at a global maximum, the search algorithm ends. As there are R replications of every key in the network, probability of a global maximum P_{gm} will be $\frac{R}{N}$. Hence the probability of a local maximum will be $P_{lm} = P - P_{gm}$.

4.3 Average Forward Move (AFM)

In order to compute the AFM, let us define Pb_i, which denotes the probability that the best possible neighbor of the current node is at HD i from the destination, given that the current node is not a maximum.

Theorem 5. *The probability that the best possible neighbor of the current node is at HD i from the destination given that current node is not a maximum, is,*

$$Pb_i = \left(1 - \sum_{k=0}^{i-1} P_k\right)^M - \left(1 - \sum_{k=0}^{i} P_k\right)^M$$

Proof: That the best possible neighbor of the current node is at HD i indicates that all the neighbors are at HD $\geq i$ with at least one neighbor at HD i. We know that the probability of a node being at HD i is P_i. Then the probability of all M neighbors being at HD $\geq i$ will be $(1 - \sum_{k=0}^{i-1} P_k)^M$. But this also includes the case where all the neighbors are at HD $> i$ with no neighbor at HD i. This can occur with probability $(1 - \sum_{k=0}^{i} P_k)^M$. We need to exclude this probability from the answer. Hence Pb_i can be given as,

$$Pb_i = \left(1 - \sum_{k=0}^{i-1} P_k\right)^M - \left(1 - \sum_{k=0}^{i} P_k\right)^M$$

Theorem 6. *The expected value of an average forward move (AFM) is given by the equation:*

$$AFM = \sum_{hd=1}^{n} \left(hd * \sum_{i=hd}^{n} (P_i * Pb_{i-hd})\right)$$

Proof: When a node is not a maximum, it has a neighbor which can reduce the HD further by value between 0 and n. We call this reduction in HD as ΔHD. Then the AFM can be computed as,

$$AFM = \sum_{hd=1}^{n} (hd * P(\Delta HD = hd))$$

Consider a case when the best neighbor reduces the HD by 1 (i.e. $\Delta HD = 1$). This is possible when the current node is at HD 1 and the best neighbor is at HD 0 or the current node is at 2 and the best neighbor is at HD 1 and so on. So the probability of $\Delta HD = 1$ is,

$$P(\Delta HD = 1) = \sum_{i=1}^{n} (P_i * Pb_{i-1})$$

In general we can say that the probability of $\Delta HD = hd$ is,

$$P(\Delta HD = hd) = \sum_{i=hd}^{n} (P_i * Pb_{i-hd})$$

Substituting value of $P(\Delta HD = hd)$ in AFM formula, we can say that

$$AFM = \sum_{hd=1}^{n} (hd * P(\Delta HD = hd))$$

4.4 Average Backward Move (ABM)

In order to compute ABM, let us define Pf_{i+hd}, which denotes the probability that the farthest possible neighbor of the current node is at HD i from the destination, given that the current node is a maximum.

Theorem 7. *The probability that the farthest possible neighbor of current node is at HD $i + hd$ from the destination, given that the current node is a maximum and at HD i is,*

$$Pf_{i+hd} = \left(\sum_{k=i+1}^{i+hd} P_k\right)^M - \left(\sum_{k=i+1}^{i+hd-1} P_k\right)^M$$

Proof: This proof is quite similar to the proof of theorem 5.

Theorem 8. *The expected value of the average backward move (ABM) is given by the equation:*

$$ABM = \sum_{hd=1}^{n} \left(hd * \sum_{i=1}^{n-hd} (P_i * Pf_{i+hd})\right)$$

Proof: This proof is quite similar to the proof of theorem 6.

We performed various experiments to verify all these formulae. Details of these experiments will be discussed later in the performance evaluation section.

4.5 Analysis of Random Search

In the case of random search, a node chooses the neighbor to forward the query at random and when the query reaches the node having the destination key replica, the algorithm ends. We can derive a formula for the expected number of hops for the random search. A similar formula has been derived in [1]. We are presenting the result for the completeness of the article.

Theorem 9. *Let R be the number of replications of the key of a node in the network and N be the number of nodes in the network. Then the expected number of hops taken by the random search is given as*

$$Hops_{ran} = \frac{R}{N} * \left(\sum_{i=0}^{\infty} \left(i * \left(1 - \frac{R}{N}\right)^i\right)\right)$$

Proof: Let $P_{ran}(i)$ be the probability that the destination key replica is found at the ith hop. Then $Hops_{ran}$ can be given as,

$$Hops_{ran} = \sum_{i=0}^{\infty} (i * P_{ran}(i))$$

Given the key replication value R and the network size N, the probability that a node has the destination key replica is $\frac{R}{N}$. Now probability that the query finds the destination key replica in one hop is $\left(1 - \frac{R}{N}\right) * \frac{R}{N}$ as it is the case when first node doesn't have the destination key (probability of this being $\left(1 - \frac{R}{N}\right)$) and the second one has it (probability being $\frac{R}{N}$). Similarly the probability that the query finds the destination key replica in two hops is $\left(1 - \frac{R}{N}\right)^2 * \frac{R}{N}$. In general case, we can say that the probability of finding the destination key at ith hop ($P_{ran}(i)$) is,

$$P_{ran}(i) = \left(1 - \frac{R}{N}\right)^i * \frac{R}{N}$$

Substituting this into the formula of $Hops_{ran}$, we get

$$Hops_{ran} = \frac{R}{N} * \left(\sum_{i=0}^{\infty} \left(i * \left(1 - \frac{R}{N}\right)^i\right)\right)$$

5 Performance Evaluation

This section presents the results of the performance evaluation of the hapto-search algorithm. The experimental setup as well as the specific experiments which are performed to evaluate the performance are noted next.

5.1 Experimental Setup

We ran the algorithm (both simple and restricted versions) on E-R networks. For each network, and for different values of key replication, we ran the algorithm for randomly chosen 2000 source-destination pairs. The key size is considered as 32 for all the cases. We were mainly interested in the following aspects:

(i.) Experimental verification of the theoretical prediction made for the simple version of the hapto-search algorithm
(ii.) Performance analysis of the restricted version. The performance is measured in terms of the average number of hops. We compared the performance of the restricted version of the algorithm with state-of-the-art algorithm:
  - the random search in unstructured p2p networks
  - DHT-based search in structured p2p networks
(iii.) Understanding the dependence of the performance parameters upon the network size N for the restricted version of the algorithm.

5.2 Simple Version - Experimental Results and their Prediction by Theoretical Analysis

We ran the simple version of the hapto-search on E-R networks with different values of N (number of nodes) and R (key replication). We measured the values of AFM,

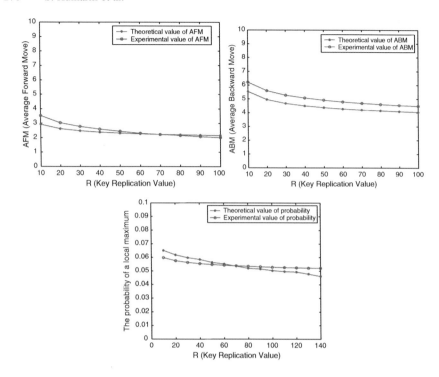

Fig. 2. Comparison of theoretical and experimental values of the parameters, Average Forward Move (AFM), Average Backward Move (ABM) and the probability of a local maximum (P_{lm}). The first figure shows AFM on the y-axis vs key replication value on the x-axis for the E-R network of 10K nodes, M=10. The second figure shows the plot of ABM on the y-axis vs key replication value on the x-axis where as the third figure plots the probability of local maxima for the same configuration. As one can observe, the theoretical and the experimental values match for all these parameters. Also interesting to note is that the parameter values except in the initial phase are almost independent of the replication value R

ABM, average hops and the number of dead ends. The number of dead ends are computed by a method called Steepest Ascent Graph (SAG)[4]. Figure 2 shows the comparison between theoretical and experimental values of the parameters, AFM, ABM and the probability of a local maximum. It can be seen that for all the plots, theoretical and experimental values match.

Figure 3 shows the same comparison for the average number of hops. Here we observe that the average number of hops (both experimentally and theoretically) in the case of the simple version is insensitive to the changes in the key replication value. One can also observe that there is a slight difference between the theoretical and experimental values for this plot. This is due to the practicalities of the

---

[4] Steepest Ascent Graph (SAG): SAG is constructed for a given network, destination and the key distribution. Every node maintains only a link with the best neighbor and drops the rest of the links. All the nodes having a self-loop are maxima.

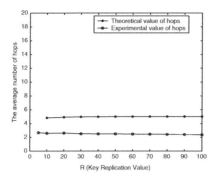

Fig. 3. Comparison of theoretical and experimental values of the average number of hops. The figure depicts the plot of the probability of a local maximum on the y-axis vs key replication value on the x-axis for the E-R network of 10K nodes, M=10. As one can observe, there is a slight mismatch between the theoretical and the experimental values. The average number of hops is largely insensitive to the key replication value

implementation of the simple version. As no path vector is maintained, the query tends to go in a loop most of the times. To avoid this, a TTL (Time To Live) value is used to restrict the number of hops. The search is considered as a failure if the query doesn't reach the destination within TTL hops. We haven't considered the failure cases during the plot. It was experimentally observed that only 5-10% of the queries turn out to be successful. The successful queries are the lucky queries, which find the destination in a shorter path. Hence the experimental values of the average number of hops tend to be smaller than the theoretical values.

In the next section, we analyze the performance of the restricted version of the hapto-search, which can also be considered as a more practical version.

5.3 Performance of the Restricted Version

In this section we compare the performance of the restricted hapto-search with the random search algorithm and DHT-based search algorithm. Moreover, the experimental testing of the scalability of the algorithm is reported.

Comparison with Random Search

Figure 4 shows the comparative plot of the average number of hops needed for the random search and the hapto-search algorithm for an E-R network with 10000 number of nodes and average degree 10. One can see that the hapto-search algorithm performs far better than random search. The average number of hops taken by the random search algorithm at key replication value 140 is 276.6 whereas for the hapto-search algorithm even at the key replication value of 10, the number of average hops needed is 98.4. This shows that the gradient-based guided search is far more efficient than the random search.

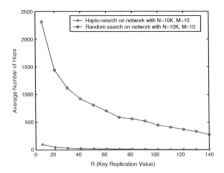

Fig. 4. Comparative plot of the average number of hops on the y-axis vs key replication value on the x-axis for the random search and the hapto-search algorithm on an ER network with N=10000 and average degree (M)=10

Comparison with DHT and Scalability Analysis

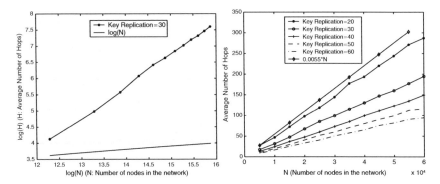

Fig. 5. Plot of average number of hops vs the number of nodes (N). The first figure is a log-log plot of the average number of hops (H) required to reach the destination with key replication value = 30 on the y-axis and the number of nodes on the x-axis. It also shows log-log plot of $\log N$. It can be seen that the slope of the curve $\log N$ is much less than that of the average number of hops, which shows that hapto-search does not scale to $\log N$. In the second plot, a E-R network of degree 10 is considered with different key replication values = 20, 30, 40, 50 and 60 respectively. It can be seen that the average number of hops varies linearly with N (number of nodes in the network) unlike the formula for the average hops for the simple version of the algorithm, which is independent of N. As the key replication value increases, the slope of the curve decreases

We know that the performance bound on DHT-based search algorithms is $\log N$. Therefore we tried to compare the performance of the hapto-search with $\log N$. The first plot in figure 5 shows this comparison. It is a log-log plot of the average number of hops vs the number of nodes in the network. $\log N$ is also plotted for the comparison. It can be seen that the performance of the hapto-search does not scale to $\log N$.

We then tried to evaluate the effect of the key replication on the average number of hops. The second plot in the figure 5 shows the average number of hops vs the network size for E-R networks of degree 10 and for different values of R (key replication value) ranging from 20 to 60. Previously we saw that the formula for the average number of hops for the simple version of the algorithm is not dependent upon the network size N. Moreover, we found that the performance is also largely insensitive to R. But in the restricted version, we see that the average number of hops increases linearly with N. Also the performance is dependent on R. As the value of R increases, the slope of the curve for the average number of hops decreases.

We are still working towards formulating a theory behind the dependence upon N in the restricted version. As a first step, we have made some important observations regarding the behavior of the parameters, AFM and ABM in the restricted version. This is noted next.

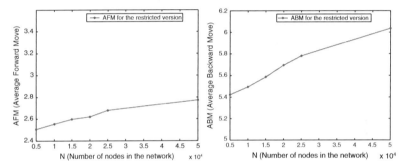

Fig. 6. Plot of the parameters Average Forward Move (AFM) and Average Backward Move (ABM) vs the network size (N). The first figure shows AFM on the y-axis vs N on the x-axis for the E-R network with key replication value 10, M=10 for the restricted version whereas the second figure plots ABM on the y-axis vs the network size N on the x-axis for the same configuration

Dependence of AFM and ABM upon N

Figure 6 shows the plot of AFM and ABM vs the network size N for the restricted version. The first figure shows the plot of AFM vs the network size N whereas the second figure shows the plot of ABM vs the network size N. Here one can observe that the values of AFM and ABM increase linearly with N. The rate of change in AFM and ABM are different.

6 Conclusion and Future Work

In this paper, we have discussed a new bio-inspired algorithm for location search in an unstructured peer to peer network. It is a key-based search algorithm, which ensures that the query always moves "closer" and ultimately reaches the destination.

We demonstrated that the algorithm is far more efficient than plain random walk. We presented two versions of the algorithm, a simple version as well as a restricted version. A rigorous theoretical analysis to explain the simple version is reported. The theoretical predictions and the experimental results match to a high degree of precision. We observed that the performance parameters such as the average number of hops, the probability of a local maximum, AFM and ABM is independent of the network size N.

While studying the experimental results of the restricted version, we identified that these performance parameters are also dependent on N. Hence we have realized that the theory proposed here needs to be suitably modified in the future to explain the experimental results of the restricted version of the hapto-search. Moreover, all the results presented here are from the experiments on the random networks. Future work will also explore other interesting and realistic topologies like the small-world and power law networks.

References

1. Q. Lv, P. Cao, E. Cohen, K. Li, and S. Shenker. *Search and Replication in Unstructured Peer-to-peer Networks*. In Proceedings of the 16th International Conference on Supercomputing, pages 84-95. ACM Press, 2002.
2. B. Alberts, A. Johnson, J. Lewis, M. Raff, K. Roberts, and P. Walter. *Molecular Biology of the Cell*, fourth edition, Garland Science, New York, 2002
3. A. Deutsch, and S. Dormann. *Cellular Automaton Modeling of Biological Pattern Formation*, Birkhauser, Boston, 2005
4. N. Ganguly, G. Canright, and A. Deutsch. *Design of an Efficient Search Algorithm for p2p Networks Using the Concepts from Natural Immune Systems*, In proceedings of PPSN VIII: The 8th International Conference on Parallel Problem Solving from Nature, 18-22 September 2004, Birmingham, UK.
5. G. Canright, A. Deutsch, and T. Urnes. *Chemotaxis-Inspired Load Balancing*, In the proceedings of the European Conference on Complex Systems, November 2005.
6. O. Babaoglu, G. Canright, A. Deutsch, G. Di Caro, F. Ducatelle, L. Gambardella, N. Ganguly, M. Jelasity,, R. Montemanni, A. Montresor. *Design Patterns from Biology for Distributed Computing* ACM Transaction of Autonomous and Adaptive Systems, Vol 1 Issue 1, September 2006.
7. T. Urnes. *An Analysis of the Skype Peer-to-Peer Protocol* Internal Publication, Telenor R&D, May 2006.
8. J. Rosenberg, J. Weinberger, C. Huitema and R. Mahy. *STUN: Simple Traversal of User Datagram Protocol (UDP) Through Network Address Translators (NATs)*, RFC 3489, IETF, Mar. 2003.
9. *http://www.gnutella.com*
10. Open Source Community. The free network project - rewiring the internet. In *http://freenet.sourceforge.net/, 2001*.

Ant Colony Optimization and its Application to Regular and Dynamic MAX-SAT Problems

Pedro C. Pinto[1,2], Thomas A. Runkler[1] and João M. C. Sousa[2]

[1] Siemens AG, Corporate Technology
Information and Communications, CT IC 4
Otto-Hahn-Ring 6, 81730 Munich - Germany
pinto.pedro.ext@siemens.com, thomas.runkler@siemens.com

[2] Technical University of Lisbon, Instituto Superior Técnico
Department of Mechanical Engineering, GCAR - IDMEC
Avenida Rovisco Pais, 1049-001 Lisbon - Portugal jmsousa@ist.utl.pt

Summary. In this chapter we discuss the ant colony optimization meta-heuristic (ACO) and its application to static and dynamic constraint satisfaction optimization problems, in particular the static and dynamic maximum satisfiability problems (MAX-SAT). In the first part of the chapter we give an introduction to meta-heuristics in general and ant colony optimization in particular, followed by an introduction to constraint satisfaction and static and dynamic constraint satisfaction optimization problems. Then, we describe how to apply the ACO algorithm to the problems, and do an analysis of the results obtained for several benchmarks. The adapted ant colony optimization accomplishes very well the task of dealing with systematic changes of dynamic MAX-SAT instances derived from static problems.

1 Introduction

In this chapter we discuss the ant colony optimization (ACO) meta-heuristics [1] and its application to constraint satisfaction optimization problems, in particular the dynamic variant of the maximum satisfiability problem, or MAX-SAT [2]. By applying ACO to the dynamic variant of MAX-SAT we are solving a problem with different characteristics from regular MAX-SAT, especially due to the adaptation required during the optimization process, to which meta-heuristical approaches are particularly suited. The original ACO applied to the static regular form of MAX-SAT was validated by using the results published by Roli, Blum and Dorigo in [3] as a benchmark. The adapted ACO accomplishes very well the task of dealing with systematic changes of dynamic MAX-SAT instances derived from regular problems. This chapter begins with an introduction to Ant Colony Optimization, useful if the reader is not familiar with the subject. The second part concerns satisfaction problems, particulary regular and dynamic MAX-SAT. This is followed by an overview of how to apply ACO to both problems. The last section concerns the results obtained and an analysis of some relevant algorithm parameters and how these affect the results of the optimization process. The chapter ends with a conclusion about applying ACO to dynamic SAT, and an outline of future work.

2 Ant Colony Optimization

Ant Colony Optimization is a bio inspired meta-heuristic, one of a few which attempt to emulate behaviors characteristic of social insects. Meta-heuristics are a set of approximate algorithms that essentially try to combine basic heuristic methods of local search in higher level frame-works aimed at exploring the search space of a problem in an efficient way. Usually, the meta-heuristic is used to guide an underlying, more problem specific heuristic, with the goal of increasing its performance [4], [5], [6]. Besides the bio inspired methods, several others have been proposed and implemented successfully in many occasions. The most popular are perhaps simulated annealing [7], tabu search [8], iterated local search [9] and GRASP [10].

Heuristic algorithms are necessary because many theoretical and real world optimization problems are unsuitable or unfeasible for analytical or exact approaches, that is, it is strongly believed that the optimal solution of the problems cannot be found within reasonable polynomial computational time. This problem occurs in dynamic environments, distributed systems, and in general in problems with a large number of variables and sources of uncertainty. Heuristics solve this kind of problems through the use of approximate methods that usually yield good enough solutions after a reasonable computational time.

2.1 Meta-heuristics Inspired in Social Insects

Some meta-heuristic algorithms try to derive optimization procedures from the natural behavior of social insects. Social insects live in organized communities (nests or colonies) where the members depend on each other to survive. The dynamic assignment of tasks in those communities to the community members, such as foraging and brood care, and the efficient way a colony achieves goals without a conscious planning effort from any of its agents, serve as inspiration source for optimization methods.

There are four basic families of social insects - ants, wasps, bees and termites, of which three are successfully explored as meta-heuristics. Wasp algorithms [11] have been used to optimize scheduling and logistic processes, and more recently regular constraint satisfaction problems [12] and dynamic constraint satisfaction problems [13]. Honey-Bees have been used as well in constraint optimization [14]. We have no knowledge of practical termites based heuristics.

Ant colony optimization algorithms [1] are multi-agent approaches where the behavior of each agent mimics the behavior of real life ants and how they interact with each other. Several different aspects of the behavior of real ants can be exploited to coordinate populations of artificial agents in order to solve computational problems. Examples are foraging, division of labor, brood sorting and cooperative transport. In all these examples, ants coordinate their activities via stigmergy, a form of indirect communication through modification of the environment [15]. ACO is used to optimize a wide range of problems, such as the satisfiability problem [3], [16], the traveling salesman problem [17], supply-chain logistics [18], [19] and routing problems [17], among others.

2.2 Ant Colony Optimization

When an ant is searching for the nearest food source and comes across with several possible trails, it tends to choose the trail with the largest concentration of pheromone τ, with a certain probability p. After choosing the trail, it deposits a certain quantity of pheromone, increasing the concentration of pheromones on this trail. The ants then return to the nest, depositing another portion of pheromone on the way back. When two ants at the same location choose two different trails at the same time, the pheromone concentration on the shortest trail will increase faster than the other. The ant that chooses the faster trail will deposit more pheromones in a smaller period of time.

The ant k at node i will choose one of the possible trails (i, j) connecting the actual node to one of other possible positions $j \in \{1, \cdots, n\}$, with probability

$$p_{ij}^k = f(\tau_{ij}) \qquad (1)$$

where τ_{ij} is the pheromone concentration on the path connecting i to j. If a whole colony of thousands of ants follows this behavior, soon the concentration of pheromone on the shortest path will be much higher than the concentration on other paths. Then the probability of choosing any other way will be very small, and only very few ants in the colony will fail to follow the shortest path. There is another phenomenon related with the pheromone concentration. Since it is a chemical substance, it tends to evaporate in the air, so the concentration of pheromones vanishes over time. In this way, the concentration of the less used paths will be much lower than that on the most used ones, not only because the concentration increases on the other paths, but also because their own concentration decreases. The pheromone level on a trail changes according to:

$$\tau_{ij}(t+1) = \tau_{ij}(t) \cdot \rho + \delta_{ij}^k \qquad (2)$$

where δ_{ij}^k is the pheromone released by the ant k on the trail (i, j) and $\rho \in [0, 1]$ is the evaporation coefficient. The system is continuous, so the time acts as the performance index, since the shortest paths will have the pheromone concentration increased in a shorter period of time.

This is a mathematical model of a real colony of ants. However, the artificial ants that mimic this behavior can be uploaded with more characteristics, e.g. memory and ability to see. The pheromone expresses the *experience* of the colony in the job of finding the shortest path. Memory and ability to see express useful *knowledge* about the problem the ants are solving. In this way, the function f in (1) can be defined accordingly as:

$$p_{ij}^k(t) = \begin{cases} \dfrac{\tau_{ij}{}^\alpha \cdot \eta_{ij}{}^\beta}{\sum_{r \notin \Gamma}^{n} \tau_{ir}{}^\alpha \cdot \eta_{ir}{}^\beta} & \text{if } j \notin \Gamma \\ 0 & \text{otherwise} \end{cases} \qquad (3)$$

where η_{ij} is a *heuristic function* and Γ is a *tabu list*. In this case, the heuristic expresses the capability of seeing which is the nearest node j to travel towards the food source. Γ is a list that contains all the trails that the ant has already passed and must not be chosen again (artificial ants can go back before reaching the food source). This acts as the memory of an ant. As an example, if the traveling salesman problem (TSP) [20] is the problem to be solved, the heuristic function can be the inverse of the distance from city i to city j expressed in a matrix d_{ij} [1]. Then $\eta_{ij} = 1/d_{ij}$ and the tabu list Γ is the list of cities that the ant has already visited. Finally, each ant deposits a pheromone δ_{ij}^k to the chosen trail:

$$\delta_{ij}^k = \tau_c \qquad (4)$$

where τ_c is a constant. In the artificial ants framework, due to the time factor, equation (2) is not sufficient to mimic the increasing pheromone concentration on the shortest path. With real ants, time acts as a performance index, but the artificial ants use all the same time to perform the task, whether they choose a short path or not. For solving the TSP with artificial ants, for example, it is the length l of the paths they have passed that quantifies the quality of the solution. Thus the best solution should increase even more the pheromone concentration on the shortest trail. To do so, (2) is changed to:

$$\tau_{ij}(t+n) = \tau_{ij}(t) \cdot \rho + \Delta\tau_{ij} \qquad (5)$$

where $\Delta\tau_{ij}$ are pheromones deposited onto the trails (i, j),

$$\Delta\tau_{ij} = \sum_{k=1}^{q} \delta_{ij}^k \cdot f\left(1/z_k\right) \qquad (6)$$

and z_k is the performance index. Once more, in the TSP case z_k can be the length $l_k = \sum d_{ij}$ of the path chosen by ant k. This way, the global update is biased by the solution found by each individual ant. The paths followed by the ants that achieved the shortest paths have their pheromone concentrations increased. Notice that the time interval taken by the q ants to do a complete tour is $t + n$ iterations. A *tour* is a complete route between the nest and the food source and an *iteration* is a step from i to j done by all the ants. The algorithm runs N_{max} times, where in every N^{th} tour, a new *ant colony* is released. The total number of iterations is $\mathcal{O}(N_{max}(n \cdot m + q))$. The general algorithm for the ant colonies is described in Algorithm 2.1.

Algorithm 2.1 Ant Colonies Optimization Algorithm.

Initialization:
  Set for every pair (i, j): $\tau_{ij} = \tau_0$
  define a N_{max}
  Place the q ants
  For $N = 1, ..., N_{max}$
Build a complete tour
    For $i = 1$ to n
      For $k = 1$ to m
        Choose the next node with probabilities p_{ij}^k from (3)
        Update locally $\tau_{ij}(t)$ using (4)
        Update the **tabu list** Γ
  Analyze solutions
    For $k = 1$ to q
      Compute performance index z_k
      Update globally $\tau_{ij}(t+n)$ using (5)

3 Constraint Satisfaction

The general constraint satisfaction problem (CSP) consists of finding an assignment to a set of variables that satisfies a set of constraints over the domains of these variables [21].

In recent years, constraint satisfaction has come to be seen as the core problem in many applications, for example temporal reasoning [22], resource allocation [23] and scheduling [24], [25], [26].

3.1 General Constraint Satisfaction Problem

The general constraint satisfaction problem (CSP) [21] consists of finding an assignment to a set of variables that satisfies a set of constraints over the domains of these variables. In a more formal way

Definition 1. *A CSP is a triple (X,D,C) where* X = $\{x_1, ..., x_n\}$ *is the set of variables,* D = $\{D_1, ..., D_n\}$ *is the set of domains which defines the values each variable can assume and* C = $\{C_1, ..., C_m\}$ *is the set of constraints among the variables. The defined triple is called an* $instance$ *and represented by* $\Phi = (X, D, C)$.

A solution S of the CSP is a complete assignment of the variables satisfying all the constraints. Depending on the domains and the type of constraints, several classes of CSPs are obtained. In this work, we focus our attention on finite discrete CSPs, that is, CSPs in which each domain is discrete and finite, and on dynamic discrete CSPs, in which the constraints of the CSPs vary with time.

A constraint on a set of variables is a restriction on the values that they can take simultaneously. Conceptually, a constraint can be seen as a set that contains all the legal compound labels for the subject variables. In practice constraints can be represented in many other ways, such as functions, inequalities, matrices, etc. The variables of a constraint are the variables of the members of the constraint.

The MAX-CSP is a variant of the CSP where the problem may be unsatisfiable, and the goal is then to find the solution that satisfies the maximum number of constraints possible.

3.2 Constraint Satisfaction Optimization Problems

In applications such as industrial scheduling, some solutions are better than others, or the constraints of the problem are so tight that it is impossible to find a solution to satisfy all of them, and a decision has to be made on which are more important to satisfy. The task in such problems is to find optimal solutions, where optimality is defined in terms of some application-specific functions. These problems are called Constraint Satisfaction Optimization Problems (CSOP) to distinguish them from the standard CSP.

Definition 2. *A **CSOP** is defined as a CSP together with an optimization function f which maps every solution tuple to a numerical value: X, D, C, f where (X,D,C) is a CSP, and if S is the set of solution tuples of (X,D,C) then $f : S \rightarrow$ numerical value. Given a solution tuple* T, *$f(T)$ is the f-value of* T.

The task in a CSOP is then to find the solution tuple with the optimal (minimal or maximal) f-value with regard to the application dependant optimization function f.

3.3 Regular MAX-SAT Problem

The regular, or static, maximum satisfiability problem (MAX-SAT) is a $\mathcal{NP}$-complete CSP [2],[27] in which the constraints are Conjunctive Normal Form (CNF) clauses, i.e., disjunctions between literals, where the domain of each variable is true or false, or equivalently $X \in \{0,1\}^n$.

In MAX-SAT, given a set of clauses, each of which is the logical disjunction of $K \geq 2$ variables, the problem is to find an assignment that satisfies the largest number of clauses or, in the weighted version, the assignment that maximizes the sum of weights of the satisfied clauses. The f-function returns then either the number of unsatisfied clauses, or the weight of those unsatisfied clauses.

The 3-satisfiability decision problem, or 3-SAT, is a special case of the k-satisfiability (k-SAT) CSP problem with boolean variables, where each clause contains at most $K = 3$ literals. It is known that 3-SAT experiences dramatic transitions from easy to difficult and then from difficult back to easy when the ratio of the number of clauses to the number of variables increases [28].

The difficulty of a SAT problem depends on the ratio of the number of constraints to number of variables [29],[30]. The critical value of this order parameter for 3-SAT is around 4.13. A 3-SAT is almost always satisfiable when the clause to variable ratio is below this critical value and is almost always unsatisfiable beyond it, making a sharp transition from satisfiability to unsatisfiability. The computational cost is also low when the probability of satisfiability is close to one or zero, being the highest around the 4.13 ratio.

MAX-3SAT follows the same pattern of 3-SAT to enter the computationally difficult region, but being a CSOP it follows an easy-hard pattern as the clause to variable ratio increases instead of easy-hard-easy. In other words, while 3SAT follows an easy-hard-easy pattern, MAX-3SAT follows an easy-hard pattern as the clause to variable ratio increases. When a constraint problem is overconstrained, a small subset of the problem is very likely to be overconstrained as well, so the problem can be declared unsatisfiable when such an overconstrained subproblem is detected unsatisfiable. The more constrained the problem is, the more quickly the decision process can conclude that no solution exists. However, in an overconstrained case, finding an optimal solution to minimize the total number of violated constraints is typically hard since every possible variable assignment can be a candidate of an optimal solution.

3.4 Dynamic Constraint Satisfaction Optimization Problems

In practice, many applications require not the optimization of a single CSOP but instead the optimization of a constantly changing problem. Changes can happen either because additional information is added to the model, or because of the intrinsic changing nature in the problem. As an example of the first, there is the planning of a network that has to be prepared to handle not only the current system but be robust to future conditions as well (it is likely money and time consuming to change the network configuration), and as a straightforward example of the second there is the Global Positioning System (GPS) navigation computer, that has to recalculate the route when the car does not follow the proposed route.

The changed CSOP problem is often similar to the old problem, only differing in a few constraints or variables. While it is possible for a small change in the problem to create a significant change in the optimal solution, in most practical cases the optimal solutions of such similar problems do not differ a lot. Continuing the previous example, in the GPS system when a car goes forward instead of turning left, the initial positions are almost the same and it may be enough to take the next turn.

A number of algorithms have been proposed for solving dynamic constraint problems. They can be divided into traditional methods and methods based on meta-heuristics. While the complete methods perform better on harder static problems, the more reactive nature of the incomplete methods based on meta-heuristics makes them more suited for easier static problems and for tracking changes in dynamic problems [31],[32],[33].

3.5 Dynamic MAX-SAT Problem

The conventional MAX-SAT problem can be generalized in different ways to allow for dynamic changes in the problem over time [34]. One possibility is to start from a given SAT instance and allow clauses to be dynamically added to or retracted from this instance. The motivation behind this definition is that of modeling a system which is subject to different constraints at different points in time. These constraints can reflect the state of the environment or of a subsystem, or the input by a user who controls the system.

The notion of a dynamic MAX-SAT instance is captured by the following formal definition:

Definition 3. *An **instance of the dynamic MAX-SAT problem** over a set V of proportional variables is given by a function* $\Psi : N \mapsto CNF(V)$, *where N is the set of nonnegative integers, and CNF(V) is the set of all propositional formula in conjunctive normal form which only use the variables in V.*

For practical purposes, Ψ mentions only a finite number of clauses.

Another way of defining dynamic MAX-SAT is to use a fixed set of clauses but allow certain propositional variables to be set to true or false at different points in time. In a practical problem, this can happen when a user forces a certain input. Returning to the GPS example, the user may require the passage through a particular city during the trip. This can be formalized in the following way:

Definition 4. *An instance of dynamic MAX-SAT over a set V of proportional variables is given by a CNF formula F over V and a second-order function* $\Psi : N \mapsto CNF(V \mapsto \{true, false, free\})$, *where N is the set of positive integers.*

For each time n, $\Psi(n)$ determines for each variable appearing in F whether it is fixed to true, fixed to false or not fixed.

It is not hard to see that Definitions 3 and 4 are equivalent in the sense that each dynamic SAT instance according to Definition 3 can always be transformed into an equivalent dynamic SAT instance according to Definition 4 and vice versa, and thus results obtained for one case are valid for both [34].

Definition 3 is conceptually simpler and a slightly more obvious generalization from conventional SAT from a theoretical point of view. This makes it slightly more adequate for theoretical considerations. Definition 4, on the other hand, reflects actual dynamic systems in a more direct way, and is easier to apply directly to existent regular MAX-SAT problems without changing their characteristics by adding out of place clauses. For both this reasons, we focused on dynamic CSOP problems formalized as according to Definition 4.

As an example, we can formulate a simple problem. Let us have a regular MAX-SAT problem defined by a set of clauses with a fixed number of variables being set forcibly to true or false in every stage, which means that for the MAX-SAT problem the variable set X is defined in instance $\Phi(i)$ as $x \in \{0, 1, free\}$, where $free$ means that the variable is unassigned. In this example, the problem is composed of an instance Φ with 3 clauses and 5 variables

$$\Phi = (x_1 \vee \bar{x}_2 \vee x_4) \wedge (x_3 \vee x_4 \vee \bar{x}_5) \wedge (x_1 \vee x_3 \vee \bar{x}_4) \tag{7}$$

Ψ is the function that determines which variables of X are fixed for each instance and their values, from the set of existing variables: $X_1 = \Psi(X_1)$. In this example, it fixes two variables randomly at each stage

$$Stage\ 1: X = [0, free, 1, free, free] \tag{8}$$

In order to assure the variability of the problem, at least one of the variables fixed at stage i cannot have been fixed in stage $i - 1$.

$$Stage\ 2: X = [0, free, free, free, 1] \tag{9}$$
$$Stage\ 3: X = [free, 1, free, free, 1]$$
$$\vdots$$

Stage 1 can then be formulated as a modified set of constraints and variables. The new constraints are $x_1 = 0$ and $x_3 = 1$, and the variables are x_2, x_4 and x_5. The assignment $x_2 = 1$, $x_4 = 1$ and $x_5 = 0$ solves the problem, since

$$(0 \vee 0 \vee 1) \wedge (1 \vee 1 \vee 1) \wedge (0 \vee 1 \vee 0) = 1. \tag{10}$$

Therefore, the solution of the first stage of the dynamic problem is $S = \{0, 1, 1, 1, 0\}$. In this case, the cost of the solution is 0 (no clauses are left unsatisfied) and the solution is not unique, which is not a concern since the problem here consists only in finding one solution at all.

4 Ant Colony Optimization for the MAX-SAT Problem

Ant Colony Optimization (ACO) was adapted to the regular MAX-SAT problem in [3] in the form of an ant colony system (ACS), and studies on the viability of ACO applied to CSP have been made in [16] and [35]. In ACS each ant is initially placed on a randomly chosen node of the construction graph $\mathcal{G} = (\mathcal{C}, \mathcal{L})$, defined such that nodes (components) are the pairs $(x_i \in \mathcal{X}, d_i \in \mathcal{D})$ and edges (connections) fully connect the nodes. A solution is a sequence of n nodes, being n de number of variables. Ants construct a solution by probabilistically choosing a node in their feasible neighborhood. The feasible neighborhood of an ant m is the set of pairs (x_i, d_i) such that the variable has not yet been assigned a value. The neighborhood chosen implements therefore the problem constraints $\mathcal{C}$ that say that nodes associated to the same variable are not in a same solution.

The solution the ants obtain is built from the artificial pheromone matrix and a set of heuristical rules. The artificial pheromones can either be placed on the components (set of nodes) or the connections (edges), and are represented by a matrix. The pheromones are updated both step-by-step, by every solution built by the ants, and globally, by the best solution found so far by adding a quantity of pheromone that depends on the quality of that solution.

4.1 Ant Colony Optimization for the Regular MAX-SAT Problem

Three different variants of ACS for MAX-SAT, namely ACS-comp, ACS-conn and ACS-conn+, are presented in [3]. In the first, the pheromones represent the components, while for the other two the pheromones represent the connections. In [3] it is

Algorithm 4.1 ACO applied to regular MAX-SAT

Initialization:
    Set for every pair (i,j): $\tau_{ij} = \tau_0$
    define the number of iterations N_{max}
    Place the q ants
    For $N = 1, ..., N_{max}$
Build a complete tour
        For $i = 1$ to n
            For $k = 1$ to q
                Generate new solution S_i based on heuristics η and pheromone matrix τ using (12)
                Update locally τ using (13)
    Analyze solutions
        For $k = 1$ to q
            Compute the *cost* or *utility* of S_k
            Update best solution *cost* or *utility*

---

concluded that the results obtained with ACS-comp and ACS-conn+ are very similar and slightly better than for ACS-conn. These values are used as a benchmark for our algorithm, before we extend it to the *dynamic* MAX-SAT problem.

The ACO proposed here is represented in Algorithm 4.1. It is a single ant algorithm ($m = 1$) where the ant builds a new solution at each iteration based on heuristics and pheromones. The solution is built in two phases:

(i.) Selection of the next variable to assign from the variables that are currently unassigned. That decision can be made through systematic analysis, but satisfactory results can be obtained (for the instances considered) by choosing the variable randomly from the set of unassigned variables.

(ii.) Selection of the value to assign to the variable, from $\{0,1\}$. This is done through a balance between the heuristic information η, given by

$$\eta = \frac{1}{1 + f(s_k \cup \langle j, e \rangle) - f(s_k)} \quad (11)$$

where s_k is the partial solution of the problem, $s_k \cup \langle j, e \rangle$ is the partial solution of the problem after assigning the value e to variable j, and the information contained in the pheromone matrix. f is the cost function of the optimization problem, $f(s_k)$ being the cost of s_k. The pheromone matrix has the size $2 \times N$ (the pheromones represent the components) and contains knowledge of how good the results were when either 0 or 1 where chosen in the creation of former solutions. The probability of variable j being assigned the value e is given by

$$p_{ej} = \frac{\tau_e^\alpha \cdot \eta_e^\beta}{\tau_e^\alpha \cdot \eta_e^\beta + \tau_{\bar{e}}^\alpha \cdot \eta_{\bar{e}}^\beta} \quad (12)$$

In the unweighted variant of the MAX-SAT problem *cost* is defined as the total number of unsatisfied clauses. An unsatisfied clause is a clause whose variables are all assigned and has logical value 0. For the weighted variant, *utility* is defined as the sum of the weight of every satisfied clauses. In the case when all weights are one, the utility is identical to the cost.

The pheromone matrix τ is updated for every iteration with the last calculated solution, using:

$$\tau = (1 - \rho)(\tau_{S_i} + 1/f) \tag{13}$$

where ρ is the evaporation coefficient (constant) and f is either the number of unsatisfied clauses or the utility, depending on the instance being unweighted or weighted.

For the problems in [3] this algorithm obtained results similar to the ones published. The optimization of the algorithm parameters yielded some improvements. In the next section we modify this algorithm in order to solve *dynamic* MAX-SAT problems.

4.2 Ant Colony Optimization for the Dynamic MAX-SAT Problem

The basic algorithm for the optimization of dynamic MAX-SAT problems is presented in Algorithm 4.2.

Algorithm 4.2 ACO applied to dynamic MAX-SAT

Initialization:
   Set for every pair (i, j): $\tau_{ij} = \tau_0$
   define the number of iterations N_{max}
   for stages 1 to *Nr of Stages*
      Reset the best value of *cost* or *utility* to zero
      Place the q ants
      For $N = 1, ..., N_{max}$
Build a complete tour
        For $i = 1$ to n
          For $k = 1$ to q
            Generate new solution S_i with heuristics η and pheromone matrix τ using (12)
            Update locally τ using (13)
      Analyze solutions
        For $k = 1$ to q
          Compute the *cost* or *utility* of S_k
          Update best solution *cost* or *utility*
          Compute the *diversity* of S_i
          Update evaporation coefficient $\rho(\text{div}_i)$ using (14)

We consider three approaches in order to study how to make the algorithm adapt better to the changes in the instance structure. First, we applied the algorithm as described for the regular problem, (static ACO). For the second approach, we initialize the pheromone matrix at the beginning of each stage (algorithm restart ACO). Lastly, we introduced the algorithm adaptive ρ ACO. This algorithm is based on the adaption of the evaporation coefficient ρ, which is a constant in the static algorithm. In order to prevent the pheromone matrix from saturating, the evaporation coefficient is set to a bigger value if the solution starts to stabilize significantly around a result. This is done by defining ρ as a function of the diversity div_i of the last solution computed.

$$\rho_i = \begin{cases} \rho_0 & \text{if } \text{div}_{i-1} > \rho_0 \\ K \cdot \text{div}_{i-1} & \text{otherwise} \end{cases} \quad (14)$$

The Diversity div_i is defined as the lowest Hamming distance (number of bits which differ between two binary strings) between solution i and any solution computed previously, divided by the number of variables. The concept was used in [36] by R. Battiti and M. Protasi to adapt a reactive local search algorithm with good results.

5 Experiments and Results

This section presents the results obtained for both the regular and dynamic MAX-SAT problems. (Just as in [3]), *regular* MAX-SAT is applied to the following data sets, that can be obtained from [37] and [38]:

- Weighted problems: several instances of the family jnh, with 800, 850 or 900 clauses and 100 variables (therefore, mostly overconstrained) and weighted using uniformly distributed integer weights.
- Unweighted problems: instances of the family uuf with 250 variables and 1065 clauses.

Both the weighted and unweighted families of instances are used as benchmarks because they have a clause to instance ratio close or bigger than 4.13 and thus being past the area of the *from easy to hard* phase transition.

For the experiments with dynamic MAX-SAT, we modify the same weighted and unweighted instances that are used for the static case. 100 stages with 50 (for the unweighted problem) and 100 (for the weighted problem) iterations per stage with the procedure described in section 3.5.

5.1 Results for the Regular MAX-SAT Problem

Table 1 shows the results for the algorithm by Roli et al. as presented in [3] (ACO'= results in [3] for the ACS algorithm with pheromones on the components) and for our proposed algorithm. For the jnh instances we show the average error from the optimal solution and the average number of iterations until the best solution is found. For the uuf instances (1) we show the best average solution (in number of satisfied clauses) instead of the average error. The averages were evaluated for 100 runs, with different random initializations of the pheromone matrices, and for each run we stopped the optimization process after 100 iterations, which for the considered problems was always sufficient to yield a stable solution. The parameter settings used were: $\alpha = 2$, $\beta = 6$ and $\rho = 0.2$.

Table 1. Results for the regular MAX-SAT problems for 100 runs with 100 iterations per run

instance	nr. satisfied or error				# iterations			
	average		standard dev.		average		standard dev.	
	ACO'	ours	ACO'	ours	ACO'	ours	ACO'	ours
uuf250-01	1038.19	1047.20	2.13	2.85	63.21	88.17	24.46	9.89
uuf250-02	1031.35	1043.90	2.37	3.40	70.85	90.15	21.23	8.40
jnh1	0.558	0.490	0.084	0.150	76.530	86.610	18.193	9.128
jnh201	0.568	0.380	0.086	0.180	74.810	85.360	19.882	10.419
jnh301	0.777	0.660	0.127	0.170	81.710	86.930	16.299	12.826

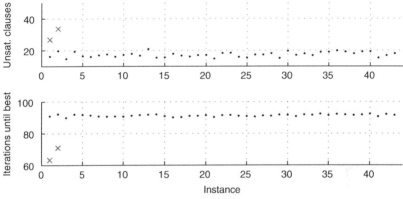

Fig. 1. Application of the proposed ACO algorithm to the set of uuf instances (x = ACO')

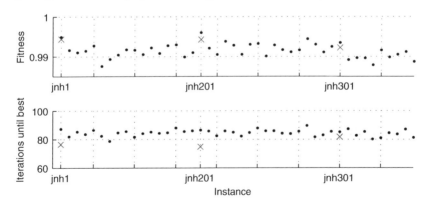

Fig. 2. Application of the proposed ACO algorithm to the set of jnh instances (x = ACO')

The quality of the solutions found with our algorithm is on average better for the problems in question, when compared to ACS, so it is reasonable to expect a good performance also for the dynamic problem. Figures 1 and 2 show the results obtained with our algorithm for 44 unweighted $uuf250$ and 44 weighted jnh instances, all averaged for 100 runs. The results of both algorithms as displayed in table 1 are included as special cases. For each instance the algorithm finds a solution with a low number of unsatisfied clauses or a high fitness, respectively, after around 80 iterations. This indicates a high effectiveness and a high robustness of the algorithm. It is also visible that the quality of the solution obtained is similar for every instance of the same family *uuf250* or *jnh*.

We also tried local search for post-optimization and obtained slightly better results. However, the difference is marginal and local search slows down the optimization process, so it was not used any further.

5.2 Results for the Dynamic MAX-SAT Problem

We applied our proposed ACO algorithms to the unweighted and weighted versions of the dynamic MAX-SAT problem.

In section 5.1 it is seen that the quality of the solution obtained for each static problem does not vary much, both for the unweighted as well as the weighted regular MAX-SAT (Figures 1 and 2, respectively). It is therefore reasonable to expect that the result of the optimization of the dynamic variant of one instance of each family of problems is representative of the whole family. This was verified to be the case.

We constructed then three unweighted dynamic MAX-SAT problems based on the static problem uuf250-n01 (representative of the *uuf250* family), and three weighted dynamic MAX-SAT problems based on the static weighted MAX-SAT problem jnh301 (representative of the *jnh* family). The characteristics of each unweighted dynamic problem are:

- 100 stages
- 50 iterations per stage
- number of fixed variables per stage set to $\psi=4$, $\psi=6$ and $\psi=10$, respectively

while the weighted dynamic problems have:

- 100 stages
- 100 iterations per stage
- number of fixed variables per stage set to $\psi=4$, $\psi=6$ and $\psi=10$, respectively

The parameter settings were: $\alpha = 2$, $\beta = 6$, $\rho_0 = 0.2$ and $K = 0.6$.

Figures 3 to 8 show the results for static ACO, for restart ACO and for adaptive ρ ACO for both weighted and unweighted problems, respectively. The results are the average for 100 runs.

For the jnh instances we show, as in the previous regular MAX-SAT problem, the average fitness (utility of the best solution of the stage divided by the utility of the theoretical optimal solution) and the average number of iterations until the best solution is found. For the uuf instances we show the average in number of unsatisfied clauses instead of the average fitness.

The results show that for both weighted and unweighted problems adaptive ρ ACO clearly outperforms both static ACO and restart ACO. The standard deviation for the fitness is similar for all three algorithms and it is too small to be of relevance in the comparison of the algorithm's results. Restart ACO always "needs" to run each stage until near its end to find the best result, which is to be expected since it is essentially the static algorithm applied at each stage. Static ACO performs adequately in the initial stages, but soon its results get progressively worse. This is observed better in the unweighted, larger problems (Figures 6-8) and corresponds to the eventual saturation of the pheromone matrix, which invalidates the exploration by the algorithm of different solutions. The adaptive ρ ACO algorithm reaches much better solutions, with less iterations needed (i.e. lower computational cost) than restart ACO.

The results are better for the three algorithms in problems with less constraints, with the fitness being higher for the problems with $\psi = 4$ and lower for the problems with $\psi = 10$. Regardless of the value of ψ, the pattern of the results obtained with each algorithm through the stages is similar, with adaptive ρ ACO performing the best, followed by restart ACO and static ACO.

If the number of iterations per stage is high enough, then it is clear that restart ACO will eventually find results equal to the results of adaptive ρ ACO, although always with a higher computational cost. After this transition point, restart ACO may reach a better solution than adaptive ρ ACO. For the class of unweighted problems studied here this number of iterations per stage is somewhere around 100, while for the weighted class of problems it is around 200. However, in this situation the algorithm is basically solving one static problem per stage, so we actually abandon the dynamic problem.

If we consider the case of a low or very low number of iterations per stage the effect seen in the figures will still occur, but the advantages of the adaptive ρ over static ACO will only happen after a very large number of stages since it will take a long time for the static ACO to saturate. Restart ACO performs worst in this situation, since it requires the largest number of iterations per stage to find a good solution.

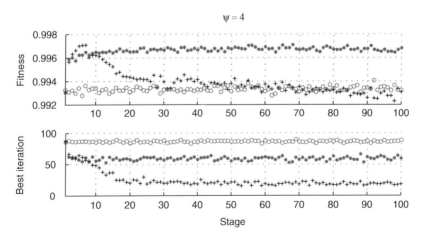

Fig. 3. The proposed ACO algorithm (*: Adaptive ρ, o: Restart, +: Static) applied to dynamic jnh301, $\Psi = 4$

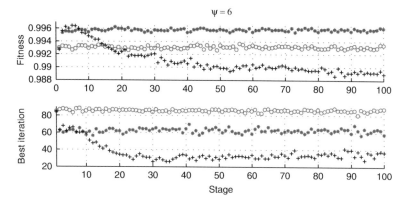

Fig. 4. The proposed ACO algorithm (*: Adaptive ρ, o: Restart, +: Static) applied to dynamic jnh301, $\Psi = 6$

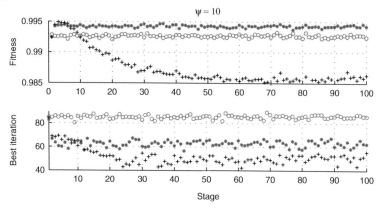

Fig. 5. ACO (*: Adaptive ρ, o: Restart, +: Static) applied to dynamic jnh301, $\Psi = 10$

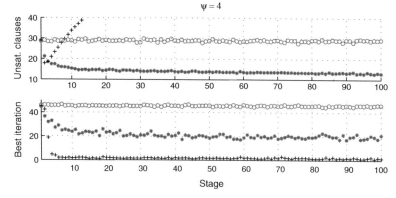

Fig. 6. The proposed ACO algorithm (*: Adaptive ρ, o: Restart, +: Static) applied to dynamic uuf250-n01, $\Psi = 4$

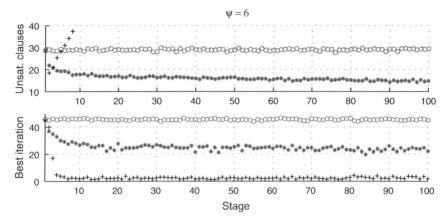

Fig. 7. The proposed ACO algorithm (*: Adaptive ρ, o: Restart, +: Static) applied to dynamic uuf250-n01, $\Psi = 6$

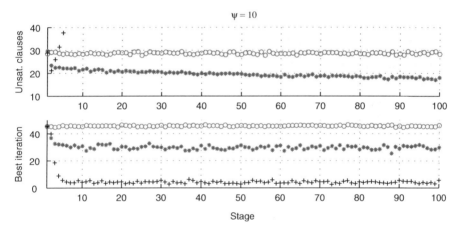

Fig. 8. The proposed ACO algorithm (*: Adaptive ρ, o: Restart, +: Static) applied to dynamic uuf250-n01, $\Psi = 10$

6 Conclusions

This chapter proposed an adaptive ρ algorithm. The algorithm works very well for dynamic MAX-SAT problems and the quality of the solutions it finds is significantly better than the quality of the solutions found with restart ACO and static ACO, for both the weighted and unweighted variants of MAX-SAT. The quality of the best solution at each stage is similar through the iterations. Adaptive ρ ACO is faster than restart ACO in all cases. For future work it is our intention to expand these results to other types of problems. Here we limited the analysis to the problems analyzed in [3] and expanded them to the dynamic problems.

How ACO compares with other algorithms used to solve the same problem is another question. Intelligent algorithms of Local Search [39] and [40], such as Walk-SAT and IROTS, are excellent solvers for SAT problems and in future work it is our intention to see how they compare in solving dynamic MAX-SAT problems.

Acknowledgments

This work is supported by the Portuguese foundation for Science and Technology (FCT) under Grant no, SFRH /BD / 17869 / 2004 and by the project POCI / EME / 59191 / 2004, co-sponsored by FEDER, Programa Operacional Ciência e Inovação 2010, FCT, Portugal.

References

1. Dorigo, M., Stützle, T.: Ant Colony Optimization. MIT Press (2004)
2. Garey, M.R., Johson, D.S.J.: Computers and Intractability: A Guide to the Theory of NP-Completeness. WH Freeman Publishers (1979)
3. Roli, A., Blum, C., Dorigo, M.: ACO for maximal constraint satisfaction problems. MIC'2001 - Metaheuristics International Conference, Porto, Portugal (2001) 187–192
4. Kennedy, J., Eberhart, R.: Swarm intelligence. Morgan Kaufmann Publishers (2001)
5. Osman, I., Laporte, G.: Metaheuristics: A bibliography. Annals of Operations Research (63) (1996) 513–628,
6. Blum, C., Roli, A.: Metaheuristics in combinatorial optimization: Overview and conceptual comparison. ACM Compututational Survey **35**(3) (2003) 268–308
7. Aarts, E.H.L., Korst, J.H.M., Laarhoven, P.J.M.V.: Simulated annealing. In: Local Search in Combinatorial Optimization. Wiley-Interscience Eds, Chichester, England (1997) 91–120
8. Glover, F., Laguna, M.: Tabu search. (1993) 70–150
9. Ramalhinho-Lourenço, H., Martin, O.C., Stützle, T.: Iterated local search. (513) (2000)
10. Feo, T., Resende, M.: Greedy randomized adaptive search procedures. Journal of Global Optimization **6** (1995) 109–133
11. Cicirello, V.A., Smith, S.F.: Wasp-like agents for distributed factory coordination. Autonomous Agents and Multi-agent systems **8** (2004) 237–266
12. Pinto, P., Runkler, T.A., Sousa, J.M.C.: Agent based optimization of the MAX-SAT problem using wasp swarms. Controlo 2006, 7th Portuguese Conference on Automatic Control (2006)
13. Pinto, P., Runkler, T.A., Sousa, J.M.C.: Wasp swarm algorithm for dynamic MAX-SAT problems. ICANNGA 2007, 8th International Conference on Adaptive and Natural Computing Algorithms (2007)
14. Abbass, H.: An agent based approach to 3-SAT using marriage in honey-bees optimization. International Journal of Knowledge-Based Intelligent Engineering Systems (2002) 1–8
15. Grasse., P.P.: La reconstruction du nid et les coordinations inter-individuelles chez bellicositermes natalensis et cubitermes sp. la theorie de la stigmergie: Essai d'interpretation du comportement des termites constructeurs. Insectes Sociaux **6** (1959) 41–81

16. Pimont, S., Solnon, C.: A generic ant algorithm for solving constraint satisfaction problems. 2nd International Workshop on Ant Algorithms (ANTS 2000), Brussels, Belgium (2000) 100–108
17. Cicirello, V.A., Smith, S.F.: Ant colony for autonomous decentralized shop floor routing. In: Proceedings of ISADS-2001, Fifth International symposium on autonomous decentralized systems. (2001)
18. Silva, C.A., Runkler, T.A., Sousa, J.M.C., da Costa, J.M.S.: Optimization of logistic processes in supply–chains using meta–heuristics. In Pires, F.M., Abreu, S., eds.: Lecture Notes on Artificial Intelligence 2902, Progress in Artificial Intelligence, 11th Portuguese Conference on Artificial Intelligence. Springer Verlag, Beja, Portugal (2003) 9–23
19. Silva, C.A., Runkler, T.A., Sousa, J.M.C., Palm, R.: Ant colonies as logistic process optimizers. In Dorigo, M., Caro, G.D., Sampels, M., eds.: Ant Algorithms, International Workshop ANTS 2002, Brussels, Belgium. Lecture Notes in Computer Science LNCS 2463, Heidelberg, Springer (2002) 76–87
20. Applegate, D.L., Bixby, R.E., Chvátal, V., Cook, W.J.: The Traveling Salesman Problem: A Computational Study. Princeton University Press (2007)
21. Tsang, E.: Foundations of Constraint Satisfaction. Academic Press (1993)
22. Gennari, R.: Temporal reasoning and constraint programming: A survey (1998)
23. Frei, C., Faltings, B.: Resource allocation and constraint satisfaction techniques. In: CP '99: Proceedings of the 5th International Conference on Principles and Practice of Constraint Programming, London, UK, Springer-Verlag (1999) 204–218
24. Cesta, A., Cortellessa, G., Oddi, A., Policella, N., Susi, A.: A constraint-based architecture for flexible support to activity scheduling. In: AI*IA 01: Proceedings of the 7th Congress of the Italian Association for Artificial Intelligence on Advances in Artificial Intelligence, London, UK, Springer-Verlag (2001) 369–381
25. Abril, M., Salido, M.A., Barber, F., Ingolotti, L.: Distributed constraint satisfaction problems to model railway scheduling problems. ICAPS 2006 Workshop on Constraint Satisfaction Techniques for Planning and Scheduling Problems, Cumbria (England) (2006)
26. Rasconi, R., Policella, N., Cesta, A.: Fix the schedule or solve again - comparing constraint-based approaches to schedule execution. In: Proceedings of the ICAPS Workshop on Constraint Satisfaction Techniques for Planning and Scheduling Problems. (2006)
27. Goodrich, M.T., Tamassia, R.: Algorithm Design - Foundations, Analysis, and Internet Examples. John Wiley & Sons, Inc. (2001)
28. Zhang, W.: Phase transitions and backbones of 3-SAT and maximum 3-SAT. In: Principles and Practice of Constraint Programming. (2001) 153–167
29. Cheeseman, P., Kanefsky, B., Taylor, W.M.: Where the really hard problems are. In: Proceedings of the Twelfth International Joint Conference on Artificial Intelligence, IJCAI-91, Sidney, Australia. (1991) 331–337
30. Mitchell, D., Selman, B., Levesque, H.: Hard and easy distributions of SAT problems. In: 10-th National Conf. on Artificial Intelligence (AAAI-92), San Jose, CA (1992) 459–465
31. Mertens, K., Holvoet, T., Berbers, Y.: The dynCOAA algorithm for dynamic constraint optimization problems. In Weiss, G., Stone, Peter, e., eds.: Proceedings of the Fifth International Joint Conference on Autonomous Agents and MultiAgent Systems. (2006) 1421–1423
32. Mertens, K., Holvoet, T., Berbers, Y.: Which dynamic constraint problems can be solved by ants. In Sutcliffe, G., Goebel, Randy, E., eds.: Proceedings of The 19th International FLAIRS Conference. (2006) 439–444

33. Mailler, R.: Comparing two approaches to dynamic, distributed constraint satisfaction. In: AAMAS '05: Proceedings of the Fourth International Joint Conference on Autonomous Agents and Multiagent Systems, New York, NY, USA, ACM Press (2005) 1049–1056
34. Hoos, H.H., O'Neill, K.: Stochastic local search methods for dynamic SAT - an initial investigation. AAAI-2000 Workshop "Leveraging Probability and Uncertainty in Computation", Austin, Texas (2000) 22–26
35. Solnon, C.: Ants can solve constraint satisfaction problems. IEEE Transactions on Evolutionary Computation **6** (2002) 347–357
36. Battiti, R., Protasi, M.: Reactive search, a history-based heuristic for MAX-SAT. ACM Journal of Experimental Algorithmics **2** (1997)
37. Johnson, D., Trick, M.: Cliques, coloring, and satisfiability: Second DIMACS implementation challenge. DIMACS Series in Discrete Mathematics and Theoretical Computer Science (1996)
38. Hoos, H.H., Stützle, T.: SATLIB: An online resource for research on SAT. SAT 2000 (2000) 283–292
39. Hoos, H.H., Stützle, T.: Local search algorithms for SAT: an empirical evaluation. Journal of Automated Reasoning, special Issue "SAT 2000" (1999) 421–481
40. Smyth, K., Hoos, H.H., Stützle, T.: Iterated robust tabu search for MAX-SAT. Lecture Notes in Computer Science, Springer Verlag **44** (2003) 279–303